CW00500731

Metal Failures: Mechanisms, Analysis, Prevention

Metal Failures:

Mechanisms, Analysis, Prevention

Arthur J. McEvily, Professor Emeritus
Department of Metallurgy and Materials Engineering
The University of Connecticut

A WILEY-INTERSCIENCE PUBLICATION
JOHN WILEY & SONS, INC.

This book is printed on acid-free paper. ♾

Published simultaneously in Canada.

Library of Congress Cataloging-in-Publication Data:

McEvily, A. J.
Metal failures : mechanisms, analysis, prevention / Arthur J. McEvily.
p. cm.
"A Wiley-Interscience publication."
Includes index.
ISBN 0-471-41436-0 (cloth)
1. Metals—Fracture. 2. Fracture mechanics. I. Title.
TA460.M382 2001
620.1'66—dc21 2001026653

Printed in the United States of America.

10 9 8 7 6 5 4 3 2 1

To
Geoff, Allysha, Keith, Joey, Ryan, Kyle, and Courtney

Contents

Preface

This book is intended for use by senior engineering students in a one-semester course, as well as by graduate engineers who are seeking further information concerning the analysis of failures. The book is the outcome of teaching a 14-week course to both undergraduates and graduates for more than ten years. By dealing with a wide scope of types of failures, the book provides the information usually found in a course on mechanical metallurgy. A large number of case studies, often based upon the author's experience of over 50 years, are used to illustrate many of the basic principles involved, both in metallurgy and in failure analysis. These case studies are intended to demonstrate how basic principles are applied to real-world situations. There are 14 chapters, of varying lengths, and it is expected that an instructor will balance the time spent on each to cover the material in a semester as suits his or her interests and those of the students.

My appreciation is expressed to my colleagues Mark Aindow, Martin Blackburn, Steven Boggs, Maurice Gell, Jorge Gonzalez, Yoshiyuki Kondo, Iain Le May, Gary Marquis, John Morral, Yukitaka Murakami, Nitin Padture, and Leon Shaw for their helpful comments and encouragement.

Arthur McEvily
Storrs, CT
September 2001

Metal Failures:
Mechanisms, Analysis, Prevention

1

Failure Analysis

I. INTRODUCTION

Despite the great strides forward that have been made in technology, failures continue to occur, often accompanied by great human and economic loss. This text is intended to provide an introduction to the subject of failure analysis. It cannot deal specifically with each and every failure that may be encountered, as new situations are continually arising, but the general methodologies involved in carrying out an analysis are illustrated by a number of case studies. Failure analysis can be an absorbing subject to those involved in investigating the cause of an accident, but the capable investigator must have a thorough understanding of the mode of operation of the components of the system involved, as well as a knowledge of the possible failure modes, if a correct conclusion is to be reached. Since the investigator may be called upon to present and defend opinions before highly critical bodies, it is essential that opinions be based upon a sound factual basis and reflect a thorough grasp of the subject. A properly carried out investigation should lead to a rational scenario of the sequence of events involved in the failure as well as to an assignment of responsibility, either to the operator, the manufacturer, or the maintenance and inspection organization involved. A successful investigation may also result in improvements in design, manufacturing, and inspection procedures, improvements that preclude a recurrence of a particular type of failure.

The analysis of mechanical and structural failures might initially seem to be a relatively recent area of investigation, but upon reflection, it is clear the topic has been an active one for millenia. Since prehistoric times, failures have often resulted in taking one step back and two steps forward, but often with severe consequences for the designers and builders. For example, according to the Code of Hammurabi, which was written in about 2250 BC (1):

> If a builder build a house for a man and do not make its construction firm, and the house which he has built collapse and cause the death of the owner of the house, that builder shall be put to death. If it cause the death of a son of the owner of the house, they shall put to death a son of that builder. If it destroy property, he shall restore what ever it destroyed, and because he did not make the house which he built firm and it collapsed, he shall rebuild the house which collapsed at his own expense.

The failure of bridges, viaducts, cathedrals, and so on, resulted in better designs, better materials, and better construction procedures. Mechanical devices, such as wheels and axles, were improved through empirical insights gained through experience, and these improvements often worked out quite well. For example, a recent program in India was directed at improving the design of wheels for bullock-drawn carts. However, after much study, it was found that improvements in the design over that which had evolved over a long period of time were not economically feasible.

An example of an evolved design that did not work out well is related to the earthquake that struck Kobe, Japan, in 1995. That area of Japan had been free of damaging earthquakes for some time, but had been visited frequently by typhoons. To stabilize homes against the ravages of typhoons, the local building practice was to use a rather heavy roof structure. Unfortunately, when the earthquake struck, the collapse of these heavy roofs caused considerable loss of life as well as property damage. The current design codes for this area have been revised to reflect a concern for both typhoons and earthquakes.

The designs of commonplace products have often evolved rapidly to make them safer. For example, consider the carbonated soft-drink bottle cap. At one time, a metal cap was firmly crimped to a glass bottle, requiring a bottle opener for removal. Then came the easy-opening, twist-off metal cap. These caps were made of a thin, circular piece of aluminum that was shaped by a tool at the bottling plant to conform to the threads of the glass bottle. If the threads were worn, or if the shaping tool did not maintain proper alignment, then the connection between cap and bottle would be weak and the cap might spontaneously blow off the bottle, for example, on the supermarket shelf. Worse than that, there were a number of cases where, during the twisting-off process, the expanding gas suddenly propelled a weakly attached cap from the bottle and caused eye damage. To guard against this danger, the metal caps were redesigned to have a series of closely spaced perforations along the upper side of the cap, so that as the seal between the cap and bottle was broken at the start of the twisting action, the gas pressure was vented, and the possibility of causing an eye injury was minimized. The next stage in the evolution of bottle cap design has been to use plastic bottles and plastic caps. In a current design, the threads on the plastic bottle are slotted, so that, as in the case of the perforated metal cap, as the cap is twisted the CO_2 gas is vented, and the danger of causing eye damage is reduced.

Stress analysis plays an important role both in design and in failure analysis. Ever since the advent of the industrial revolution, concern about the safety of structures has resulted in significant advances in stress analysis. The concepts of stress and strain developed from the work of Hooke in 1678, and were firmly established by Cauchy and Saint-Venant early in the nineteenth century. Since then, the field of stress analysis has grown to encompass strength of materials, and the theories of elasticity, viscoelasticity, and plasticity. The advent of the high-speed computer has led to further rapid advances in the use of numerical methods of stress analysis by means of the finite element method (FEM), and improved knowledge of material behavior has led to advances in development of constitutive relations based upon dislocation theory, plasticity, and mechanisms of fracture. Design philosophies such as safe-life and fail-safe have also been developed, particularly in the aerospace field.

In a safe-life design, a structure is designed as a statically determinant structure that is intended to last without failure for the design lifetime of the structure. To guard against premature failure, the component should be inspected at intervals during its in-service lifetime.

In the fail-safe approach, the structure is designed such that if one member of the structure were to fail, there would be enough redundancy built into the structure that an alternate load path would be available to support the loads, at least until the time of the next inspection. (The use of both suspenders and a belt to support trousers is an example of a fail-safe, redundant approach.) Consideration must also be given to the spectrum of loading that a structure will be called upon to withstand in relation to the scatter in the ability of materials to sustain these loads. As indicated in Fig. 1-1, danger of failure is present when these two distributions overlap.

In addition, new fields such as fracture mechanics, fatigue research, corrosion science, and nondestructive testing have emerged. Important advances have also been made in improving the resistance of materials to fracture. In the metallurgical field, these advances have been brought about through improvements in alloy design, better control of alloy chemistry, and improvements in metal processing and heat treatment. The failure analyst often has to determine the nature of a failure; for example,

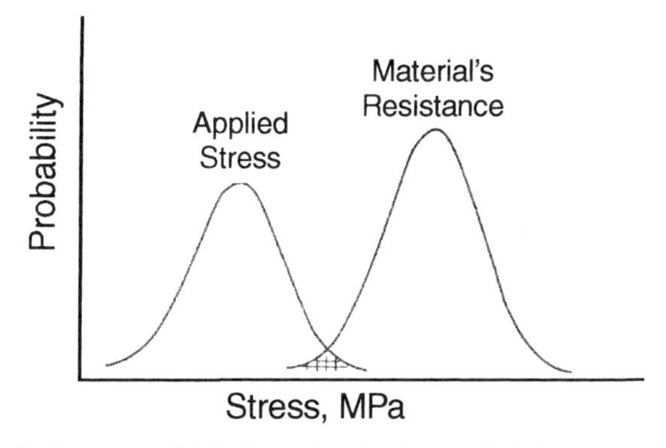

Fig. 1-1. Schematic frequency distributions showing the applied stresses and the resistance of the material.

was it due to fatigue or to an overload? In many cases, a simple visual examination may suffice to provide the answer. In other cases, however, the examination of a fracture surface (fractography) may be more involved and may require the use of laboratory instruments such as the light microscope, the transmission electron microscope, and the scanning electron microscope.

Many of today's investigations are quite costly and complex, and require a broad range of expertise as well as the use of sophisticated laboratory equipment. In some instances, the investigations are carried out by federal investigators, as in the case of the TWA Flight 800 disaster (center fuel tank explosion), where both the Federal Bureau of Investigation (FBI) and the National Transportation Safety Board (NTSB) had to determine if the cause of the failure was due to a missile attack, sabotage, mechanical failure, or an electrical-spark-ignited fuel tank explosion. The case of the Three Mile Island accident (faulty valve) involved the Nuclear Regulatory Commission (NRC), and the Challenger space shuttle disaster (O-ring) involved the National Aeronautics and Space Administration (NASA). Many investigations are also carried out by manufacturers to ensure that their products perform reliably. In addition, a number of companies now exist for the purpose of carrying out failure analyses to assist manufacturers and power plant owners, as well as to aid in litigation. The results of many of these investigations are made public, and thus provide useful information as to the nature and cause of failures. Unfortunately, the results of some investigations are sealed as part of a pretrial settlement to litigation, and the general public is deprived of an opportunity to learn that certain products may have dangers associated with them. A company may decide on the basis of costs versus benefits that is cheaper to settle a number of claims rather than to issue a recall. This policy can sometimes be disastrous, as in the case of the recent rash of tire failures. Another example involved a brand of cigarette lighter that repeatedly malfunctioned and caused serious burn injuries. It was only after some fifty of these events had occurred and the cases had been settled that the dangers associated with this item were brought to light in a public trial.

An important outcome of failure analyses has been the development of building codes and specifications governing materials [the American Society for Testing and Materials (ASTM)], manufacturing procedures [the Occupational Safety and Health Administration (OHSA)], design [the American Society of Mechanical Engineers (ASME) Boiler and Pressure Vessel Codes, the Federal Aviation Administration (FAA), NASA, American Petroleum Institute (API)], construction (state and municipal codes), and operating codes (NASA, NRC, FAA). These codes and standards have often been developed to prevent a repetition of past failures, as well as to guard against potentially new types of failure, as in the case of nuclear reactors. Advances in steel making, nondestructive examination, and analytical procedures have led to a reduction of the material design factor (safety factor) for power boilers and pressure vessels from 4 to 3.5 (2). (Allowable stress values based upon the tensile strength are obtained by dividing the tensile strength by the material design factor.) Today, the reliability of engineered products and structures is at an all-time high, but this reliability often comes with a high cost. In fact, in the nuclear industry, compliance with regulations intended to maximize safety may be so costly as to warrant the taking of a reactor out of service. It is also important for manufacturers to be aware of the state of the art as well as the latest standards. The number of manufacturers of small planes has dwindled because of product liability losses incurred when it was shown that their manufacturing procedures did not meet the current state-of-the-art safety standards. To guard against product failures, a number of firms now are organized in such a way that failure analysis is a line function rather than a staff function, and a member of the failure analysis group has to sign off on all new designs before they enter the manufacturing stage.

II. EXAMPLES OF CASE STUDIES IN FAILURE ANALYSIS

A. Problems with Loads and Design

1. Problems with Wind Loadings The Tay Bridge was a 10,300 foot long single track railroad bridge built in 1878 to span the Firth of Tay in Scotland (3). A portion of the bridge consisted of 13 wrought iron spans, each 240 feet in length and 88 feet above the water, which were supported by cast iron piers. On the fateful day of December 28, 1879, a gale developed with wind speeds up to 75 mph. That evening a passenger train, while making a scheduled crossing, plunged into the Firth, together with the 13 center spans, and 75 passengers and crew members lost their lives.

The subsequent investigation revealed that a major cause of the disaster was that the gale force winds produced lateral forces on the passenger cars that were transmitted to the bridge structure and led to its collapse. Such wind loading had not been properly taken into account in the design stage. This disaster underscored the obvious fact that all potential loading conditions must be considered in order to design safe and reliable structures.

Today, we are much more aware of the importance of wind loading in structural design. Nevertheless, from time to time, problems still arise. For example, the Citicorp Tower in New York City was built in 1977 in accord with the building code, which required calculations for winds perpendicular to the building faces. However, this was a unique structure in that a church occupies one corner of the building site, and the CiticorpTower is built over and around it. In 1978, it was discovered that the building was unstable in the presence of gale-force quartering winds, that is, winds that come in at a 45° angle and hit two sides of the building simultaneously. The building was quickly reinforced to insure its safety in the event of all types of wind loading, and a potential disaster was averted.

An instance where wind loading did result in a spectacular failure was that of the Tacoma Narrows suspension bridge, which failed in 1940 after only four months of service. The bridge, which connected the Olympic peninsula with the mainland of Washington, had a narrow, two-lane center span over a half mile in length. The design was unusual in that a stiffened-girder, which caught the wind, was used, rather than a deep open truss, which would have allowed the wind to pass through. The design resulted in low torsional stiffness and so much flexibility in the wind that the bridge was known as "Galloping Gertie." As the wind's intensity increased to 42 mph, the bridge's rolling, corkscrewing motion also increased, until it finally tore the bridge apart. The ultimate cause of the failure was the violent oscillations, which were attributed to forced vibrations excited by the random action of turbulent winds as well as to the formation and shedding of vortices created as the wind passed by the bridge.

2. Comet Aircraft Crashes In the early 1950s, the Comet aircraft was the first jet transport introduced into commercial passenger service. The plane was so superior to propeller-driven transports that it soon captured a large share of the market for future transport planes. However, not long after coming into service, two planes of the Comet fleet, on climbing to cruise altitude, underwent explosive decompressions of the fuselage (as shown by subsequent investigation), which resulted in the loss of the planes as well as the lives of all aboard. Intensive investigation revealed that these crashes were due to fatigue cracking of the fuselage at regions of high stress adjacent to corners of more-or-less square (rather than round) windows, as shown in Fig. 1-2. The fatigue loading was due to the pressurization and depressurization of the cabin, which occurred in each takeoff and landing cycle. The presence of fatigue cracking was confirmed through study of the fracture surfaces of critical parts of the wreckage. These surfaces were found to contain fractographic markings, which are characteristic of fatigue crack growth (4).

The results of these crashes were significant. First of all, the Comet fleet was grounded and orders for new aircraft were canceled. Secondly, the crashes drew attention to the importance of fatigue crack growth in aircraft structures. Thirdly, it was realized that pressurized fuselages had to be designed so as to avoid catastrophic depressurization in the presence of damage such as fatigue cracking or penetration by debris should an engine explode. As a result of these crashes, significant steps to improve the reliability of aircraft structures were taken in terms of design philosophy, consideration of the effects of fatigue crack growth, and inspection procedures.

As underscored by the Comet crashes, fatigue must be an important consideration in the design of aircraft. Certain components such as turbine blades, which may experience 10^{10} stress cycles over their lifetimes, are designed such that the stresses are well below the fatigue strengths of the materials. The design objective for such components is that fatigue cracks never develop within the design lifetime, for if a crack were to form in a turbine blade, it would rapidly grow to critical size, and hence periodic inspection would not detect it in time to avert disaster. The situation with respect to the aircraft structure is different. Here cycles are accumulated at a slower rate than in engine components, and if a fatigue crack were to form, the critical size for fracture would be measured in terms of centimeters rather than millimeters, as in the case of a small turbine blade. This means that with proper inspection it is possible to detect fatigue cracks in a structure before they have grown to critical size.

Aluminum alloys are widely used in the construction of aircraft structures. Their high strength-to-density ratio makes them attractive for this application. However, these alloys are characterized by relatively low fatigue strengths. If an aircraft struc-

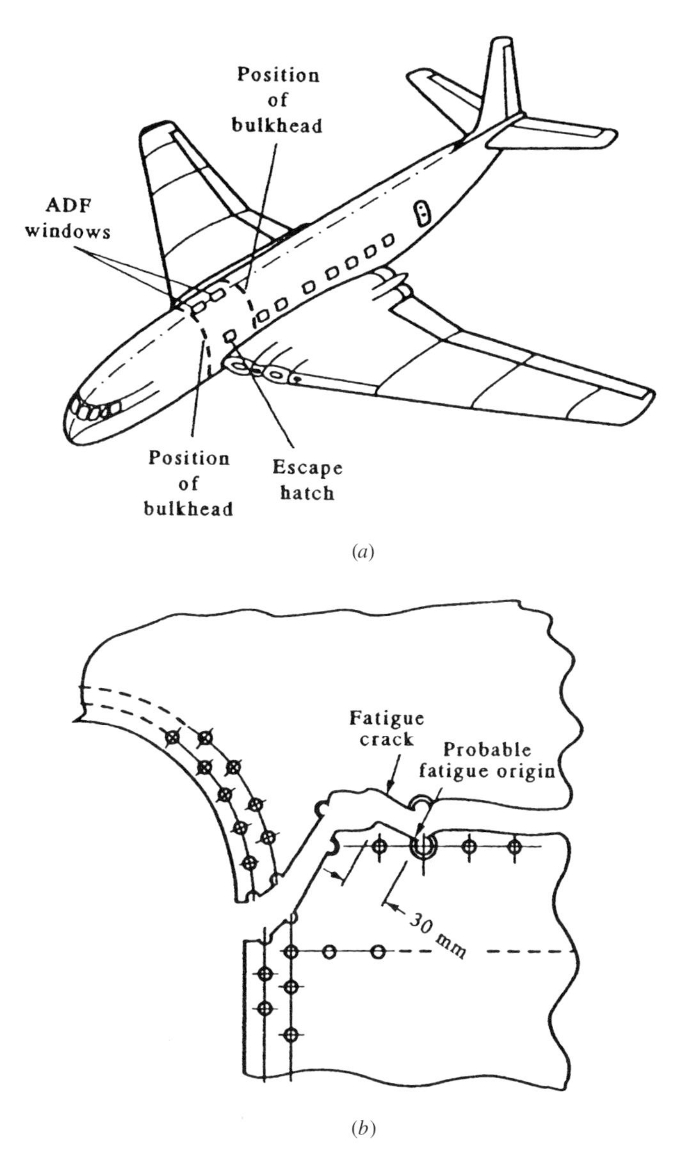

Fig. 1-2. (*a*) The Comet aircraft. (*b*) The location of fatigue cracking near an aft corner of the ADF (automatic direction finder) window. (After Jones, 3, reprinted by permission.)

ture were to be designed such that all repeated stresses were below the fatigue limit, the aircraft would be too heavy for economical flight. To reduce the weight of the structure, the design cyclic stresses are set at levels above the fatigue strength in what is referred to as the finite life range. This means that if the cyclic stresses are repeated often enough, fatigue cracks would eventually develop. Because of the statistical variation in fatigue lifetimes, as well as uncertainty with respect to the actual loading conditions, the designer must consider the possibility that fatigue cracks may appear within the lifetime of the structure. If cyclic tests are carried out on full-scale prototypes, the results will provide some knowledge of the fatigue strength of the structure as well as information about where fatigue cracks are likely to be located. However, actual structures in service may experience different cyclic loading conditions than the prototype, and in addition, as in the case of aging aircraft, long-time effects associated with corrosion and fretting-corrosion may take place, effects that would not have been reflected in the prototype tests.

As mentioned earlier, two different design approaches have been developed in order to deal with the problem of fatigue cracking in aircraft structures. When the structure is designed to be statically determinant, a safe-life design approach is used. In this approach, the components of the structure are designed to have sufficient fatigue life to exceed the design lifetime of the aircraft, but inspections for fatigue cracks are required to insure the safety of the aircraft structure. The other approach is known as fail-safe. In this approach, there is sufficient redundancy in the structure such that if a structural component failed, other structural members would have enough strength to carry the redistributed load. Further, these now more highly stressed surviving members should themselves not be in danger of failing prior to the next scheduled inspection. In principle this approach is more reliable than safe-life, but it entails a weight penalty.

3. Dan Air Boeing 707 Crash (5) The following case study illustrates an instance where the fail-safe approach did not work out as planned. In 1977, a Boeing 707-300C aircraft on a scheduled cargo flight from London to Zambia was preparing to land when the right horizontal stabilizer and elevator separated in flight, causing the aircraft to pitch rapidly nose down and dive into the ground about two miles short of the runway. The pilot, copilot, and flight engineer were killed. This plane was the first off the B-707-300C series convertible passenger/freighter production line, and had accumulated a total of 47,621 airframe hours and had made a total of 16,723 landings. It had made 50 landings since its last inspection. The horizontal stabilizer, as well as other components of this aircraft, had been designed using the fail-safe approach, but full-scale fatigue testing of the B-707-300C stabilizer had not been done.

However, a fail-safe design is only fail-safe if after the failure of one component the remaining components have sufficient residual strength to support the applied loads. A singly redundant structure (as in this case) is only fail-safe while the primary structure is intact. Once this has failed, the principle of safe-life obtains, and it becomes necessary to find the failure in the primary structure before the fail-safe members themselves can be weakened by fatigue, corrosion, or any other mechanism. Because the strength reserves in the fail-safe mode are usually well below those of the intact structure, this means that, in practice, the failure must be found and appropriate action taken within a short time compared with the normal life of the structure. In order to maintain the safety of a fail-safe structure, an adequate inspection program must be an integral part of the total design to insure that a failure in any part of the primary structure is identified well before any erosion of the strength of the fail-safe structure can occur.

Postaccident examination of the detached stabilizer revealed a failure of the top chord of the rear spar of the stabilizer due to the growth of a fatigue crack from a fastener hole. (The word chord has two different meanings in aircraft structural terminology. It is defined as the straight line joining the leading and trailing edges of an airfoil, and also as either of the two outside members of a truss connected and braced by web members. The latter definition is applicable here, i.e., the chords of the stabilizer ran in the span-wise direction.) The rear spar consisted of a top chord, a middle chord, and a bottom chord, which were joined by an aluminum web. The purpose of the nominally unstressed middle chord was to act as a crack arrestor in the event that a fatigue crack propagated in the rear spar web from the top chord. There was evidence that the fracture of the web between the upper chord and the center chord had also failed prior to the crash. There was some fatigue cracking of the center chord, and both the center chord and lower chord had failed due to overload. This was not an isolated case, for a survey of 521 B-707 aircraft equipped with this type of horizontal stabilizer revealed that 7% had rear spar cracks of varying sizes.

The investigation was directed at the establishment of (a) the reason for and age of the fatigue failure, and (b) the reason why the fail-safe structure in the rear spar had failed to carry the flight loads once the top chord had fractured as a result of fatigue. The examination indicated that the total number of flights between the initiation of the fatigue crack and final failure of the upper chord was on the order of 7200. The study concluded that additional fatigue crack growth had occurred after the top chord failure, and that there were probably up to 100 flights between top chord failure and stabilizer separation.

The recommended time to be spent in inspecting the horizontal stabilizer was of such a duration, 24 minutes, as to suggest that a visual inspection rather than a more detailed examination was intended. The rear top and bottom spar chords had been designed to permit them to be inspected externally, and the recommended inspection should have been adequate to detect a crack in the top chord provided the crack was reasonably visible. It was known from those cracks detected as a result of the postaccident fleet inspection that partial cracks on the top chord, although visible to the naked eye when their precise location was known, were for all practical purposes undetectable visually. The recommended inspection could not therefore detect the crack in the spar chord unless the inspection occurred during the interval between top chord severance and total spar failure, which was not so in this case.

The investigators concluded that following the failure of the stabilizer rear spar top chord, the structure could not sustain the flight loads imposed upon it long enough to enable the failure to be detected by the then existing inspection schedule. Although the manufacturer had designed the horizontal stabilizer to be fail-safe, in practice it was not, because of the inadequacy of the inspection procedure. The inspections were not adequate to detect partial cracks in the horizontal stabilizer rear spar top chord, but would have been adequate for the detection of a completely fractured top chord.

Horizontal stabilizers remain prone to fatigue. A British Concorde was recently grounded when a growing fatigue crack in the left rear wing spar had propagated to 76 mm (6).

4. Hartford Coliseum Roof Collapse The roof of this three-year-old structure collapsed at 4:00 am on January in 1978 during a freezing rainstorm after a period of snow. A triangular lattice steel space grid, 360 feet by 300 feet, supported on four reinforced concrete pylons giving spans of 270 feet and 210 feet, was used to support the roof. Smith and Epstein (7) concluded that the interaction of top chord compression members and their bracing played an important role in the redistribution of load and the eventual collapse. They noted that certain compression members were braced against buckling only in one plane. As loads increased, these members buckled out of plane and redistributed load to other members. Over a period of time, more chords buckled and fewer and fewer members carried the load. This situation worsened until the remaining members were unable to withstand the added stress due to the loads present that night, and the final, sudden collapse took place.

This is an instance primarily of inadequate structural design.

5. Kansas City Hyatt Regency Walkways Collapse (8) On July 20, 1981, two suspended walkways within the atrium area of the Hyatt Regency Hotel in Kansas City, MO, collapsed, leaving 113 people dead and 186 people injured. In terms of loss of life and injuries, this was the most devastating structural collapse ever to take place in the United States. The second floor walkway was suspended from the fourth floor walkway, which was directly above it. In turn this fourth walkway was suspended from the atrium roof framing by three pairs of hanger rods. In the collapse, the second and fourth floor walkways fell to the atrium floor, with the fourth floor walkway coming to rest on top of the lower walkway. Most of those killed or injured were either on the first floor level of the atrium or on the second floor walkway.

As originally approved for construction, the plans for the walkways called for the hanger rods to pass through the fourth floor box beams and on through the second floor box beams. The box beams were made up of a pair of 8-inch steel channels with the flanges welded toe-to-toe. The beams were to rest on hanger-rod washers and nuts below each set of beams, Fig. 1-3*a*. Under this arrangement, each box beam would separately transfer its load directly into the hanger rods.

However, during construction, drawings were prepared by the steel fabricator that called for discontinuous rather than continuous hanger rods, Fig. 1-3*b*. In this modified design, three pairs of hanger rods extended from the fourth floor box beams to the roof framing, and three pairs of hanger rods extended from the second floor box beams to the fourth floor box beams. Under this arrangement, all of the second floor walkway load was first transferred to the fourth floor box beams, where both that load and the fourth floor walkway load were transmitted through the box beam hanger rod connections to the ceiling hanger rods. This change essentially doubled the load to be transferred by the fourth floor box beam-hanger rod assembly connections.

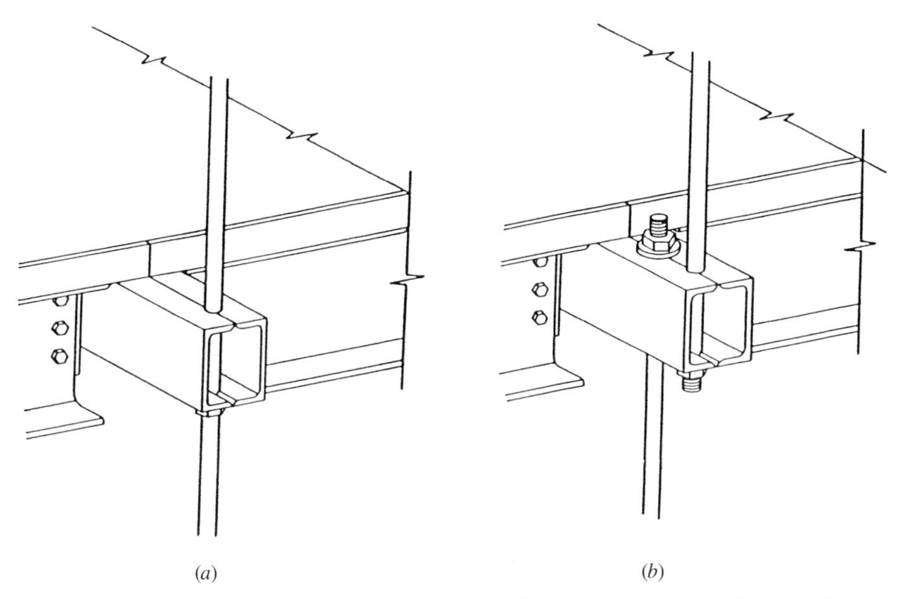

Fig. 1-3. A comparison of the Kansas City Hyatt walkway connectors. (*a*) As originally designed. (*b*) As built. (From National Bureau of Standards, 8.)

Postcollapse failure analysis indicated that the failure of the walkway system initiated in one of the box beam-hanger rod connections. In this instance, the fabricator, structural engineer, and the architect, each of whom had approved the design change, had not appreciated the consequences of the design change.

B. Problems with Inspection, Maintenance, and Repair

6. Mianus River Bridge Failure Demers and Fisher (9) provide a description of the collapse of a portion of this bridge. A six-lane interstate highway supported by six sets of piers, which are skewed to run parallel to the river, spans the Mianus River in Greenwich, CT. The bridge is composed of a number of individual spans, each supported on the outer edges by longitudinal girders. The bridge had been in service for 24 years when, on June 28, 1983, in the early hours of the morning—fortunately an hour when traffic was light—one of the eastbound spans completely separated from the bridge and fell to the river below, causing several fatalities. The span that failed had been suspended, as indicated in Fig. 1-4*a*, between adjacent spans that were cantilevered out from supporting piers. The failed span was a statically determinant structure, which meant that the failure of one main structural member of a span would lead to collapse of that span. Recall that a redundant structure is one in which failure of a structural component leads to a redistribution of loads to other members, but not complete collapse. The span that failed employed pin-and-hanger assemblies at its eastern corners to connect the girders of adjacent

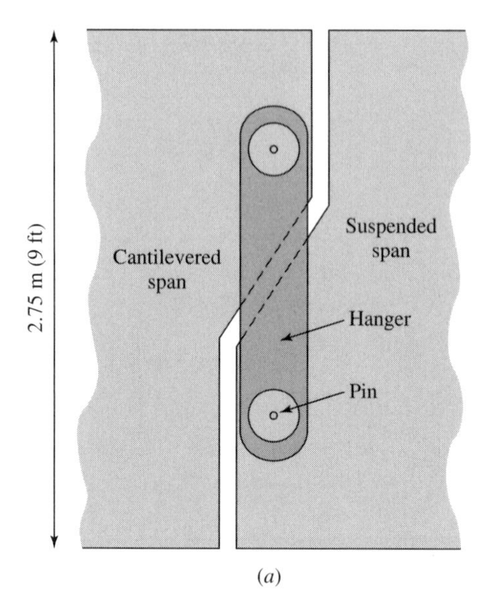

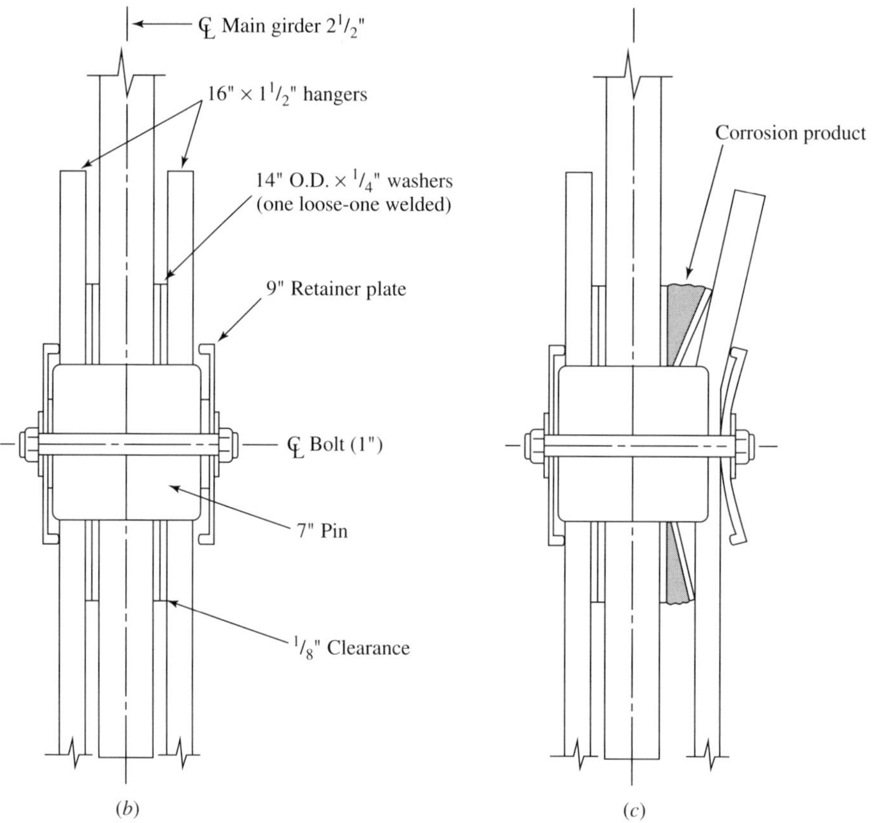

Fig. 1-4. The Myannis River bridge hangers. (*a*) Method of supporting failed suspended span. (*b*) Pin-hanger assembly as built. (*c*) Pin-hanger assembly after 24 years of service. (Fig. 1-4 b and c, after Demers and Fisher, 9.)

components of the bridge. Collapse started at the southeast corner as deduced from the postfailure position of the span. After the collapse, the southeast corner inside hanger was found to be straight and attached to the upper pin, whereas other connectors had been severely deformed. It was concluded that since the inside hanger was straight, the lower pin had separated from it prior to the collapse and had moved in the direction of the outer hanger. This unloading of the inner hanger doubled the load on the outer hanger. The resulting high bearing pressure at the upper surface of the upper pin led to the formation of a fatigue crack in the pin, which caused a portion of the upper pin to separate from the pin, thereby allowing the hanger to slip off the upper pin to bring about the final collapse of the span.

Postaccident inspection revealed that the bearing surface of the southeast corner inside hanger at the lower hole was severely corroded. The inside end of the lower pin was severely corroded and tapered, and the bottom edge had broken off. Movement of the lower pin required the failure of the restraining bolt through the pin. Extensive corrosion packout (compare Figures 1-4*b* and 1-4*c*) between the outer washers was found on the outer side of both upper and lower assemblies, which resulted in plastic deformation of the retainer plate and high tensile stresses in the bolt, which led to its fracture. It was concluded that failure was the result of a progressive process that occurred over a period of time, and that corrosion packout was primarily responsible for the hanger displacement on the pin, which led to the collapse.

This failure underscored the importance of maintaining effective corrosion prevention and inspection programs to maintain the integrity of such structures.

7. Aloha Airlines Boeing 737-200 Accident (10) In 1988, a Boeing 737-200 operated by Aloha Airlines, while en route from Hilo to Honolulu, HI, experienced an explosive decompression and structural failure as the plane leveled at 24,000 feet. Approximately 18 feet of the cabin skin and structure aft of the cabin entrance door and above the passenger floor had separated from the airplane, Fig. 1-5. There were 89 passengers and 6 crewmembers on board. One flight attendant was swept overboard and seven passengers and one flight attendant received serious injuries. An emergency landing was made on the island of Maui. As a result of the accident the airplane was damaged beyond repair and was dismantled and sold for scrap.

The B-737 involved had been manufactured in 1969. At the time of the accident it had acquired 35,496 flight hours and 89,680 flight cycles (landings), the second highest number of cycles in the worldwide 737 fleet. Due to the short distance between destinations on some Aloha Airlines routes, the full pressurization of 52 kPa (7.5 psi) was not reached on every flight. Therefore, the number of full pressure cycles was significantly less than 89,680. The plane had also been exposed to warm, humid, maritime air, which promoted corrosion.

Failure was found to have initiated along a fuselage skin longitudinal lap joint that had been "cold bonded." The cold bonding process utilizes an epoxy-impregnated woven "scrim" cloth to join the longitudinal edges of the single-thickness 0.036-inch skin panels together. In addition, the joint contained three rows of

Fig. 1-5. General view showing the damage sustained by the Aloha Airlines 737. (From NTSB, 10.)

countersunk rivets. Fuselage hoop loads were intended to be transferred through the bonded joint, rather than through the rivets, allowing for thinner skin with no degradation in fatigue life. However, early service history with production B-737 airplanes revealed that difficulties were encountered with the bonding process, and it was discontinued after 1972. In order to safeguard those B-737 planes that had been "cold bonded," Boeing issued a number of service bulletins over a period of time directing the attention of operators to the problem of disbonding and providing information on how to check for disbonding using the eddy current nondestructive examination (NDE) method. In 1987, the FAA issued an airworthiness directive (AD) requiring that eddy current inspections of the bonds and repairs, if needed, be carried out in compliance with the Boeing service bulletins. Some of the bonds had low environmental durability, with susceptibility to corrosion. Some areas of the lap joints did not bond at all, and moisture and corrosion could contribute to further disbonding. When disbonding did occur, the hoop load transfer though the joint was borne by the three rows of countersunk rivets. However, the countersinking extended through the entire thickness of the 0.036-inch sheet, which resulted in a knife edge being created at the bottom of the hole, which concentrated stress and promoted fatigue crack nucleation, Fig. 1-6. For this reason, fatigue cracking would be expected to begin in the outer layer of the skin along the lap joint along the upper, more highly stressed, row of rivet holes.

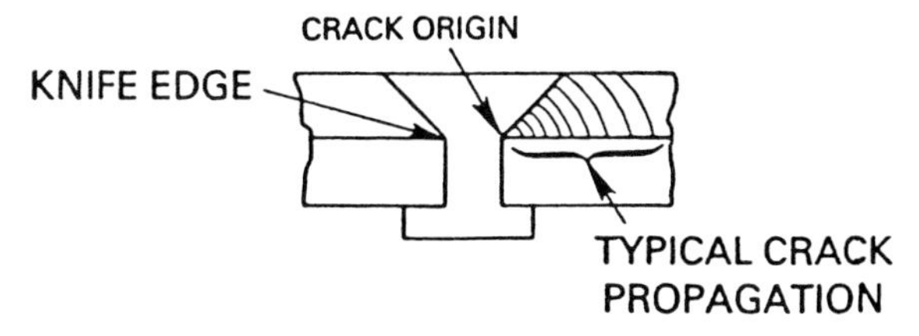

Fig. 1-6. A sketch of a countersunk rivet and an associated fatigue crack observed in Aloha Airlines 737. (From NTSB, 10.)

The NTSB believed that the top rivet row was cracked at the critical lap joint before the accident flight takeoff, and determined that the probable cause of the accident was the failure of the Aloha Airlines maintenance program to detect the presence of the significant disbonding and fatigue damage that ultimately led to the failure of the lap joint and the separation of the fuselage upper lobe. This accident was significant in that it brought attention to some of the corrosion and fatigue problems that could develop in *aging aircraft*. It also focused attention on the problem of multiple-site damage (MSD), that is, the formation and possible linking up of fatigue cracks formed at adjacent rivet holes.

8. Chicago DC-10-10 Crash (11) On May 25, 1979, as American Airlines Flight 191, a McDonnell-Douglas DC 10-10 aircraft, was taking off from the Chicago-O'Hare International Airport, the left engine and pylon assembly separated from the aircraft, went over the top of the wing, and fell to the runway. The plane continued to climb to about 325 feet off the ground and then rolled to the left and crashed. The aircraft was destroyed in the crash and subsequent fire, and the 271 persons on board were killed, as were two others on the ground, the worst loss of life in U.S. aviation history. The accident aircraft had entered service in 1972. It had accumulated a total of 19,871 flight hours, 341 of which had come since a maintenance procedure in Tulsa, OK.

The cause of the separation of the engine from the wing was found to be cracking of the aft bulkhead of the pylon, a problem created during the maintenance procedure in Tulsa. Figure 1-7 shows the pylon assembly. Note that the upper spar is attached to a flange (not shown) on the forward side of the aft bulkhead. This flange turned out to be a critical element in the accident sequence. McDonald-Douglas had issued a service bulletin calling for the replacement of the upper and lower spherical bearings that attached the pylon to the wing. In this procedure, McDonald-Douglas indicated that the 13,477-lb engine was to be removed from the 1865-lb pylon before the pylon was removed from the wing. Procedures for accomplishing this maintenance were also described. However, in contrast to the maintenance procedure advocated by McDonald-Douglas, American Airlines decided to lower and raise the engine and pylon assembly using a forklift-type supporting device, since this procedure would save about 200 man-hours per aircraft and would reduce the

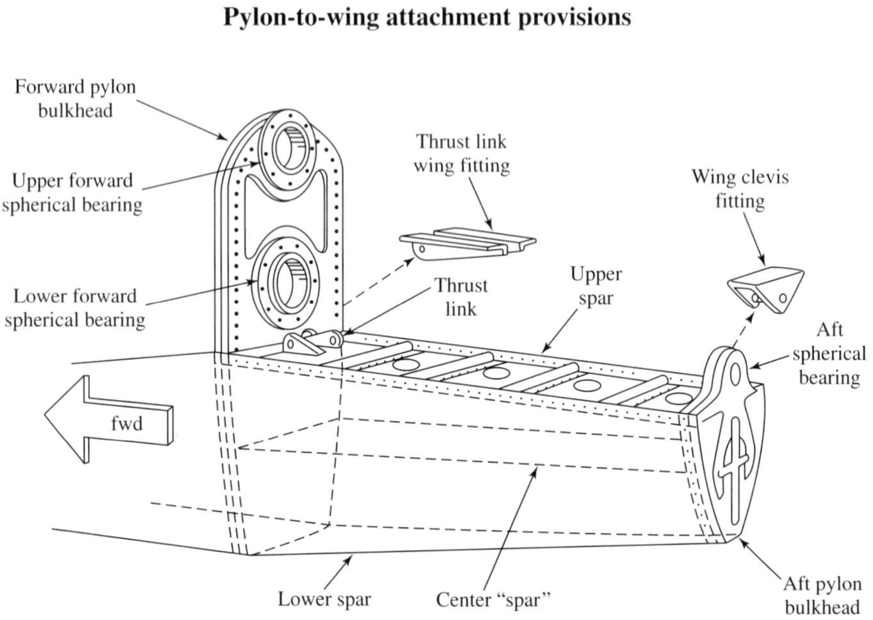

Pylon-to-wing attachment provisions

Fig. 1-7. The pylon assembly of a DC-10. The upper spars and sheet metal are attached to the critical forward flange of the aft bulkhead. (From NTSB, 11.)

number of disconnects from 79 to 27. An engineering change order (ECO) was issued by American Airlines in 1978 prescribing this maintenance procedure, and in March 29 through 31, 1979, the accident aircraft underwent the spherical bearing modification using this procedure. It is noted that McDonald-Douglas had discouraged the use of this procedure because of the risk involved in remating the engine-pylon assembly to the wing attach points, but lacked the authority to either approve or disapprove the maintenance procedures of its customers. Also, members of the American Airlines engineering department did not witness the removal of the wing to pylon attachment assemblies, and consequently, they were not aware of difficulties such as controlling the forklift accurately.

Postaccident investigation revealed that a portion of the upper forward flange of the aft bulkhead had been fractured by overload in the inboard-outboard direction just forward of the radius between the flange and the bulkhead plane. The fracture had been initiated by a downward bending moment at the center section of the flange just forward of the fracture plane due to contact between the clevis and the flange. As a result of this contact, the aft fracture surface of the upper flange was deformed into a crescent shape that matched the shape of the lower end of the wing clevis. The length of this overload fracture was 10 inches. Fatigue cracking was present at both ends of the overload fracture, and the total length of the crack due to both overload and fatigue was 13 inches.

In postaccident inspections of the DC-10 fleet, four American Airlines planes and two Continental Airlines planes were found to have cracked upper flanges on the pylon aft bulkheads, with the longest of these cracks being 6 inches. In addition, it was discovered that two Continental Airlines DC-10s, one in December 1978 and the other in February 1979, had had fractures on their upper flanges. These two flanges had been damaged during this same maintenance operation, but they had been repaired and returned to service. McDonald-Douglas had been informed of these problems, but neither the FAA nor other airlines had been informed because the events were considered to have been maintenance errors.

An examination of the maintenance procedure disclosed numerous possibilities for the upper flange of the aft bulkhead to be brought into contact with the wing-mounted clevis. A fracture-producing load could be applied during or after removal of the attaching hardware in the aft bulkhead fitting. Because of the close fit between the pylon to wing attachments and the minimal clearance between the structural elements, maintenance personnel had to be extremely cautious when they detached or attached the pylon. A minor mistake by the forklift operator could easily damage the aft bulkhead and its upper flange.

The structural separation of the pylon was caused by a complete failure of the forward flange of the aft bulkhead after its residual strength had been reduced by the fracture induced during the maintenance operation as well as by additional fatigue crack growth in service. It is also clear the poor communications between engineering and maintenance personnel, and between the FAA, the manufacturer, and the airlines contributed to this accident.

9. Japan Airlines Boeing 747SR, Crash 1985 (12) In August of 1985 a Japan Airlines Boeing 747SR (short range) jet aircraft was on a flight from Tokyo to Osaka. On climbing through 24,000 feet, the rear pressure bulkhead failed, and as a result, there was an explosive decompression, which led to loss of hydraulic power and of the pilot's ability to control the aircraft. Thirty minutes later the aircraft crashed into a mountain. This was the worst single-plane accident in aviation history, for of the 524 people on board, only 4 survived.

This aircraft had been in a takeoff mishap in June of 1978 in which the tail section struck the runway, causing damage to the lower half of the rear pressure bulkhead. This bulkhead is in the shape of a hemisphere and is made of thin-gauge aluminum alloy sheets. At a joint between sheets the sheets overlap, and an additional piece of sheet material, known as a doubler, spans the riveted joint to provide extra strength. To repair the damage after the 1978 accident, a new lower half bulkhead was riveted to the upper half. However, the two halves were not properly spliced together. On the upper side of the joint, there was a doubler and a stiffener on the inner side of the bulkhead, Fig. 1-8. On the lower side of the joint there was a doubler, but the doubler was not continuous with the upper doubler, so that a gap existed between the doublers, with only the sheet material carrying the load. In addition, the centroid of the load-bearing material was now on the inner side of the bulkhead. Therefore, the load on the sheet spanning the gap consisted not only of that due to the hoop tension, but also that due to bending because of the eccentric loading condition created by the doublers and the stiffener. Each time the cabin was

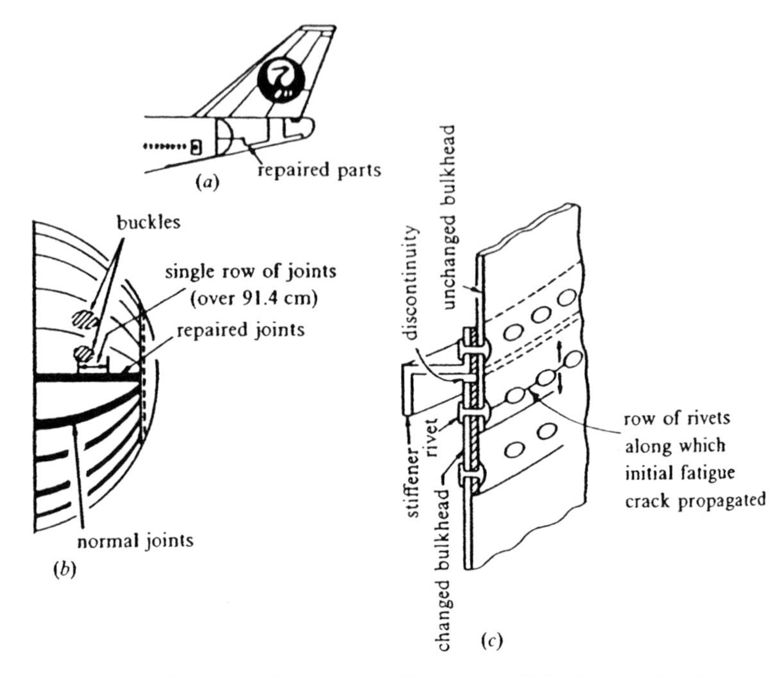

Fig. 1.8. The rear bulkhead of the JAL 747 SR. (After Kobayashi, 12.)

pressurized, there was an increase in stress in the aluminum sheet spanning the gap over that expected. As a result of this stress increase, fatigue cracks were formed at each of the rivet holes on the lower half of the bulkhead just below the gap, another example of MSD. These fatigue cracks eventually linked up, and the resultant long crack led to the explosive decompression.

C. Other Problems

10. Air France Concorde Crash, July 25, 2000 The crash of Air France Flight 4590, a supersonic Concorde (SST), moments after takeoff from the Charles de Gaulle Airport near Paris, resulted in the deaths of 109 people aboard the jet and 5 people on the ground. This accident is currently under investigation, but the preliminary evidence, as well as a past history of similar, but fortunately not catastrophic, events, indicate that the bursting of a tire on the left side of the plane while the plane was accelerating during its takeoff run was critical. A 16-inch piece of metal that fell to the runway from the engine of a plane that had taken off shortly before the Concorde may have caused the tire to burst. The metal strip matched a gash found in one of the Concorde's left tires, and it is probable that this piece caused the cut. There have been 57 cases of burst tires on Concordes, and in 7 instances these bursts have led to the rupture of fuel tanks, the severing of hydraulic lines, and the damaging of engines. French investigators believe that in the Paris crash, after the tire burst, an 8-lb piece of rubber penetrated the fuel tank, thereby releasing a plume of fuel that, being in close proximity to the two left side engines, ignited. Both British and French aviation authorities have grounded the Concordes pending further assessment of their fail-safe capabilities in the event of a tire burst. The addition of fuel-tank liners to minimize damage related to tire bursts is being considered

11. TWA Flight 800 Crash In July of 1996, TWA Flight 800, a Boeing 747, crashed into the Atlantic off Long Island, NY, killing 230 people. A leading theory as to the cause of this accident is that fuel vapors in the empty center fuel tank may have been explosively ignited by a spark between two elements of a terminal strip that was part of a fuel probe. Moments before the explosion, a fuel gauge behaved erratically, an indication of a wiring problem. The potential difference between the elements of the terminal strip was 170 volts, which ordinarily would not be a cause for concern. However, the plane was 25 years old and over time a semiconducting sulfur compound had built up on the terminal strip, which allowed current to flow between the elements. As a result, the compound may have "burned" and triggered a spark that caused the explosive ignition of the fuel vapors.

III. SUMMARY

These eleven examples indicate the range of design, maintenance, environmental, and inspection problems that can arise and endanger the integrity of structures. In the following chapters, the failure mechanisms and investigative procedures are discussed in greater detail, and additional case studies are presented.

REFERENCES

(1) R. F. Harper, The Code of Hammurabi, University of Chicago Press, 1904.

(2) D. A. Canonico, Adjusting the Boiler Code, Mechanical Engineering, vol. 122, no. 2, Feb. 2000, pp. 54–57.

(3) D. R. H. Jones, Engineering Material 3, Material Failure Analysis, Pergamon Press, Oxford, UK, 1993, pp. 291–314.

(4) British Ministry of Transport and Civil Aviation, Civil Aircraft Accident: Report of the Court of Inquiry into the Accidents to Comet G-ALYP on 10th January 1954 and Comet G-ALYY on 8th April 1954, HMSO, London, 1955.

(5) Aviation Week and Space Technology, Sept. 17 and Sept. 24, 1979.

(6) Aviation Week and Space Technology, Aug. 7, 2000, p.31.

(7) E. A. Smith and H. I. Epstein, Hartford Coliseum Roof Collapse, Civil Engineering-ASCE, Apr. 1980, pp. 59–62.

(8) Investigation of the Kansas City Hyatt Regency Walkways Collapse, National Bureau of Standards Building Science Series 143, 1982.

(9) C. E. Demers and J. W. Fisher, A Survey of Localized Cracking in Steel Bridges, 1981 to 1988, ATLSS Report No. 89-01, Lehigh University, Bethlehem, PA, 1989.

(10) Aloha Airlines Flight 243, National Transportation Safety Board Aircraft Accident Report, NTSB AAR-89/03, 1989.

(11) American Airlines Flight 101, National Transportation Safety Board Aircraft Accident Report, NTSB-AAR-79-17, 1979.

(12) H. Kobayashi, On the Examination Report of the Crashed Japan Airlines Boeing 747 Plane; Failure Analysis of the Rear Pressure Bulkhead (in Japanese), J. Japan Soc. Safety Eng., vol. 26, 1987, pp. 363–372.

2

Elements of Elastic and Plastic Deformation

I. INTRODUCTION

In a failure analysis, the determination of the nature and magnitude of the stresses that had been developed in a failed component is often a significant aspect of the investigation. This chapter briefly reviews some basic definitions, constitutive relations, principal stresses, Mohr circles, plane stress and plane strain, yield criteria, and the state of stress within a plastic zone ahead of a blunt notch.

II. DEFINITIONS OF ENGINEERING AND TRUE STRESS AND STRAIN; POISSON'S RATIO (1–3)

Figure 2-1 depicts a tetrahedron with a stress σ acting on the inclined face whose area is A_0. With reference to this figure, the stress σ is defined as the load P acting parallel to σ divided by A_0:

$$\sigma = \frac{P}{A_0}, \tag{2-1}$$

the normal stress σ_n acting on the plane A_0 is defined as

$$\sigma_n = \frac{P_n}{A_0}, \tag{2-2}$$

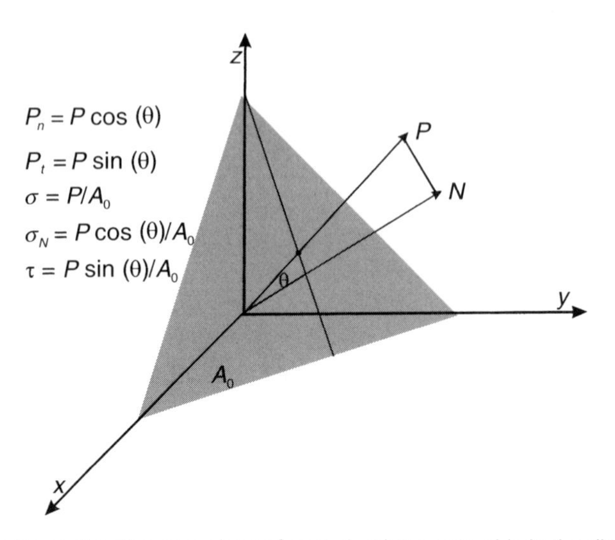

Fig. 2-1. Force P acting on a plane of area A_0 whose normal is in the direction N.

and the tangential (shear stress) τ is defined as

$$\tau = \frac{P_t}{A_0}. \tag{2-3}$$

The true normal stress σ_{tn} takes into account the change in area with strain and is defined as

$$\sigma_{tn} = \frac{P_n}{A}, \tag{2-4}$$

where A is the actual area. For a strain of 10%, the true stress exceeds the engineering stress by about 5%, and therefore, the difference between the two for strains less than 10% is not significant.

Figure 2-2 shows the stresses acting on the face of a cube. The stresses shown are all positive in the sense that they act parallel to the positive direction of an axis on a positive face of the cube and in the reverse direction on a negative face. The positive face is that one of a pair of parallel faces that intersects the axis at the more positive value. The first subscript denotes the axis intersected by the face on which a stress acts, and the second subscript denotes the direction in which the stress acts.

The engineering normal strain ε is defined as

$$\varepsilon = \frac{\Delta l}{l_0}, \tag{2-5}$$

where l_0 is the original length and Δl is the change in length.

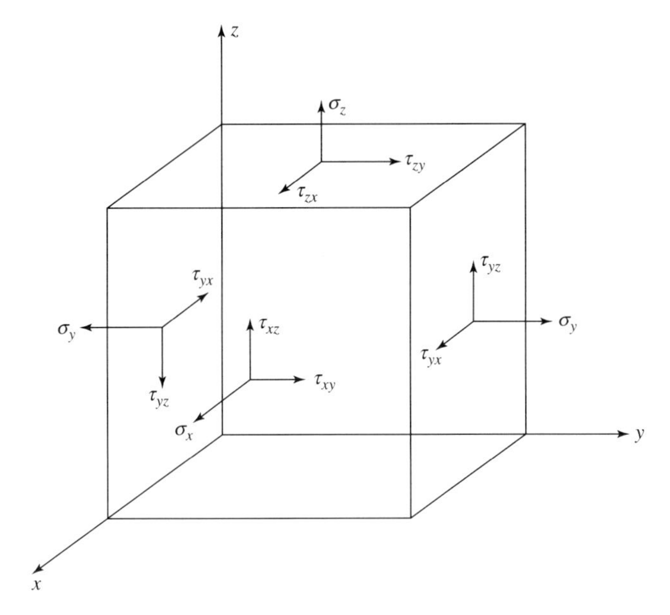

Fig. 2-2. Stresses acting on a cubic element.

The true normal strain ε_t is defined as

$$\varepsilon_t = \ln \frac{l}{l_0} = \ln(1 + \varepsilon_{eng}). \tag{2-6}$$

For $\varepsilon_{eng} = 0.1$, $\varepsilon_t = 0.095$, and as with the definitions of stress, there is relatively little difference in the definitions of strain, for strains of less than 10%.

The shear strain γ is defined as

$$\tan \gamma = \frac{\Delta}{h}, \tag{2-7}$$

where Δ and h are defined in Fig. 2-3. Since γ is a small angle, $\tan \gamma$ can be replaced by γ, where γ is in radians, that is,

$$\gamma = \frac{\Delta}{h}. \tag{2-7a}$$

Poisson's ratio μ is defined for unidirectional loading as minus the ratio of the transverse strain to the longitudinal strain, that is,

$$\mu = -\frac{\varepsilon_{lat}}{\varepsilon_{long}}, \tag{2-8}$$

where ε_{lat} is a transverse strain and ε_{long} is the longitudinal strain.

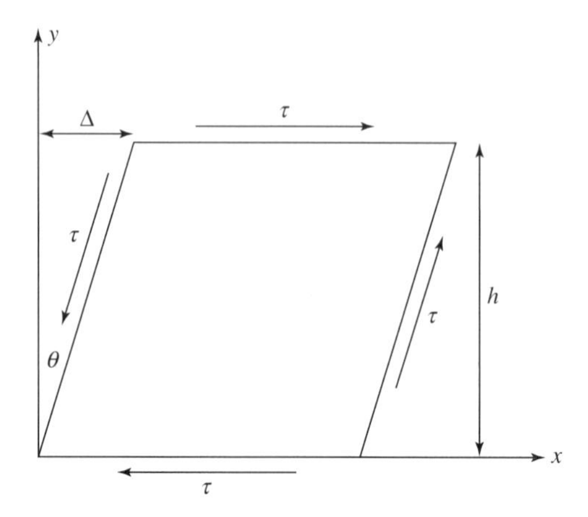

Fig. 2-3. Shear displacement Δ and shear strain Δ/*h*.

III. THE NUMBER OF INDEPENDENT STRESS AND STRAIN COMPONENTS

In Fig. 2-2, there appear to be a total of nine different stresses. However, if the moment of forces is taken about the x-axis, for example, and the result set equal to zero for equilibrium, it can be shown that $\tau_{yz} = \tau_{zy}$. Therefore, the order of the subscripts can be interchanged, and the number of independent stress components is reduced to six. This is also true of the number of independent strain components.

IV. CONSTITUTIVE RELATIONSHIPS FOR ISOTROPIC, HOMOGENEOUS, AND CONTINUOUS MEDIA

$$\text{Young's modulus} = E = \frac{\sigma}{\varepsilon}, \tag{2-9}$$

$$\text{the bulk modulus} = K = \frac{p}{\Delta} = \frac{E}{3(1-2\nu)}, \tag{2-10}$$

where p is the hydrostatic pressure and Δ is the dilatation, which is given by

$$\Delta = \frac{\Delta V}{V} = \varepsilon_x + \varepsilon_y + \varepsilon_z, \tag{2-11}$$

where V is the volume.

$$\text{The shear modulus} = \text{G} = \frac{\tau}{\gamma} = \frac{E}{2(1+\nu)}. \tag{2-12}$$

Note that these relationships involve only two independent material constants, E and ν.

The hydrostatic component of stress σ_{hyd} is given by

$$\sigma_{hyd} = \frac{\sigma_x + \sigma_y + \sigma_z}{3}. \tag{2-13}$$

The total stress tensor can be separated into two components: one is the hydrostatic component, which is related to the change in volume, and the other is the distortional component, also known as the stress deviator tensor, which is related to the change in shape. The following example shows how a simple tensile stress can be divided into hydrostatic and distortional components:

$$\begin{pmatrix} \sigma_{xx} & 0 & 0 \\ 0 & 0 & 0 \\ 0 & 0 & 0 \end{pmatrix} = \begin{pmatrix} \frac{\sigma_{xx}}{3} & 0 & 0 \\ 0 & \frac{\sigma_{xx}}{3} & 0 \\ 0 & 0 & \frac{\sigma_{xx}}{3} \end{pmatrix} + \begin{pmatrix} \frac{2\sigma_{xx}}{3} & 0 & 0 \\ 0 & -\frac{\sigma_{xx}}{3} & 0 \\ 0 & 0 & -\frac{\sigma_{xx}}{3} \end{pmatrix}. \tag{2-14}$$

V. PRINCIPAL STRESSES

There are three principal stresses σ_1, σ_2, and σ_3, and they act on planes on which $\tau = 0$. In designating these stresses, the usual convention is to take $\sigma_1 > \sigma_2 > \sigma_3$ in the algebraic sense, that is, if the three principal stresses were of magnitude $+10$, $+20$, and -30, then σ_1 would equal $+20$, σ_2 would equal $+10$, and σ_3 would equal -30.

The maximum value of the shear stress acts on a plane whose normal is at 45° to the σ_1 and σ_3 principal stress directions. The magnitude of this shear stress is given by $(\sigma_1 - \sigma_3)/2$. On the plane whose normal is at 45° to the σ_2 and σ_1 principal stress directions, the shear stress is given by $(\sigma_2 - \sigma_1)/2$, and on the plane whose normal is at 45° to the σ_3 and σ_2 directions, the shear stress is given by $(\sigma_3 - \sigma_2)/2$. Note also that to obtain the normal stresses acting on the 45° planes, simply replace the minus sign in the above relations with a plus sign. These relationships are easily visualized by means of Mohr circles, which are discussed in the next section.

VI. MOHR CIRCLES

The state of stress at a point can be conveniently depicted by means of a Mohr circle. The axes for the Mohr circle plot are the shear stress τ (ordinate) and the normal stress σ (abscissa). Recall that the angle between two directions in the plot is twice the actual angle in the stressed body. Examples of Mohr circles for tension,

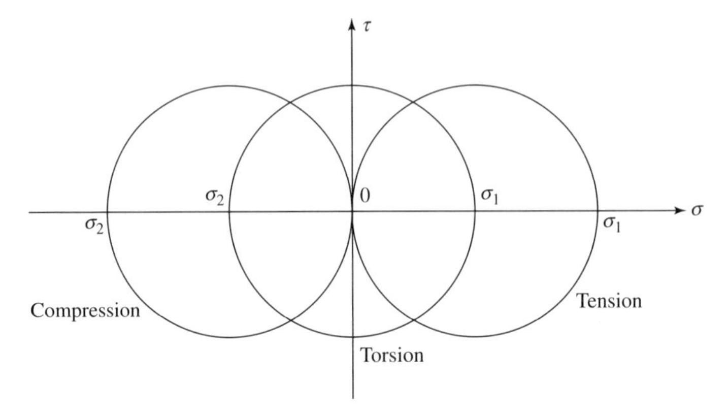

Fig. 2-4. Mohr circles for tension, compression, and torsion.

compression, and torsional loading are given in Fig. 2-4. The stresses acting on an element under pure shear (torsion) are shown in Fig. 2-5. The shear stresses shown are defined as positive when they act in the positive direction on a positive face and in the negative direction on a negative face (the face further along the positive direction of the x- or y-axis is the positive face). Since all the shear stresses shown are positive, a question arises as to how to plot them on the Mohr circle diagram. Note that the shear stress τ_{yx} acts on a plane whose normal is 45° counterclockwise from the first principal direction σ_1. On the Mohr circle, this shear stress is therefore plotted at 90° counterclockwise from σ_1, and the shear stress τ_{xy} is plotted at 90° clockwise from σ_1.

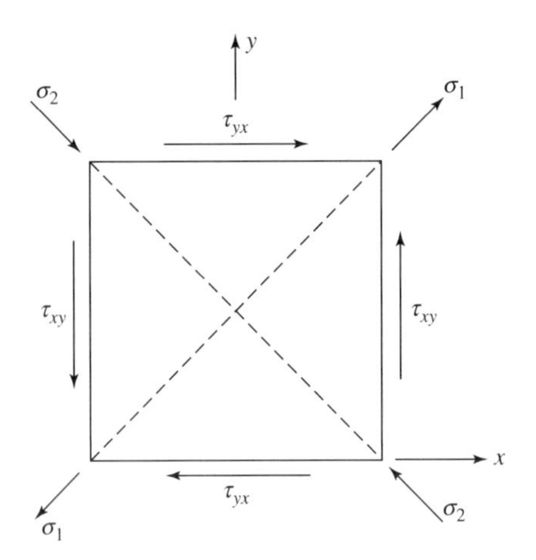

Fig. 2-5. Stresses in pure shear.

VII. PLANE STRESS AND PLANE STRAIN

Under tensile loading, the thickness of an unnotched or uncracked plate has no effect on the state of stress in the plate. However, when a notch or crack is present in the plate, the state of stress is affected by the thickness, as will be discussed.

It is clear that, if the number of independent stress components can be reduced from six to a smaller number, the analysis will be simpler. For example, if the independent components of stress lie only in one plane, there are only three independent stresses, that is, two normal stresses and one shear stress. There are two types of in-plane situations: one is known as plane stress, and the other is known as plane strain. In the analysis of these in-plane situations, the x- and y-axes will be taken in the plane, and the z-axis will therefore be perpendicular to the plane.

In *plane stress*, $\sigma_{zz} = \tau_{xz} = \tau_{yz} = 0$, that is, the components of stress in the direction normal to the plane are zero. The three remaining components of stress are σ_{xx}, σ_{yy}, and τ_{xy}. In *plane strain*, $\varepsilon_{zz} = \gamma_{xz} = \gamma_{yz} = 0$, that is, the components of strain in the z-direction are zero.

In plane strain, there are also three independent components of stress in the plane. However, a σ_{zz} stress is also present because of a constraint on both elastic and plastic deformation in the z-direction. Consider a plate containing a hole of radius a, which is small compared to the width of the plate. The plate is elastically loaded in tension with the y-axis in the direction of loading, and the x-axis transverse to the direction of loading. At the notch root (coordinates 0, a), σ_{yy} is the principal stress σ_1 and is equal to $K_T\,\sigma$, where K_T is the stress concentration factor, in this case equal to 3.0, and σ is the applied stress. Because of the Poisson effect, the material at the notch root will try to contract in the z- (thickness) direction. However, the adjacent, less heavily stressed material surrounding the root of the notch will not contract as much and will exercise a constraint on the ability of the most highly stressed material to contract. This constraint leads to a tensile stress σ_{zz} being developed in the z-direction. This stress must be zero at the surface of the plate but quickly rises up to a constant value.

Similarly, material in the x-direction at the root of the notch will try to contract, but again a constraint will be developed in the x-direction, which results in a tensile stress σ_{xx}. In this case, the σ_{xx} stress must be zero at the notch, but rises up to a maximum in the x-direction before falling off. In the case of a hole in a plate, the maximum value of this stress is 0.375 of the applied stress at a distance measured from the center of the hole equal to $\sqrt{2}$ times the radius. It is noted that the solutions for in-plane stresses in plane-stress and plane-strain problems do not depend upon the material constants E and ν. The σ_{xx} stress also leads to a constraint being developed in the z-direction, and the total resultant stress in the z-direction is given by $\nu(\sigma_{xx} + \sigma_{yy})$.

These z- and x-direction constraints transform the uniaxial tensile stress into a triaxial tensile stress state at the notch. It is important to note that the hydrostatic stress ahead of the notch is higher in plane strain than it is in plane stress, because this state of stress has a deleterious effect on fracture resistance.

VIII. ELASTIC STRESS-STRAIN RELATIONS

The following expressions are used to determine the strains when the stresses are known:

$$\varepsilon_x = \frac{1}{E}[\sigma_x - \nu(\sigma_y + \sigma_z)],$$

$$\varepsilon_y = \frac{1}{E}[\sigma_y - \nu(\sigma_2 + \sigma_x)],$$

$$\varepsilon_z = \frac{1}{E}[\sigma_z - \nu(\sigma_x + \sigma_y)]. \quad (2\text{-}15)$$

The following expressions are used to determine the stresses when the strains are known:

$$\sigma_x = \frac{\nu E}{(1+\nu)(1-2\nu)}(\varepsilon_x + \varepsilon_y + \varepsilon_z) + \frac{E}{1+\nu}\varepsilon_x,$$

$$\sigma_y = \frac{\nu E}{(1+\nu)(1-2\nu)}(\varepsilon_x + \varepsilon_y + \varepsilon_z) + \frac{E}{1+\nu}\varepsilon_y,$$

$$\sigma_z = \frac{\nu E}{(1+\nu)(1-2\nu)}(\varepsilon_x + \varepsilon_y + \varepsilon_z) + \frac{E}{1+\nu}\varepsilon_z. \quad (2\text{-}16)$$

For plane stress, with $\sigma_z = 0$, the values of σ_x and σ_y are given by

$$\sigma_x = \frac{E}{1-\nu^2}(\varepsilon_x + \nu\varepsilon_y),$$

$$\sigma_y = \frac{E}{1-\nu^2}(\varepsilon_y + \nu\varepsilon_x). \quad (2\text{-}17)$$

IX. PLASTIC DEFORMATION (1–3)

The basic entity involved in plastic deformation is the dislocation, which is also known as a line defect. There are two types of dislocations, the edge and the screw. In metals, plastic deformation results from the motion of dislocations in response to shear stresses. These dislocations move on specific crystallographic planes known as slip planes, and the resultant direction of motion on a slip plane is known as the slip direction. In Fig. 2-6, two parallel slip planes are shown; with the open circles representing atoms in an upper plane and the solid circles representing atoms in the next lower plane. As indicated in Fig. 2-6, all dislocation loops consist of edge and screw components. In this figure, the edge components of the loop run horizontally

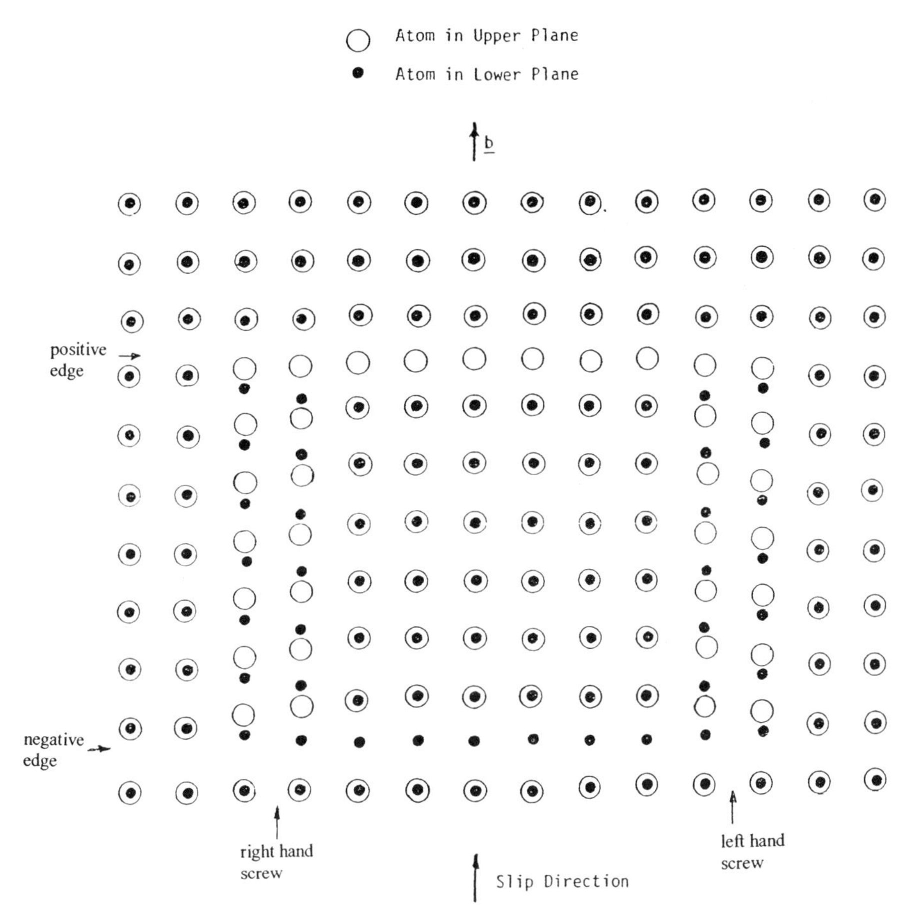

Fig. 2-6. A schematic of a glissile dislocation loop. RHS is the right-hand screw dislocation portion of the loop; LHS is the left-hand screw dislocation portion of the loop.

near the top and bottom of the figure, and the screw components run vertically at the sides of the figure. (An edge dislocation as viewed along the dislocation is also shown in the upper left-hand corner of this figure.) Note that the sense of the upper edge dislocations is opposite to that of the lower, and that the sense of one screw dislocation component of the loop is also opposite to the other. Under the action of a shear stress applied such that the upper plane is sheared upward on the page and the lower plane downward, the dislocation loop will expand. In this process, the atoms along the dislocation loop in the upper and lower planes change their positions slightly. If the direction of shearing were reversed, the dislocation loop would shrink and eventually annihilate itself.

The Burger's vector of a dislocation loop is constant everywhere on the loop, and it is defined by making a circuit around the dislocation at any location. To make this circuit (a) start at an upper atom adjacent to the loop and move a fixed number of atoms across the dislocation line, then (b) go down one atom to the lower plane and count off the same fixed number of atoms in the bottom plane going in the opposite direction as in (a). Then return to the upper plane. The distance and direction then needed to return to the atom at which the circuit began is known as the Burger's vector. The slip direction is parallel to the Burger's vector, and when a dislocation loop has progressed completely through a crystal, the upper plane will have been shifted by one Burger's vector with respect to the lower plane.

The crystallographic planes and directions are identified by their Miller indices. The Miller index for a plane in the cubic system is given as the reciprocals of the intercepts of the plane with the crystallographic axes, reduced to the smallest integers and written in parentheses. A (111) plane is shown in Fig. 2-7. A direction is specified in terms of its components along the cube axes, again reduced to the smallest whole numbers and written in brackets. A $[1\ \bar{1}\ 0]$ direction is also shown in Fig. 2-7. The combination of a slip plane and a slip direction is known as a slip system, and in face-centered cubic (fcc) crystals there are 12 of them, that is, four planes with three directions per plane, each of the $(111)[1\ 0\ \bar{1}]$ type. The four planes comprise the faces of a regular tetrahedron, with the slip directions being the edges of the tetrahedron. It is generally considered that at least five slip independent slip systems must be available in order to carry out complex plastic deformation processes such as the formation of a neck during tensile straining. The fact that 12 independent systems are available in fcc crystals is in accord with the generally high ductility of such crystals, even at very low temperatures. However, ionic single crystals have only two independent slip systems. Such crystals may extend plastically by as

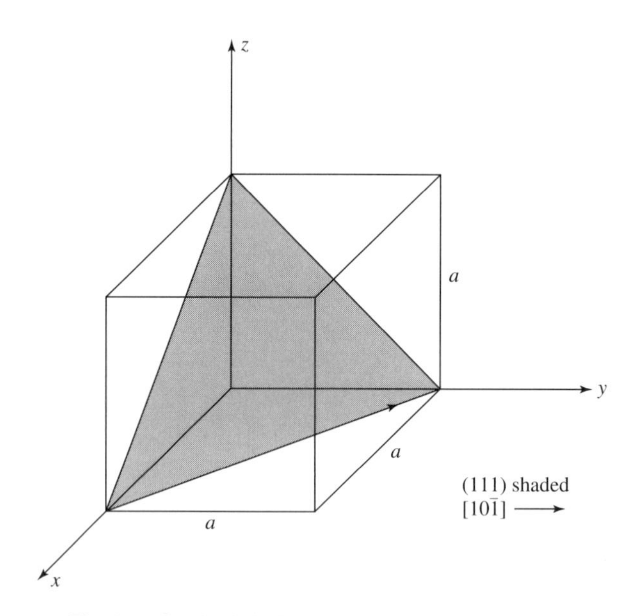

Fig. 2-7. Octahedral plane (shaded) and slip direction.

much as 10%, but when failure occurs, it does so in a brittle manner by cleavage on {100} planes. The low-temperature brittle behavior of other materials, such as zinc crystals and beta-nickel-aluminides, is similarly related to the availability of less than five independent slip systems. However, at higher temperatures, slip in these materials can take place on additional crystallographic planes, and the materials then behave in a ductile fashion, and a brittle-to-ductile transition temperature can be defined.

The resolved shear stress in a crystal is the shear stress acting on the slip plane in the slip direction of the crystal. The onset of plastic deformation in a tensile test of a single crystal often occurs at a critical value of the shear stress that is independent of the orientation of the crystal with respect to the tensile axis. For a cylindrical single crystal of cross-sectional area, if ϕ is the angle between the specimen axis and the slip-plane normal, then the area of the slip plane is given as $A/\cos\phi$. If λ is the angle between the specimen axis and the slip direction, then the component of axial load P acting in the slip direction is $P\cos\lambda$. The resolved shear stress τ_{res} is given as

$$\tau_{res} = \frac{P\cos\lambda}{A/\cos\phi} = \sigma\cos\phi\cos\lambda, \tag{2-18}$$

where σ is the tensile stress. In a tensile test of a single crystal, when plastic deformation occurs, slip steps are created on the surface due to the sliding action of adjacent parallel planes. Usually, a group of closely spaced, parallel planes are involved, so that instead of a single step, a group of steps, referred to as a slip band, is created. A number of these slip bands will be present on the surface of the crystal, with the number increasing with increase in strain.

Miller indices are not only useful in describing crystallographic planes and directions, they are also useful in carrying out calculations involving single crystals. For example, the spacing d between crystallographic planes in the cubic system is given as

$$d = \frac{a}{\sqrt{h^2 + k^2 + l^2}},$$

where a is the lattice parameter, that is, the length of the cube edge of the unit cell, and h, k, and l are the Miller indices of the plane. As another example, the cosine terms in Eq. 2–18 can be expressed in terms of Miller indices as follows: Let:

$[h_1k_1l_1]$ represent the Miller indices of the direction of loading,
$[h_2k_2l_2]$ represent the Miller indices of the direction of the slip plane normal, and
$[h_3k_3l_3]$ represent the Miller indices of the slip direction.

Then

$$\cos\phi = \frac{1}{\sqrt{h_1^2 + k_1^2 + l_1^2}\sqrt{h_2^2 + k_2^2 + l_2^2}}(h_1h_2 + k_1k_2 + l_1l_2),$$

and

$$\cos \lambda = \frac{1}{\sqrt{h_1^2 + k_1^2 + l_1^2}\sqrt{h_3^2 + k_3^2 + l_3^2}}(h_1 h_3 + k_1 k_3 + l_1 l_3).$$

Such relationships are important in analyzing the behavior of single-crystal turbine blades, for example. However, in dealing with plastic deformation of polycrystalline alloys, a macroscopic, continuum approach is usually used in which the behavior of the many individual crystals involved is averaged out. The tensile test provides basic data for characterizing a material's mechanical behavior. The information to be obtained in such a test is the Young's modulus, the yield strength, the ultimate tensile strength, the percentage of elongation in a specified gauge, and the percentage of reduction in area. In a test of a low-carbon steel, an upper and lower yield level will be evident, and it is usual to take the lower yield level in specifying the yield strength. The phenomenon of an upper and lower yield strength in low-carbon steel is due to the interaction of carbon with dislocations. This interaction prevents dislocation motion until the upper yield stress level is reached, whereupon the dislocations break away and new dislocations are created free of the restraint of the carbon atoms. Plastic deformation is usually initiated at a strain of the order of 0.001 at the shoulders of the specimen because of slightly higher stresses in this region due to stress concentrations. The plastic deformation then spreads at the lower yield level into the remainder of the test section in a series of parallel bands known as Lüder's bands. At a strain of the order of 0.01, the test section will be completely filled with these parallel bands, and the end of the lower yield level behavior is reached. Further straining results in the interaction of dislocations, and as a result, strain hardening develops. The strain hardening is manifested by an increase in the stress required for further deformation until the ultimate tensile stress is reached. Thereafter, the stress drops until the breaking stress is reached, and the specimen fractures. It is noted that three instabilities develop during the course of a tensile test of a low-carbon steel. One occurs at the transition from the upper to the lower yield point, the second occurs at the maximum stress, and the third occurs at the breaking stress. The breaking stress is not a true material property, but is a reflection of the testing machine stiffness that determines its ability to shed load in the latter part of tensile test. Under dead weight loading, for example, a type of loading corresponding to zero stiffness, the breaking stress would coincide with the ultimate tensile stress. On the other hand, in a very stiff machine, it is possible to decrease the breaking stress to very low value.

The upper and lower yield levels characteristic of low-carbon steel are not observed in most other alloys. In order to define a yield strength in such cases, an offset method is used in which a line is drawn parallel to the elastic loading line but offset from it by a specified amount of strain such as 0.002 (0.2%). The yield strength is defined as the point at which this line intersects the stress-strain curve.

In the mathematical treatment of plastic deformation, certain simplifications are made that reduce the complexity of the stress analysis. One basic assumption is that the volume remains constant, that is $\Delta V = 0$. This is a reasonable assumption since the plastic deformation process is a shearing process involving the sliding of one element of material over another without a change in volume being involved. The dilatation Δ is given as

$$\Delta = \varepsilon_{xx} + \varepsilon_{yy} + \varepsilon_{zz}. \tag{2-19}$$

For plastic deformation, Δ is zero, since $\Delta V = 0$. In plane strain with ε_{zz} equal to zero, it follows that $\varepsilon_{xx} = -\varepsilon_{yy}$. There is only one independent component of strain in plane strain plastic deformation, since the diameter of the Mohr circle is given as the difference between these two strains. Similarly, if work hardening is neglected, by setting the maximum shear stress in the plastic region equal to k, a material constant, there remains only one independent stress.

The value of Poisson's ratio for purely plastic deformation can also be evaluated making use of the assumption that Δ is equal to zero. In a tensile test, the lateral strain $\varepsilon_y = \varepsilon_z = -\nu\varepsilon_x$; hence $\Delta V = \varepsilon_x - 2\nu\varepsilon_x$. For constancy of volume, $\Delta V = 0$, and therefore $\nu = \frac{1}{2}$. (Note that in the elastic range where $\nu < \frac{1}{2}$, there is a volume expansion in tension and a volume contraction in compression.)

Therefore, in plane-strain plastic deformation,

$$\sigma_z = \tfrac{1}{2}(\sigma_x + \sigma_y), \tag{2-20}$$

and

$$\sigma_{hyd} = \tfrac{1}{3}[\sigma_x + \sigma_y + \tfrac{1}{2}(\sigma_x + \sigma_y)] = \tfrac{1}{2}(\sigma_x + \sigma_y) = \sigma_z. \tag{2-21}$$

X. YIELD CRITERIA FOR MULTIAXIAL STRESS (1–3)

An excessive amount of plastic deformation can of itself be considered to be a type of failure. For simple tension loading, it is only necessary to keep the tensile stress below the yield stress to avoid problems associated with plastic deformation. However, many components are subjected to multiaxial stresses, and criteria for yielding under such stress states are needed for purposes of design and analysis. This section deals with the two most commonly used yield criteria.

A. Distortional (Shear) Energy Criterion

In simple tension, the distortional energy is given by

$$\frac{(\sigma_1 - \sigma_3)^2}{8G} + \frac{(\sigma_2 - \sigma_1)^2}{8G} + \frac{(\sigma_3 - \sigma_2)^2}{8G} = \frac{\sigma_1^2}{4G}, \quad \text{since } \sigma_2 \text{ and } \sigma_3 = 0. \tag{2-22}$$

At yield in simple tension, $\sigma_1 = \sigma_y$, where σ_y is the yield strength. For a multiaxial stress state, the distortional energy criterion predicts that yielding will occur when the distortional energy of the combined stress state equals the distortional energy at yield in simple tension, that is,

$$\frac{(\sigma_1 - \sigma_3)^2}{8G} + \frac{(\sigma_2 - \sigma_1)^2}{8G} + \frac{(\sigma_3 - \sigma_2)^2}{8G} = \frac{\sigma_y^2}{4G}, \tag{2-23}$$

or (von Mises)

$$\sigma_y = \sqrt{\tfrac{1}{2}[(\sigma_1 - \sigma_3)^2 + (\sigma_2 - \sigma_1)^2 + (\sigma_3 - \sigma_2)^2]} = \overline{\sigma}. \tag{2-24}$$

The quantity under the square root sign is designated as $\overline{\sigma}$, and is known as the equivalent stress for the combined stress state, and is equal to the yield stress in simple tension. Note that, since $\overline{\sigma}$ is based upon shear stresses, it has no hydrostatic component. $\overline{\sigma}$ is also referred to as the flow stress in dealing with plastic deformation that exceeds the yield strain where strain hardening is important, that is, the yield strength is a function of the amount of plastic strain.

In pure shear, $\sigma_1 = -\sigma_3 = k$, where k is the shear stress at yield in pure shear, and $\sigma_2 = 0$. Therefore, $\overline{\sigma} = \sqrt{3}k$, or $k = \overline{\sigma}/\sqrt{3}$. This leads to an expression for the equivalent shear stress:

$$\overline{\tau} = \sqrt{\tfrac{1}{6}[(\sigma_1 - \sigma_3)^2 + (\sigma_2 + \sigma_1)^2 + (\sigma_3 - \sigma_2)^2]}. \tag{2-25}$$

An equivalent strain,

$$\overline{\varepsilon} = \sqrt{\frac{1}{2(1+\nu)^2}[(\varepsilon_1 - \varepsilon_3)^2 + (\varepsilon_2 - \varepsilon_1)^2 + (\varepsilon_3 - \varepsilon_2)^2]}, \tag{2-26}$$

can also be defined. For plastic deformation where $\nu = \frac{1}{2}$, the equivalent strain becomes

$$\overline{\varepsilon} = \sqrt{\tfrac{2}{9}[(\varepsilon_1 - \varepsilon_3)^2 + (\varepsilon_2 - \varepsilon_1)^2 + (\varepsilon_3 - \varepsilon_2)^2]}. \tag{2-27}$$

For pure shear, $\varepsilon_1 = -\varepsilon_3 = \gamma/2$ and $\varepsilon_2 = 0$. Therefore, $\overline{\varepsilon} = (2/\sqrt{3})\varepsilon_1$, or $\varepsilon_1 = (\sqrt{3}/2)\overline{\varepsilon}$, or $\gamma = \sqrt{3}\overline{\varepsilon}$. This leads to the expression

$$\overline{\gamma} = \sqrt{\tfrac{2}{3}[(\varepsilon_1 - \varepsilon_3)^2 + (\varepsilon_2 - \varepsilon_1)^2 + (\varepsilon_3 - \varepsilon_2)^2]}\cdot \tag{2-28}$$

B. Maximum Shear Stress Criterion

The Tresca criterion,

$$\overline{\tau} = \frac{(\sigma_1 - \sigma_3)}{2}, \tag{2-29}$$

is a maximum shear stress criterion, and does not depend upon σ_2. In general, the von Mises criterion is favored, but the Tresca criterion, because of its simple form in terms of principal stresses, is useful for purposes of illustration.

Figure 2-8 is a plot of the von Mises and Tresca yield criteria in two-dimensional stress space, that is, one of the three principal stresses is zero. For this diagram, the yield strength in tension is taken to be the common reference point.

XI. STATE OF STRESS IN THE PLASTIC ZONE AHEAD OF A NOTCH IN PLANE-STRAIN DEFORMATION

In the elastic region, the determination of the state of stress ahead of notch is a relatively complex problem involving use of the theory of elasticity. In contrast, it is relatively simple to determine the state of stress in the plastic zone ahead of a notch under plane-strain loading conditions. The analysis of this state of stress provides insight as to why fracture processes are promoted by plane-strain loading conditions at notches and cracks.

Recall that, for plastic deformation, the volume is conserved and Poisson's ratio ν is equal to $\frac{1}{2}$. From the Mohr circle it can be seen that the normal stress acting on the plane of maximum shear stress is given as $\sigma_n = \frac{1}{2}(\sigma_1 + \sigma_3)$. Under plane-strain conditions, this normal stress is also equal to σ_2, which is the hydrostatic stress σ_{hyd} (see above). In Fig. 2-9, this hydrostatic stress is shown acting in the

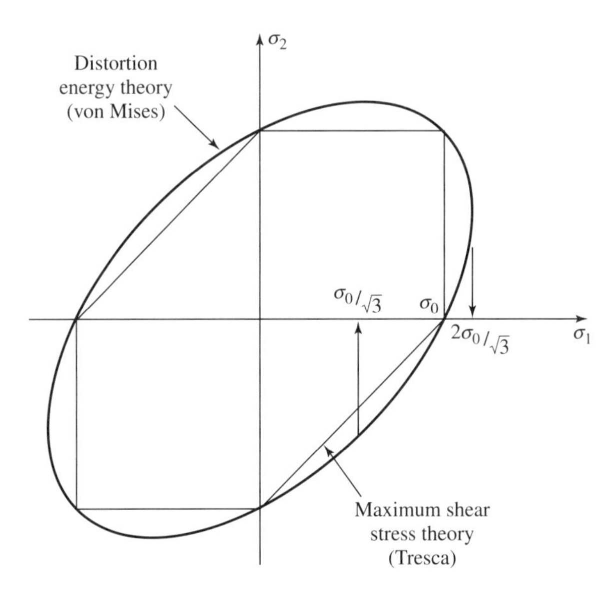

Fig. 2-8. The Tresca and von Mises yield criteria for biaxial stress.

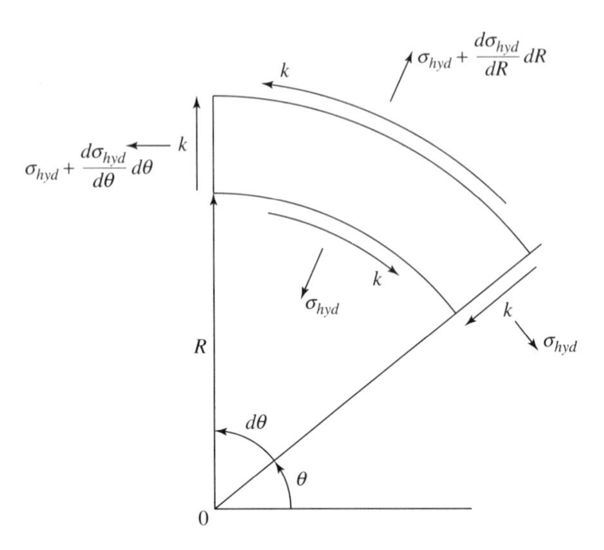

Fig. 2-9. The stress components acting on an element within the plastic range. The maximum shear stress required for plastic deformation is designated as *k*.

radial and tangential directions of the curved element, that is, perpendicular to the planes of maximum shear stress. The material is considered to be an ideally rigid plastic material, that is, strain hardening is neglected, and in plastic regions, the traces of the planes of maximum shear stress are known as slip lines. In a plastic zone, the magnitude of the maximum shear stress is indicated by k, which is the yield stress in shear (torsion) under plane-strain conditions. The value of k depends upon the yield criterion when the tensile yield strength is taken as the standard of reference:

$k = 0.577\sigma_y$, as seen from a plot of the two-dimensional von Mises relation ($\sigma_2 = \frac{1}{2}\sigma_1$, $\sigma_3 = 0$).

$k = 0.5\sigma_y$, as seen from a plot of the two-dimensional Tresca relation.

In Fig. 2-9, the value of θ is taken to be positive in the counterclockwise direction, and by taking moments about the origin, the following equation is obtained:

$$\sum M_0 = \frac{d\sigma_{hyd}}{d\theta}(\mathrm{d}\theta)dR\left(R + \frac{dR}{2}\right) + k(R + dR)d\theta(R + dR) - kR\, d\theta\, R = 0, \quad (2\text{-}30)$$

which leads to

$$\frac{d\sigma_{hyd}}{d\theta} + 2k = 0, \quad (2\text{-}31)$$

or

$$d\sigma_{hyd} + 2k\,d\theta = 0, \tag{2-32}$$

which can be integrated to yield

$$\sigma_{hyd} + 2k\theta = \text{constant} \tag{2-33}$$

on a curved plane of maximum shear stress.Note that hydrostatic stress is independent of R, and depends only on θ, the amount of rotation of the tangent to a slip line on going from one position to another. If the slip line were straight, there would be no change in θ, and hence no change in the hydrostatic stress.

As can be seen from the Mohr circle for plane-strain conditions, σ_{hyd} is at the center of the circle, $\sigma_1 = \sigma_{hyd} + k$, and $\sigma_3 = \sigma_{hyd} - k$.

From Eq. 2-33, we have $\sigma_{hyd} + 2k\theta =$ constant. The constant can be evaluated by noting that, at the free surface of the notch, Fig. 2-10, $\sigma_3 = 0$. If we set $\theta = 0$ along a slip line at this point, then the unknown constant is equal to σ_{hyd}, which in turn is equal to k at this point. (Check with Mohr circle). We can then write:

$$\sigma_{hyd} = k - 2k\theta, \tag{2-34}$$

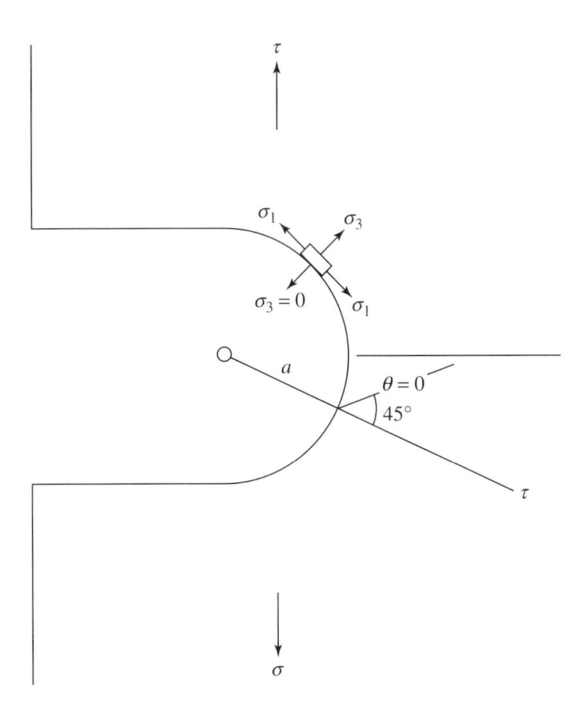

Fig. 2-10. The principal stresses acting at the free surface of a notch subjected to tensile loading.

Note that the hydrostatic component of stress changes with change in θ. Therefore, both σ_1 and σ_3 must also change with θ to maintain the state of stress at the yield level. The maximum value for the hydrostatic stress in this case (θ is negative) is

$$\sigma_{hyd} = k + 2k\frac{\pi}{2} = k(1 + \pi), \tag{2-35a}$$

and

$$\sigma_1 = k + \sigma_{hyd} = 2k\left(1 + \frac{\pi}{2}\right) \tag{2-35b}$$

As mentioned above, the value of k depends upon the yield criterion in plane strain. For the Tresca criterion, $k = (\sigma_y/2)$, where σ_y is the yield strength in simple tension, and $2k(1 + \pi/2) = 2.57\sigma_y$. That is, triaxiality has raised the local stress for yield at a point ahead of the notch by a factor of 2.57!

For the von Mises criterion, $k = 1.15(\sigma_y/2) = 0.577\sigma_y$. Therefore, $2k(1 + \pi/2) = 2.96\sigma_y$, so that the local stress for yield has been raised by a factor of almost 3! For this reason, the local yield stress in a notched bar can be considered to be increased by a factor of three in assessing the role of a notch in low-temperature, brittle fracture of steel.

(Note that the Mohr circle at yield at the base of a notch ($\sigma_3 = 0$) would have a radius 0.577 times larger for the von Mises criterion as compared to the Tresca criterion.)

Next, we derive a general expression for the value of σ_1 along the x-axis ahead of a notch. Let x be the distance along the x-axis measured from the tip of a notch of constant radius. To develop an expression for the maximum stress acting in the y-direction as a function of the distance from the notch, use is made of the mathematical expression for the log spiral.

The distance r is the radial distance measured from the center of curvature of a notch of constant radius a. The trajectories of σ_1, the first principal stress, lie along arcs of circles that are concentric with the notch. The trajectories of σ_3 must be at right angles to these circular arcs, that is, they are radial lines. The maximum shear stress trajectories must cross both sets of lines at 45°. The logarithmic or equiangular spiral is a curve that makes a constant angle with the radius vectors (Fig. 2-11), and hence also with the trajectories of σ_1.

The curved elements in Fig. 2-9 are taken to be portions of logarithmic spirals that can be mathematically expressed as

$$\frac{r}{a} = e^{-(\cot\phi)\theta} \tag{2-36}$$

and since in this case $\cot 45° = 1$,

$$\frac{r}{a} = e^{-\theta} \tag{2-37}$$

(For a spiral in the opposite sense, $r/a = e^{\theta}$.)

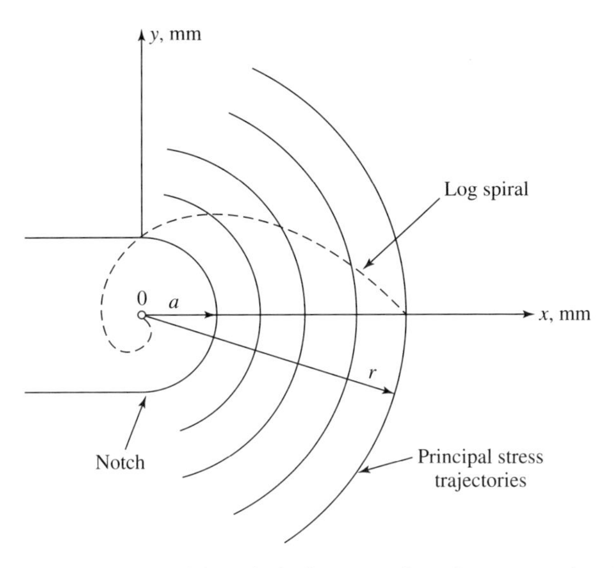

Fig. 2-11. A logarithmic spiral and the principal stress trajectories at a notch subjected to tensile loading.

Where a spiral intersects the notch, $r = a$. At this point θ for that spiral is equal to zero. The angular rotation the spiral makes in going from the intersection point on the notch surface to the x-axis, point b, is θ_b, with the counterclockwise direction taken as positive. The distance b to a point where the spiral intersects the x-axis is given as $b = ae^{-\theta_b}$. The distance b can be set equal to $a + x$, where x is the distance along the x-axis measured from the notch, so that $a + x = ae^{-\theta_b}$. Therefore

$$1 + \frac{x}{a} = e^{-\theta_b}, \qquad \text{or } \ln\left(1 + \frac{x}{a}\right) = -\theta_b. \tag{2-38}$$

To obtain an expression for σ_1 along the x-axis, this expression for θ_b is substituted into Eq. 2-35b to yield

$$\sigma_1 = 2k + 2k \ln\left(1 + \frac{x}{a}\right) = 2k\left[1 + \ln\left(1 + \frac{x}{a}\right)\right] \tag{2-39}$$

A plot of this relation is shown in Fig. 2-12. It is noted that the value of σ_1 increases from $2k$ up to its maximum value as the distance ahead of the notch increases. Once this maximum value is attained at $\theta_b = -\pi/2$ for a parallel sided notch, the stress remains constant with further increase in x until the elastic-plastic boundary is reached, whereupon the stress then decreases with further increase in x. This stress distribution within the plastic zone ahead of a notch is in marked contrast to the elastic behavior where σ_1 is a maximum at the notch root. The increase in peak

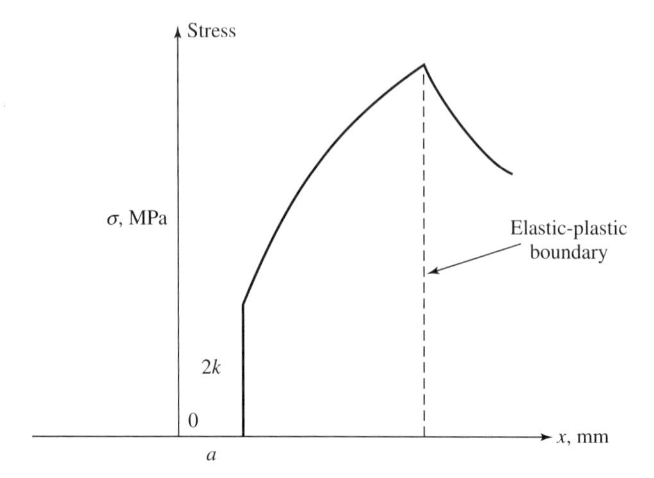

Fig. 2-12. The principal stress σ_1 perpendicular to the *x*-axis in the elastic-plastic region ahead of a notch loaded in tension.

stress within the plastic zone ahead of the notch has clear implications with respect to the site of the origin of fracture in notched components.

Note that under plane stress loading, plastic deformation is through-thickness rather than in-plane, and occurs on planes at 45° to the y-z plane. Since the slip lines are straight, the level of the hydrostatic stress is constant throughout the plastic zone. As a result, the peak stress in the plastic zone is also constant and equal to σ_Y.

XII. SUMMARY

This review of the concepts of elastic and plastic behavior should provide a sufficient basis for the understanding of the stress-strain related matters involved in many failure analyses. In some cases, a more detailed stress analysis involving, for example, the finite element method may be called for.

REFERENCES

(1) G. E. Dieter, Jr., Mechanical Metallurgy ,3rd ed., McGraw-Hill, NY, 1986
(2) A. Mendelson, Plasticity: Theory and Application, Macmillan, NY, 1968
(3) I. Le May, Principles of Mechanical Metallurgy, Elsevier, Oxford, 1981.

3

Elements of Fracture Mechanics

I. INTRODUCTION

The relatively new field known as fracture mechanics, based upon the work of Griffith and Irwin, is used to treat fracture problems involving cracks in a quantitative manner. This chapter presents the fundamentals of this field, and discusses a number of conditions that must be met in order to make valid use of fracture mechanics in a failure analysis.

II. GRIFFITH'S ANALYSIS OF THE CRITICAL STRESS FOR BRITTLE FRACTURE

Griffith (1) studied the fracture behavior of silica glass, a very brittle material. At room temperature, the stress-strain curve for this type of glass is linear up to fracture. The theoretical strength of this glass is approximately $E/10$, but in the presence of small cracks, the fracture stress is orders of magnitude below the theoretical strength of the glass. Griffith's analysis succeeded in explaining why this was so, and also provided the basis for the field of fracture mechanics.

The Griffith analysis is based upon the first law of thermodynamics, which states that in a closed system energy is conserved. Two types of energy are considered, strain energy and surface energy. Consider the two thin-sheet specimens indicated in Fig. 3-1 subjected to a tensile stress σ. One specimen has a crack, which is very small in length with respect to the width of the sheet; the other does not. If each of these specimens is loaded in tension to the same displacement, there will be a small difference in the load-extension plots, for it would take less load to extend the cracked

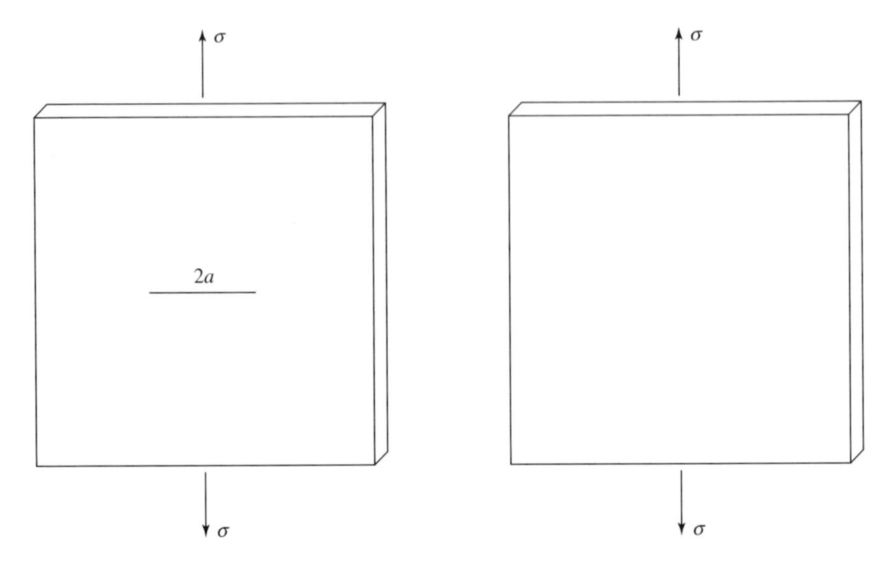

Fig. 3-1. Cracked and uncracked panels subjected to tensile loading.

specimen a given amount than the uncracked specimen. These plots are shown in Fig. 3-2, where the difference between the two cases has been exaggerated for clarity. The elastic energy stored in each specimen at a given extension Δ is given by the area under the corresponding curve and is equal to $\frac{1}{2}P\Delta$, where P for the cracked specimen is less than that for the uncracked specimen. Griffith used this difference in stored elastic energy to develop a theory for brittle fracture. He reasoned that, on going from the uncracked to the cracked state, there is not only a decrease in elastic energy, but also an increase in surface energy, due to the creation of the new crack surfaces. He calculated that if a crack of length $2a$ were to develop

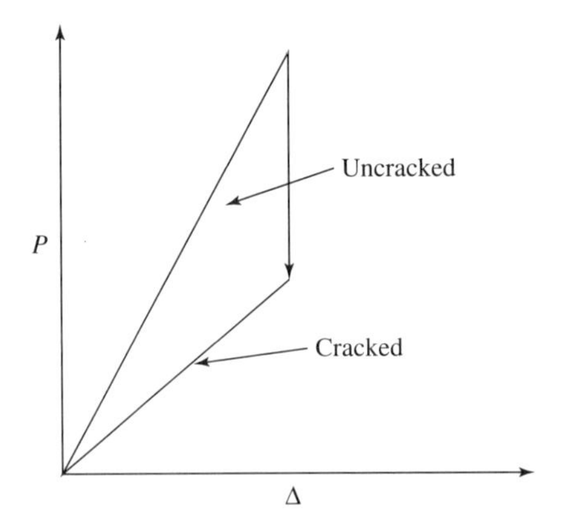

Fig. 3-2. Load deflection plots for the cracked and uncracked panels shown in Fig. 3-1.

in the uncracked body, ΔW_e, the total decrease in elastic stored energy per unit of thickness, would be given by

$$\Delta W_e = -\frac{\pi a^2 \sigma^2}{E}, \tag{3-1}$$

where a is the half-crack length and E is Young's modulus. However, there is also an increase of energy in the system due to the creation of new surface areas of energy γ_s per unit of area. The total increase of surface energy ΔW_s is given by

$$\Delta W_s = 4a\gamma_s, \tag{3-2}$$

where γ_s is the surface energy per unit area, a material constant. (The number 4 appears since the crack is of length $2a$ and it has an upper and a lower surface.)

Now comes the next step in the Griffith analysis. Suppose the specimen is already cracked; what condition must be met for the crack to propagate? Griffith reasoned that propagation would occur when the rate of strain energy release would just equal the rate at which energy was being absorbed through the creation of additional crack surface. This condition can be expressed as

$$\frac{d}{da}\left(-\frac{\pi a^2 \sigma^2}{E} + 4a\gamma_s\right) = 0, \tag{3-3}$$

or

$$-\frac{2\pi a \sigma^2}{E} + 4\gamma_s = 0, \tag{3-4}$$

which leads to

$$\sigma_c = \sqrt{\frac{2E\gamma_s}{\pi a}}, \tag{3-5}$$

where σ_c is the critical stress required for propagation of a brittle crack. Note that, for stresses less than the critical value, the crack will not propagate because the strain energy that would be released in a virtual crack advance would be less than that needed to form new surface.

Griffith tested a brittle glass and found the results to be in agreement with predictions based upon Eq. 3-5.

Equation 3-5 can also be written as

$$\sigma_c\sqrt{\pi a} = \sqrt{2E\gamma_s}, \tag{3-6}$$

where the extrinsic quantities σ_c and a are on the left-hand side of the equation, and the intrinsic quantities E and γ_s are on the right-hand side of the equation. $\sigma_c\sqrt{\pi a}$

is a common combination of terms encountered in fracture mechanics, and is designated as K_c. K in general is known as a stress intensity factor, and it depends upon component geometry, stress level, and crack length. For the Griffith geometry,

$$K = \sigma\sqrt{\pi a}. \tag{3-7}$$

The subscript c in Eq. 3-6 is used to indicate that the value of K is at the critical level for fracture, that is

$$K_c = \sqrt{2E\gamma_s}. \tag{3-8}$$

K_c is often referred to as the fracture toughness.

In the linear-elastic range, the stress ahead of a crack under tension is governed by the stress intensity factor, and is expressed as

$$\sigma_{yy} = \frac{K}{\sqrt{2\pi r}}, \tag{3-9}$$

where r is the distance measured from the crack tip.

III. OROWAN-IRWIN MODIFICATION OF THE GRIFFITH EQUATION

In later studies of the fracture of brittle steel carried out by Orowan (2) and by Irwin (3), it was found that the experimental values for γ_s were orders of magnitude higher than expected. They reasoned that plastic deformation occurred during nominally brittle fracture of steel, and that the plastic work involved greatly exceeded the surface energy γ_s. In order to preserve the linear-elastic Griffith formalism, $\sqrt{2E\gamma_s}$ was therefore written as $\sqrt{2E\gamma_p}$, where γ_p is the plastic work done per unit of area during fracture.

Equation 3-4 can be written as

$$2\gamma_p = G_c = \frac{\sigma_c^2 \pi a}{E}, \tag{3-10}$$

where G_c is referred to as the critical strain energy release rate at a crack tip. (In the treatment of the fracture of polymers, G_c rather than K_c is considered to be the fracture toughness.)

In general G can also be expressed as

$$G = \frac{\Delta W - \Delta U}{B\, da}, \tag{3-11}$$

where W is the external work, U is the internal (strain) energy, and B is the thickness. For fixed displacement conditions, $\Delta W = 0$, and so G can be written as

$$G = -\left(\frac{dU}{B\,da}\right). \tag{3-12}$$

Under plane-stress conditions, K_c and G_c are interrelated through

$$K_c = \sqrt{EG_c}, \tag{3-13}$$

and

$$G_c = \frac{K_c^2}{E}. \tag{3-14}$$

The stress intensity factors for plane stress and plane strain are the same and independent of Poisson's ratio ν. However, K_{Ic}, the plane strain value of the fracture toughness under Mode I loading, does depend upon Poisson's ratio, and is given as

$$K_{Ic} = \sqrt{\frac{EG_{Ic}}{(1 - \nu^2)}}. \tag{3-15}$$

This is because the effective Young's modulus under plane-strain conditions is equal to $E/(1 - \nu^2)$.

IV. STRESS INTENSITY FACTORS

Thus far, the only type of specimen geometry that has been discussed is the wide sheet containing a small crack. When the crack length is appreciable with respect to the width, then a finite width correction of the stress intensity factor has to be made. Figure 3-3 is a plot of the correction factor for the center-cracked plate. For this case, the stress intensity factor can be expressed as (4)

$$K = \sigma\sqrt{\pi a}\sqrt{\sec\frac{\pi a}{W}}. \tag{3-16}$$

Figure 3-4 shows a semicircular surface crack, an important type of flaw when dealing with pressure vessels. The stress intensity factor at the surface in this case is given approximately by

$$K = 1.12\left(\frac{2}{\pi}\right)\sigma\sqrt{\pi a}, \tag{3-17}$$

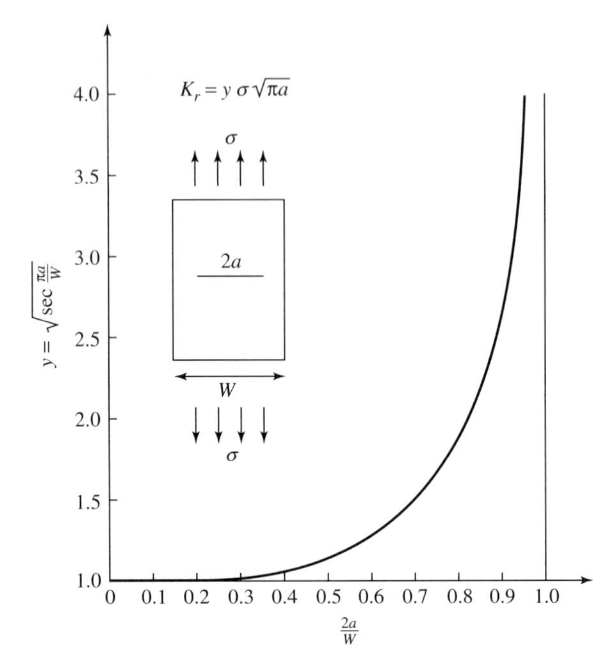

Fig. 3-3. Finite width correction of the stress intensity factor for a center cracked plate (4, 15).

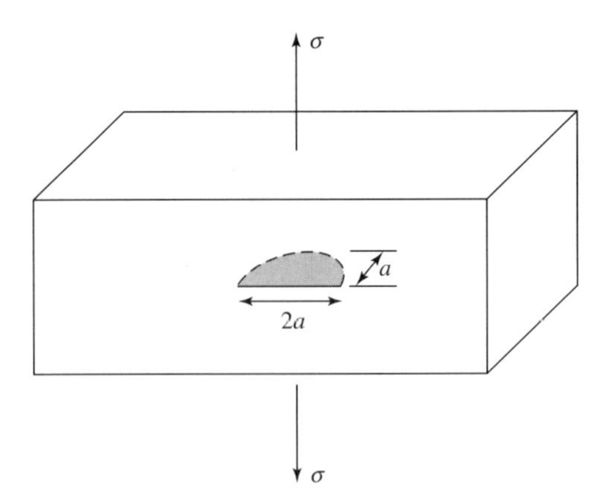

Fig. 3-4. A semi-circular surface flaw.

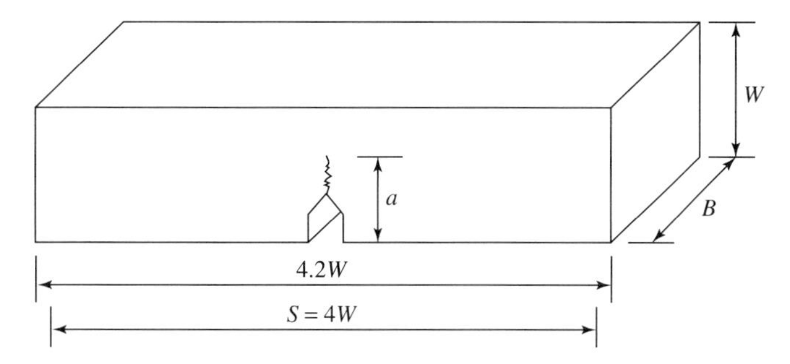

Fig. 3-5. A three-point bend specimen.

whereas at maximum depth a, the stress intensity factor is slightly less:

$$K = \frac{2}{\pi}\sigma\sqrt{\pi a}. \tag{3-18}$$

The stress intensity factor for the three-point bend specimen shown in Fig. 3-5, with a support span S, is given in ASTM Designation E 399 as

$$K = \frac{PS}{BW^{3/2}} \frac{3(a/W)^{1/2}[1.99 - (a/W)(1 - a/W) \times 2.15 - 3.93a/W + 2.7(a/W)^2]}{2[1 + 2(a/W)](1 - a/W)^{3/2}}. \tag{3-19}$$

The stress intensity factor for the compact specimen shown in Fig. 3-6 is given by ASTM E 399 as

$$K_I = \left(\frac{P}{BW^{1/2}}\right) f\left(\frac{a}{w}\right),$$

where $f(a/W)$ is given as

$$f\left(\frac{a}{W}\right) = \frac{(2 + a/W)(0.866 + 4.64a/W) - 13.32a^2/W^2 + 14.72a^3/W^3 - 5.6a^2/W^4}{(1 - a/W)^{3/2}}. \tag{3-20}$$

Further information on stress intensity factors can be found in a number of handbooks (5–8), which give them for a range of geometries and loading conditions.

V. THE THREE LOADING MODES

A component containing a crack can be loaded in three different modes. These modes are illustrated in Fig. 3-7. Mode I is the tensile opening mode, and the stress intensity factor associated with this mode is designated as K_I. Mode II is the in-plane shearing mode, and the stress intensity factor associated with this mode is desig-

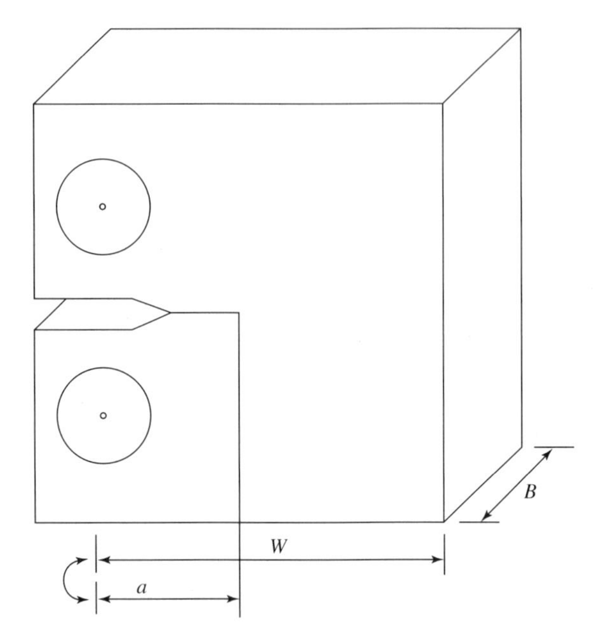

Fig. 3-6. A compact specimen.

nated as K_{II}. Mode III is the out-of-plane shearing mode, sometimes referred to as the anti-plane-strain mode. The stress intensity factor associated with Mode III is designated as K_{III}.

VI. DETERMINATION OF THE PLASTIC ZONE SIZE

Irwin has shown that the linear-elastic fracture mechanics (LEFM) approach, which was initiated by Griffith, can be used even in the presence of plastic deformation.

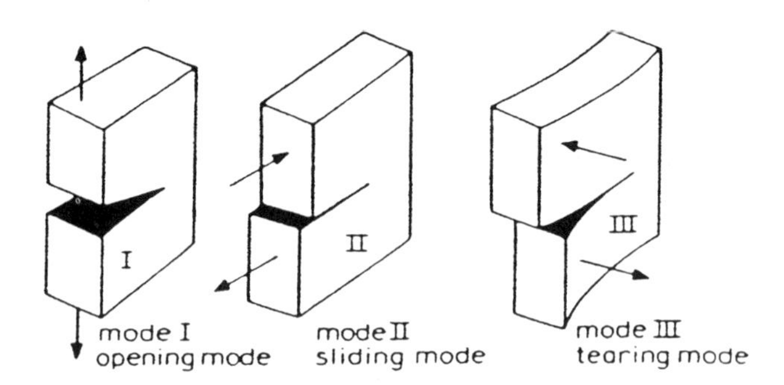

Fig. 3-7. The three loading modes.

However, limits are imposed on the extent of plastic deformation for the LEFM approach to remain valid. It is required that the size of the plastic zone that develops in a metal at a crack tip be small with respect to the crack length. The first estimate of size of the plane-stress plastic zone at a crack tip can be obtained by rearranging Eq. 3-9 to read

$$r = \frac{K^2}{2\pi\sigma_{yy}^2}. \tag{3-21}$$

Then, with σ_{yy} set equal to σ_{YS}, the yield stress, the size of the plastic zone r_{pzs} is

$$r_{pzs} = \frac{K^2}{2\pi\sigma_{YS}^2}. \tag{3-22}$$

In a more detailed analysis, Dugdale (9) derived the following expression for the plane-stress plastic zone size for a centrally cracked sheet under tension:

$$r_{pzs} = a\left[\sec\left(\frac{\pi}{2}\frac{\sigma}{\sigma_Y}\right) - 1\right]. \tag{3-23}$$

For values of σ much less than σ_Y, Eq. 3-23 can be written as

$$r_{pzs} = \frac{\pi^2}{4}\left(\frac{K^2}{2\pi\sigma_Y^2}\right), \tag{3-24}$$

about two and one-half times larger than the first estimate given by Eq. 3-22.

ASTM Designation E 399 requires that the crack length be 16 times greater than r_{pzs} to meet the conditions for a linear-elastic analysis. If it is not, then elastic-plastic fracture mechanics or some modification should be used in the analysis. Irwin (10) proposed that, in order to keep a linear-elastic framework for analysis when the crack length was small with respect to the plastic zone size, the crack length be increased by one-half of the plastic zone size. If the Dugdale value for the plastic zone size is used, then the effective crack length, a_{eff}, according to the Irwin proposal becomes

$$a_{eff} = a + \frac{1}{2}a\left(\sec\frac{\pi}{2}\frac{\sigma}{\sigma_Y} - 1\right) = \frac{a}{2}\left(\sec\frac{\pi}{2}\frac{\sigma}{\sigma_Y} + 1\right). \tag{3-25}$$

Equation 3-25 is useful in dealing with short cracks at stress levels that are high relative to the yield strength. For situations where the plastic zone is large with respect to the crack length, the stress intensity factor can be expressed as

$$K = Y\sqrt{\pi a_{eff}}\,\sigma \tag{3-26}$$

where Y is determined by the particular geometry under consideration. With the use of Eq. 3-23, Eq. 3-25 can be written as

$$K = Y\sqrt{\frac{\pi}{2}a\left(\sec\frac{\pi}{2}\frac{\sigma}{\sigma_Y} + 1\right)}\sigma \tag{3-27}$$

Irwin (10) has estimated the plane-strain plastic zone size to be one-third that of the plane-stress plastic zone size. This is consistent with the higher degree of plastic constraint in plane strain as compared to plane stress.

VII. EFFECT OF THICKNESS ON FRACTURE TOUGHNESS

The fracture toughness of a material is usually determined using a prefatigue-cracked plane specimen of the type shown in Fig. 3-6 (ASTM Designation E 399). An important factor affecting the magnitude of the fracture toughness is the thickness of the specimen. Figure 3-8 shows the typical variation of the fracture toughness as a function of thickness. For thicknesses greater than 16 times the plane-stress plastic zone size, that is, $16 \times (K^2/2\pi\sigma_Y^2) = 2.5(K^2/\sigma_Y^2)$, the specimen is in a condition of plane strain, and the fracture toughness is constant independent of thickness but at a minimum value. It is also required that the crack length a be greater than this value to insure that the crack length is long with respect to the plastic zone size to insure that a linear elastic fracture mechanics (LEFM) analysis is applicable. In addition, the length of the starter notch plus the fatigue crack is specified to be between 0.45 and 0.55 of the width W, which also fixes the size of the ligament ahead of the crack. The Mode I, plane-strain fracture toughness is designated as K_{Ic}. For lesser thickness, the fracture toughness rises due to the loss of constraint and the development of plane-stress conditions. In this range, the fracture toughness is designated as K_c.

Another method for determining the fracture toughness is to evaluate the crack tip opening displacement (CTOD) of notched and fatigue precracked specimens

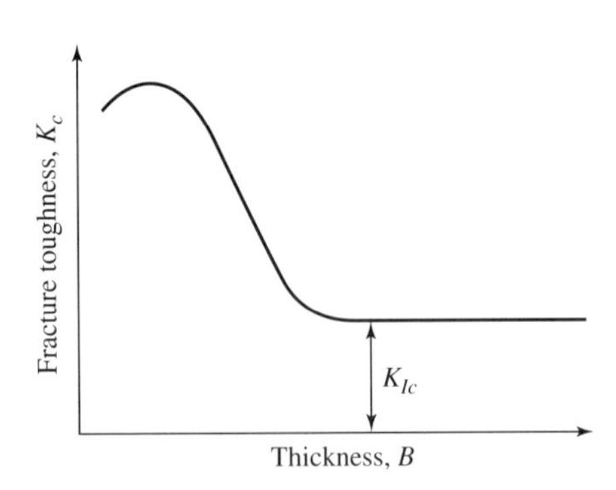

Fig. 3-8. Fracture toughness as a function of thickness.

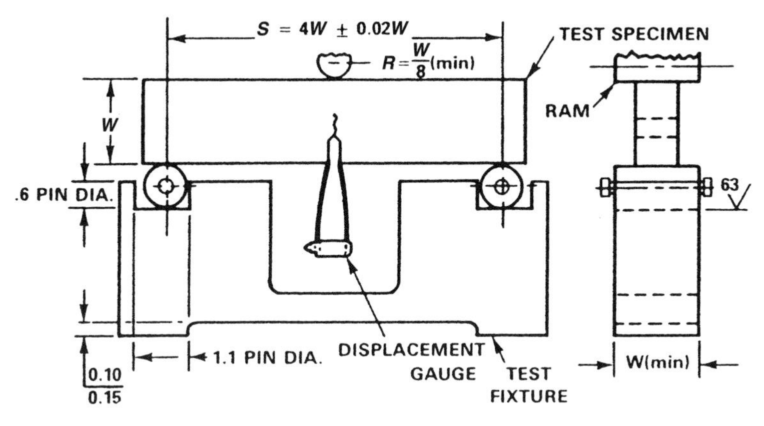

Fig. 3-9. Setup for the measurement of displacement by a clip gauge in a bend test. (From ASTM, 16. Copyright ASTM. Reprinted with permission.)

(ASTM Designation E 1290). An objective of the method is to determine the CTOD corresponding to the onset of unstable brittle crack extension. A clip gauge is mounted at the mouth of the notch, Fig. 3-9, and a load-crack mouth displacement record is obtained. The crack mouth displacement is then converted into a corresponding CTOD value, and in the case of brittle fracture, the value of the CTOD at the onset of brittle fracture can be related to the fracture toughness by means of the relation:

$$\mathrm{CTOD}_c = \frac{4}{\pi} \frac{K_{IC}^2}{\sigma_Y E}. \tag{3-28}$$

This procedure is useful when materials are too ductile or lack sufficient size to be tested in accordance with the requirements of ASTM Designation E 399.

VIII. THE *R*-CURVE

In ductile materials, particularly under plane-stress conditions, the onset of crack extension under rising-load test conditions occurs at a load level below the maximum load. There is a load range wherein a crack extends in a stable manner, that is, the resistance to unstable crack propagation increases with crack extension in the stable crack growth range. Even under plane-strain conditions of loading, there may be a small increment of stable crack growth before unstable crack growth occurs. This increased resistance is related to the increase of the plastic zone at the tip of the crack as well as to the blunting of the crack tip as it advances. This type of behavior is dependent upon specimen thickness, temperature, and strain rate. The *R*-curve is a plot of the applied stress intensity factor K versus the crack length, and the standard practice for the *R*-curve determination is given in ASTM E 561. The

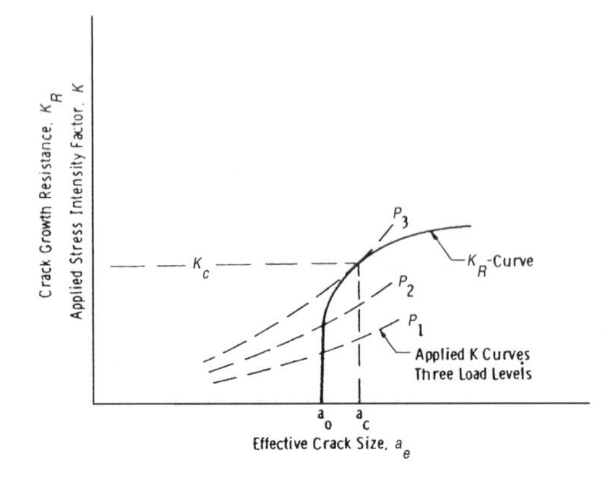

Fig. 3-10. Schematic representation of an *R*-curve and applied *K*-curves to predict instability at K_c under a load P_3, and a critical crack length a_c, for an initial crack length a_0. (From ASTM, 17. Copyright ASTM. Reprinted with permission.)

R-curve can be matched with the applied *K*-curves to estimate the load necessary to cause unstable crack propagation at K_c, as shown in Fig. 3-10. In making this estimate, *R*-curves are regarded as though they are independent of the starting crack length a_o and the specimen configuration in which they are developed. For a given material, specimen thickness, and test temperature, the *R*-curves appear to be a function of crack extension Δa only. To predict instability, the *R*-curves may be positioned as in Fig. 3-10 so that the origin coincides with the initial crack length a_o. Applied *K*-curves can be generated by assuming applied loads or stresses and calculating applied *K* as a function of crack length. The curve that develops tangency with the *R*-curve defines the critical load or stress that will cause the onset of unstable fracturing.

The unloading compliance characteristics, together with a compliance calibration curve, which is a curve that gives the compliance of the specimen as a function of crack length, can be used to determine the crack length and the corresponding *K*-level.

IX. SHORT CRACK LIMITATION

The stress intensity factor is a product of stress and the square root of the crack length. Suppose one wanted to maintain *K* at a constant value while decreasing the crack length. This would require that the stress be increased accordingly. However, if we make the crack size short enough, the value of the stress would reach levels in excess of the tensile strength of the material. Therefore, straightforward linear elastic fracture mechanics cannot be used when dealing with cracks of lengths less than one millimeter, and some modification is needed. This is a particularly important consideration in analyzing the behavior of short cracks in fatigue.

X. CASE STUDY: FAILURE OF CANNON BARRELS

When a shell is fired from a cannon, some wear of the inner surface of the cannon takes place. During World War II, when the accumulated wear after the firing of a number of rounds became excessive, the barrel was removed from service. Much of this wear was associated with the type of propellant that was used at that time, but by the time of the Vietnam War, improvements in the propellants had been made, and the rate of wear was greatly reduced. However a serious new problem arose, for the barrel could now experience many more firings in its lifetime, with each of these firings being another severe fatigue cycle. As a result, fatigue cracks formed on the inside of the barrel, and when they propagated to critical size, the gun would explode, often killing the gun crew. These explosions occurred because the critical crack length for fracture of the barrel was less than the wall thickness of the barrel because of the relatively low plane-strain fracture toughness of the steel used for the barrels. To deal with this problem, steels of higher fracture toughness were developed, so that if a fatigue crack were to propagate, it would not reach critical length while still within the wall of the barrel. It would instead penetrate to the outer side of the barrel, and any pressure buildup would be vented rather than cause an explosion. In all pressure vessels, it is desirable to have the fracture toughness sufficiently high so that penetration of the pressure vessel wall occurs before fracture. Pressure vessels that meet this criterion will "leak before burst."

The following equations review some of the relationships used in the analysis of pressure vessels.

In a thin-walled tube, the hoop stress σ_h is given by

$$\sigma_h = \frac{p_i D}{2t}, \tag{3-29}$$

where p_i is the internal pressure, D is the diameter, and t is the wall thickness.

The gun barrel is a thick-walled tube. In a thick-walled tube, the circumferential stress is given by

$$\sigma_h = \frac{D_i^2 p_i}{D_o^2 - D_i^2}\left(1 + \frac{D_o^2}{4r^2}\right). \tag{3-30}$$

The stress intensity factor at the deepest point for an internal, part-through, semielliptical, longitudinal crack of depth a and length $2c$ in a thick-walled pressure vessel of thickness t is given as (5)

$$K_I = \frac{p_i R}{t}\sqrt{\frac{\pi a}{Q}}F\left(\frac{a}{t}, \frac{a}{2c}, \frac{R}{t}\right), \tag{3-31}$$

where

$$F = 1.12 + 0.053\xi + 00055\xi^2 + (1 + 0.02\xi + 0.0191\xi^2)\frac{(20 - R/t)^2}{1400}, \tag{3-32}$$

and $\xi = a/t(a/2c)$, $Q = 1 + 1.464(a/c)^{1.65}$, a is the depth of the flaw, and $2c$ is the length of the flaw. This expression is valid for $5 \leq r/t \leq 20$ and $2c/a \leq 12$, and $a/t \leq 0.80$.

XI. THE PLANE-STRAIN CRACK ARREST FRACTURE TOUGHNESS, K_{Ia}, OF FERRITIC STEELS

As will be discussed more fully in Chapter 6, ferritic steels at a low temperature fracture in a low-toughness, brittle manner, whereas at a higher temperature they fracture in a high-toughness, ductile manner. In one type of crack-arrest test, a temperature gradient is established in a plane-strain specimen under stress, and a running crack is propagated from the low-temperature region toward the high-temperature region. The running crack will be arrested because of the increase in fracture toughness with increase in temperature. The value of stress intensity factor at the point of arrest is designated K_{Ia}, and the temperature at the point of arrest is referred to as the crack arrest temperature for that loading condition. By changing the loading conditions, a plot of K_{Ia} versus temperature can be developed. The test is of significance in the design of safe structures, since above the arrest temperature a crack whose stress intensity factor is equal to or less than K_{Ia} is not able to propagate.

The crack arrest toughness can also be obtained at constant temperature. The procedure for this test is given in ASTM Designation E 1221. In carrying out the requirements of this specification, a compact specimen is wedge loaded at a selected temperature, and a fast running crack is created. This crack is arrested because, under the constant displacement loading conditions imposed, the stress intensity factor falls off as the crack advances.

XII. ELASTIC-PLASTIC FRACTURE MECHANICS

The above linear-elastic fracture mechanics (LEFM) procedures are applicable only when the plastic deformation is limited to a small zone at the tip of the crack. One of the requirements for a valid K_{Ic} determination is that thickness of the specimen satisfy the condition

$$B \geq 2.5\frac{K_{Ic}^2}{\sigma_Y^2} \text{ inches,} \qquad \text{when } K \text{ is in ksi } \sqrt{\text{inch}} \text{ and } \sigma \text{ is in ksi,} \tag{3-33a}$$

or

$$B \geq 0.0635\frac{K_{Ic}^2}{\sigma_Y^2} \text{ m,} \qquad \text{when } K \text{ is in MPa } \sqrt{\text{m}}, \text{ and } \sigma \text{ is in MPa.} \tag{3-33b}$$

For the steels used in the nuclear reactor pressure field, K_{Ic} may have a value of 220 MPa $\sqrt{\text{m}}$ (200 ksi $\sqrt{\text{inch}}$), and σ_Y may be 345 MPa (50 ksi). The requirement on

B for linear-elastic plane-strain conditions would be 40 inches, much too large. In order to obtain a valid estimate of K_{Ic}, an elastic-plastic procedure is used to determine a quantity J_{Ic} with specimens perhaps only an inch or so in thickness. However, even in this type of test there is a requirement that to maintain plane-strain conditions,

$$B \geq 25\frac{J_{Ic}}{\sigma_Y} \text{ inches,} \qquad \text{when } J \text{ is kP-inch/inch}^2 \text{ and } \sigma_Y \text{ is in ksi,} \tag{3-34a}$$

or

$$B \geq 0.635\frac{J_{Ic}}{\sigma_Y} \text{ m,} \qquad \text{when } J \text{ is in kPa-m and } \sigma_Y \text{ is in kPa.} \tag{3-34b}$$

J_{Ic} has been shown to be experimentally the equivalent of K_{Ic}. In a J_{Ic} test, there will be large amounts of plastic deformation at the crack tip, and a plot of load versus displacement will no longer be linear. Therefore, a modification of the LEFM approach, to the determination of fracture toughness is needed. In the modified approach, it is assumed for purposes of analysis that elastic-plastic behavior can be considered to be equivalent to nonlinear elastic behavior. Figure 3-11 compares the load displacement curves for both linear and nonlinear elastic behavior for two different crack sizes, a and $a + da$. The strain energy release rate G at a fixed displacement is the area between the curves in Fig. 3-11*a*. The equivalent strain energy release rate per unit of thickness B for nonlinear behavior is designated as J, where J is given as

$$J = -\frac{dU}{B\,da}, \tag{3-35}$$

where U is the strain energy. It has been shown that J can also be mathematically expressed as a path independent integral around the crack tip (11).

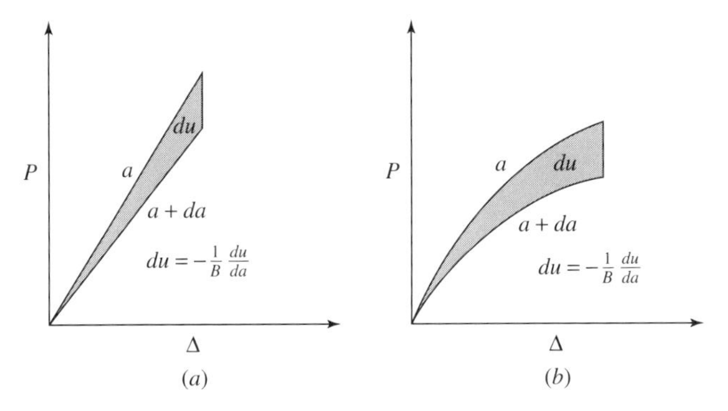

Fig. 3-11. Strain energy released due to crack advance at a fixed displacement. (*a*) In a linear-elastic material, and (*b*) In a non-linear elastic material.

ASTM Designation E 813 discusses the procedures involved in the experimental determination of J_{Ic}, the value of J near the initiation of crack growth. One procedure involves the determination of the load versus load line displacement behavior of a fatigue precracked compact specimen. The expression for J consists of its elastic and plastic components, that is,

$$J = J_{el} + J_{pl} \tag{3-36}$$

where

$$J_{el} = \frac{K^2(1 - \nu^2)}{E}, \tag{3-37}$$

and

$$J_{pl} = \frac{\eta A_{pl}}{B_N b_o}. \tag{3-38}$$

The constant η depends on the specimen type. For the compact specimen $\eta = 2 + 0.522b_o/W$, where B_N is the net specimen thickness, W is the width of the specimen measured from the load line, and b_o is the length of the uncracked ligament ahead of the fatigue crack. A_{pl} is the area shown in Fig. 3-12.

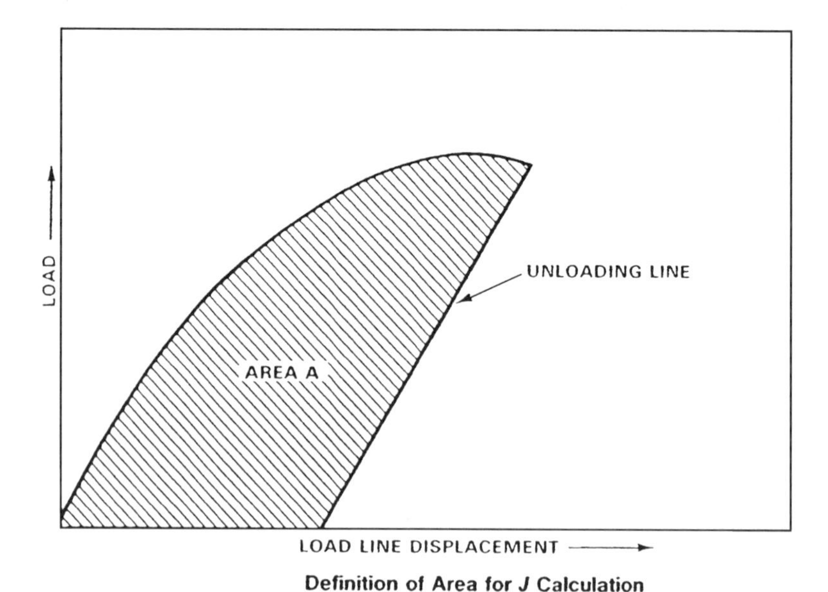

Fig. 3-12. A_{pl} in the determination of J. (From ASTM, 18. Copyright ASTM. Reprinted with permission.)

In the case of ductile metals, unstable crack growth does not occur at J_{Ic}. As in the case of the *K-R*-curve discussed above, a *J-R* curve can be constructed for elastic-plastic behavior. ASTM Designation E 1152 describes the procedures involved in establishing this type of curve. The point of unstable crack growth is reached when the crack-driving force J equals the crack-resisting force J_R.

The rate of creep crack growth has also been studied using fracture mechanics. By analogy with Eq. 3-12, J can be expressed as

$$J = -\frac{1}{B}\left(\frac{dU}{da}\right)_{\Delta}. \tag{3-39}$$

The derivative of this expression with respect to time is known as C^*, that is,

$$C^* = -\frac{1}{B}\left(\frac{d\dot{U}}{da}\right)_{\Delta}. \tag{3-40}$$

C^* has been found to be one of the more successful parameters for correlating steady-state creep crack growth data.

XIII. FAILURE ASSESSMENT DIAGRAMS

Failure assessment diagrams are plots that indicate safe and unsafe operating conditions in the presence of flaws (cracks). In these diagrams, failure is considered to occur either by general yield (collapse) or by fast fracture. From Eq. 3-26, the elastic-plastic stress intensity K_{EP} for plane stress conditions can be expressed as

$$K_{EP-P\sigma} = Y\sqrt{\frac{\pi}{2}a\left(\sec\frac{\pi}{2}\frac{\sigma}{\sigma_Y} + 1\right)}\sigma. \tag{3-41}$$

Under plane-strain conditions, with the size of the plastic zone taken to be one-third that of the plane-stress plastic zone, size K_{EP} for plane-strain conditions can be written as

$$K_{EP-P\varepsilon} = Y\sqrt{\frac{\pi}{3}a\left(\sec\frac{\pi}{2}\frac{\sigma}{\sigma_Y} + 2\right)}\sigma, \tag{3-42}$$

If $K_{EP=P\varepsilon}$ is set equal to K_{Ic}, the value of the critical crack length, a_c, is given as

$$a_c = \frac{C_I^2}{(\pi/3)(\sigma/\sigma_Y)^2[\sec(\pi/2)(\sigma/\sigma_Y) + 2]}, \tag{3-43}$$

where C_I is equal to $K_{Ic}/Y\sigma_Y$ Eq. 3-43 can be used to construct a plot of σ/σ_Y versus a_c/C_I^2, as shown in Fig. 3-13. A comparison with predictions based upon a lin-

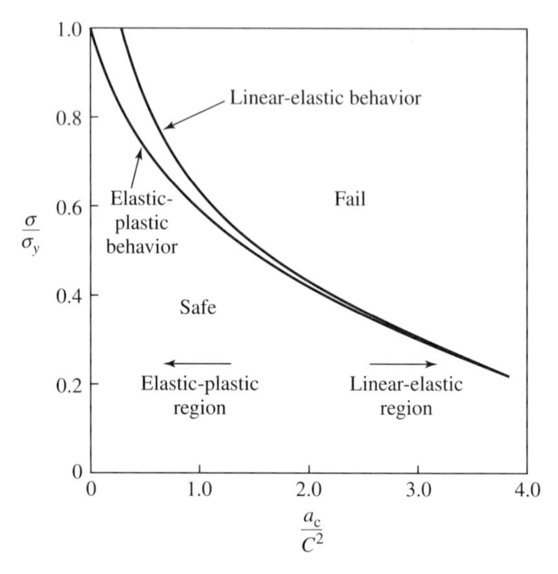

Fig. 3-13. Failure assessment diagram. $C = K_{Ic} \backslash Y\sigma_Y$ for plane strain, LE = linear elastic, $Pl\ \sigma$ = plane stress, $Pl\ \varepsilon$ = plane strain, σ_Y = yield stress.

ear elastic analysis with $K_{LE} = Y\sqrt{\pi a}\sigma$ is also shown. At large values of a_c, the linear-elastic predictions are close to the elastic-plastic predictions, but at short crack lengths, the safe region is reduced in the elastic-plastic case.

Another form of failure assessment diagram can be constructed where the coordinates are $K_{1(LE)}/K_{I(EP)}$ and σ/σ_y, as in Fig. 3-14. A somewhat similar method of analysis known as the $R6$ method has been developed in England for use in assessing the safety of nuclear power plants (12).

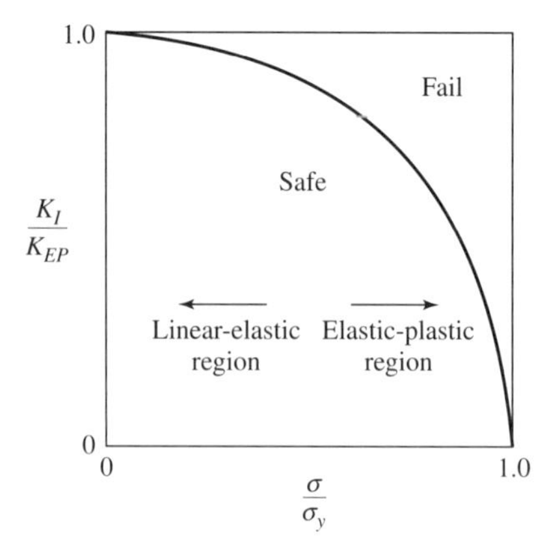

Fig. 3-14. Plane strain failure assessment diagram. σ_Y = yield stress, K_{EP} = the elastic-plastic stress intensity factor.

XIV. SUMMARY

This chapter has provided an elementary background in fracture mechanics, which will be needed in subsequent chapters. References 13–15 provide more in-depth information concerning fracture mechanics.

REFERENCES

(1) A. A. Griffith, The Phenomena of Rupture in Solids, Phil Trans. Royal Soc. London, vol. A221, 1921, pp. 163–197; The Theory of Rupture, Proc. Ist Int. Congress Appl. Mechs., 1924, pp. 55–63.

(2) E. Orowan, Energy Criteria for Fracture, Welding J., vol. 34, 1955, pp. 157s–160s.

(3) G. R. Irwin, Fracture Dynamics, in Fracturing of Metals, ASM, Materials Park, OH, 1948, pp. 147–166.

(4) C. E. Fedderson, Discussion, ASTM STP 410, 1967, pp. 77–79.

(5) H. Tada, P. C. Paris, and G. R. Irwin, The Stress Analysis of Cracks Handbook, 2nd ed., Paris Productions, Inc., St. Louis, 1985.

(6) G. C. Sih, Handbook of Stress Intensity Factors, Inst. of Fracture and Solid Mechanics, Lehigh University, Bethlehem, PA, 1973.

(7) D. P. Rooke and D. J. Cartwright, Stress Intensity Factors, Her Majesty's Stationery Office, London, 1976.

(8) Stress Intensity Factors Handbook, ed. by Y. Murakami, S. Aoki, N. Hasebe, Y. Itoh, H. Miyata, H. Terada, K. Tohgo, M. Toya, and R. Yuuki, Soc. Mater. Sci., Kyoto, Japan, 1987.

(9) D. S. Dugdale, Yielding of Steel Sheets Containing Slits, J. Mech. Phys. Solids, vol. 8, 1960, pp. 100–108.

(10) G. R. Irwin, Plastic Zone Near a Crack and Fracture Toughness, Proc. 7th Sagamore Conf., Syracuse University Press, Syracuse, NY, 1960, p. IV–63.

(11) J. R. Rice, A Path Independent Integral and the Approximate Analysis of Strain Concentrations by Notches and Cracks, J. Appl. Mech., vol. 35, 1968, pp. 379–386.

(12) I. Milne, R. A. Ainsworth, A. R. Dowling, and A. T. Stewart, Assessment of the Integrity of Structures Containing Defects, Central Electricity Generating Board Report R/H/R6-Rev 3, UK, May 1986.

(13) T. L. Anderson, Fracture Mechanics, CRC Press, Boca Raton, FL, 1991.

(14) M. F. Kanninen and C. H. Popelar, Advanced Fracture Mechanics, Oxford University Press, Oxford, UK, 1985.

(15) D. Broek, The Practical Use of Fracture Mechanics, Kluwer Academic Publisher, Boston, 1988.

(16) ASTM, 1990 Annual Book of ASTM Standards, West Conshohocken, PA, ASTM E399-83, Fig. A3.2.

(17) ASTM, 1990 Annual Book of ASTM Standards, West Conshohocken, PA, ASTM E561-86, Fig. 1.

(18) ASTM, 1990 Annual Book of ASTM Standards, West Conshohocken, PA, ASTM E813-89, Fig. A2-3.

4

Alloys and Coatings

I. INTRODUCTION

The characteristics of the materials involved in a failure are obviously of importance in a carrying out a failure analysis. This chapter reviews the microstructural features and related matters such as equilibrium and isothermal transformation diagrams of some of the more common alloys as influenced by alloying and heat treatment. The nature of the coatings used in high-temperature applications is also discussed.

II. PHASE DIAGRAMS AND SO ON

The phase diagram is a starting point in understanding microstructure-property relationships. These diagrams show the composition limits of phase fields as they exist under metastable or stable conditions. The important phase diagrams that are considered in the next sections include those for the iron-carbon system, the aluminum-copper system, the titanium-aluminum system, and the nickel-aluminum system. In addition, the time-dependent aspects of the transformation in steels from austenite to ferrite, pearlite, bainite, or martensite is also discussed.

A. Steels

For steels, the iron-carbon equilibrium diagram, Fig. 4-1 (1), together with the isothermal transformation diagrams, Fig. 4-2 (2), provide the key to understanding the multiplicity of microstructures and attendant properties that can be developed in this alloy system. Some definitions relating to these diagrams follow.

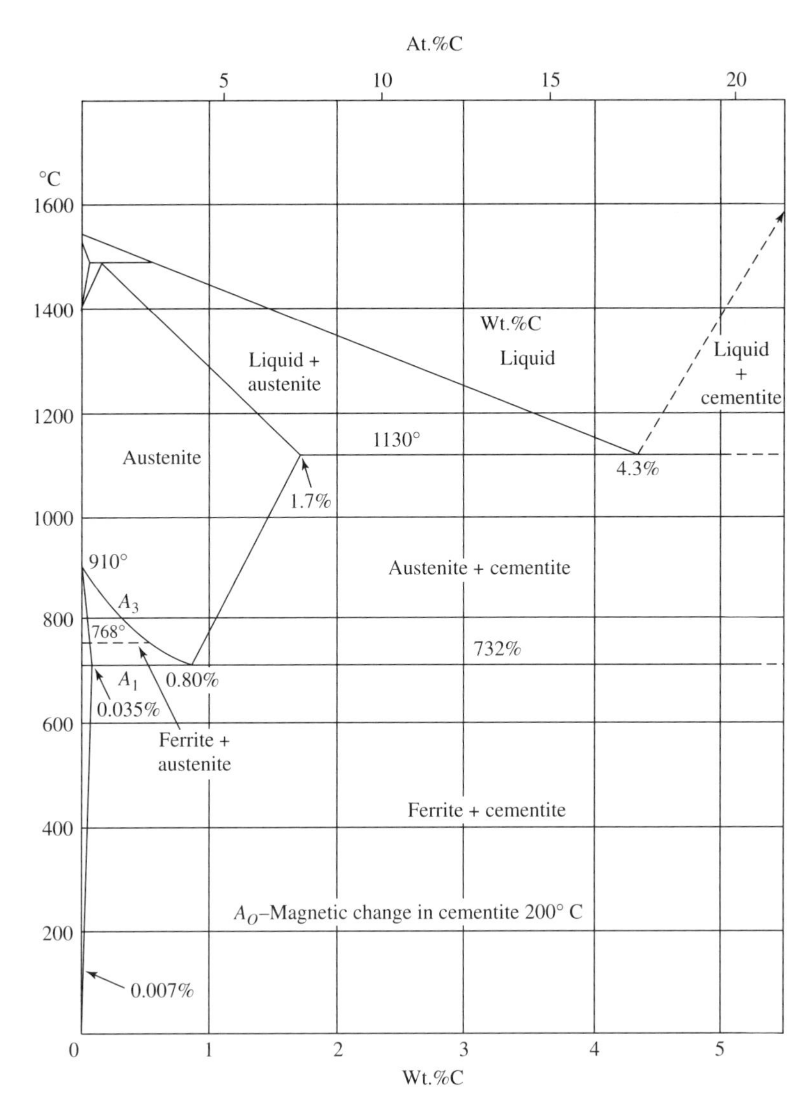

Fig. 4-1. Iron-carbon equilibrium diagram. (From Smithells Metals Reference Book, 1. Reprinted with permission.)

(a) *Ferrite:* A solid solution of one or more elements in body-centered cubic (bcc) iron. In the iron-carbon system, there are two ferrite regions separated by a region of austenite. The upper region is δ-ferrite, the lower region is a-ferrite. Acicular ferrite is a nonequiaxed ferrite that forms upon cooling by a combination of diffusion and shear in a temperature range just above that for the formation of bainite. In low-carbon steels, the size of the ferrite grains has an important effect on the yield strength; the smaller the grain size the higher the yield strength. The

1008

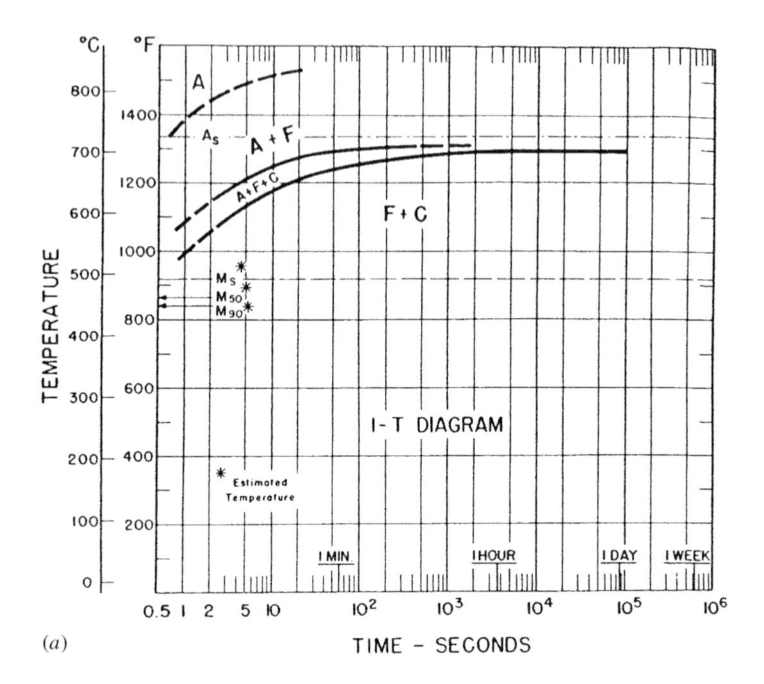

(*a*)

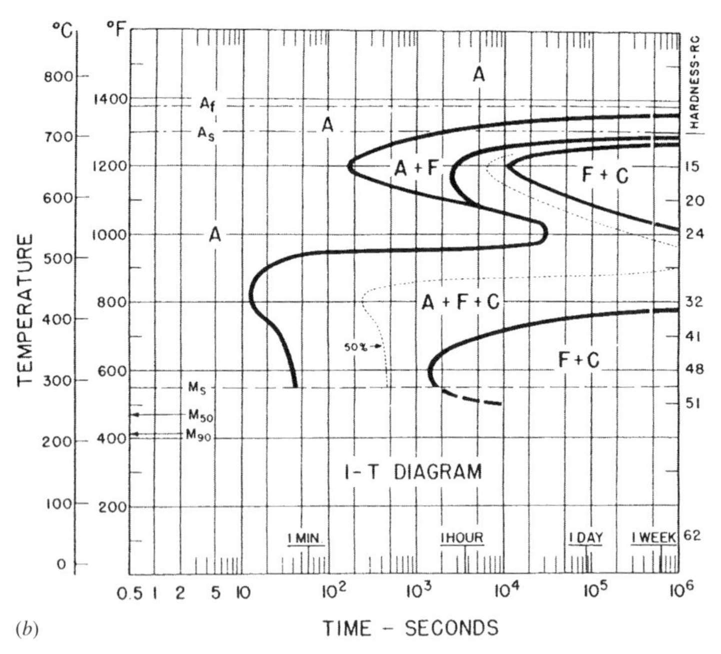

(*b*)

Fig. 4-2. Isothermal transformation diagrams (*a*) For a low carbon steel, (*b*) For a low alloy steel. (Courtesy of United States Steel, 2.)

influence of grain size d on the yield strength σ_Y is expressed through the Hall-Petch relationship:

$$\sigma_Y = \sigma_0 + k_Y d^{-1/2} \tag{4-1}$$

where σ_0 is a measure of the inherent resistance of the lattice to plastic deformation, and k_Y is a constant. For a 0.2% carbon steel, the value of σ_0 at the yield point is 70 MPa (10.1 ksi), and this value increases to 300 MPa (43.5 ksi) at a strain of 0.01, which is the maximum strain at the lower yield point. The corresponding values of k_y are 0.75 and 0.4 MPa $m^{1/2}$ (0.68 and 0.6 ksi $\sqrt{\text{inch}}$). Changes in the values of the constants with strain occur because of the increase in dislocation density with strain.

(b) *Austenite:* A solid solution of one or more elements in face-centered cubic (fcc) iron. In the iron-carbon system, it is known as the γ phase. The addition of 18% Cr and 8% Ni to low-carbon iron creates the class of alloys known as austenitic stainless steels, which are austenitic even at room temperature. They are noted for their resistance to corrosion, which is provided by a protective oxide. However, on cooling after welding some of the chromium can come out of solution to form grain boundary carbides, thus reducing the effectiveness of the protective oxide immediately adjacent to the weld and leading to "knife-line" attack, a form of intercrystalline corrosion.

(c) *Cementite:* A compound of iron and carbon having the formula Fe_3C, and also known as iron carbide.

(d) *Peritectic Reaction:* On heating, the γ phase divides into liquid and δ (bcc) phases.

(e) *Eutectoid Reaction:* On cooling, the γ phase divides into α and Fe_3C phases.

(f) *Eutectic Reaction:* On cooling, the liquid phase divides into γ and Fe_3C phases.

(g) A_1: This is the lower transformation temperature.

(h) A_3: This is the upper transformation temperature.

(i) *Transformation range:* This is the region on the phase diagram between the A_1 and A_3 lines.

(j) *Pearlite:* This is a lamellar aggregate of ferrite and cementite occurring in steel and cast iron.

(k) *Bainite:* This is a decomposition product of austenite that consists of an aggregate of ferrite and carbide and that forms at a temperature intermediate between the pearlite and martensite ranges.

(l) *Martensite:* This diffusionless transformation product of austenite that forms below the M_s (martensite start) temperature. Low-carbon martensites are bcc, and have a ductile, lathlike microstucture. High-carbon martensites are bcc tetragonal, and have a more brittle, platelike microstructure. There is a volume expansion of 1–3% when martensite forms, and the martensite is heavily deformed. As a result, martensite contains a high density of dislocations, which contributes to its high strength. The M_s temperature is dependent upon alloy content. The following equation is typical of the empirical expressions that have been developed to relate the M_s temperature and composition in carbon and low-alloy steels (3):

$$M_s\ (°F) = 930 - 570 \times \%C - 60 \times \%Mn - 50 \times \%Cr - 30 \times \%Ni - 20 \times \%Mo - 20 \times \%W, \quad (4\text{-}2a)$$

$$M_s\ (°C) = 499 - 317 \times \%C - 33 \times \%Mn - 28 \times \%Cr - 17 \times \%Ni - 11 \times \%Mo - 11 \times \%W. \quad (4\text{-}2b)$$

It is seen that carbon has a particularly strong effect on the M_s temperature.

(m) *Retained Austenite:* This is austenite that did not transform on cooling.

Some common heat treating and working operations are:

(a) *Recrystallization:* The formation of a new, strain-free grain structure from that existing in worked metal.

(b) *Hot Working:* The deforming of a metal at a sufficiently high temperature that dynamic recrystallization occurs so that there is no stain hardening.

(c) *Cold Working:* The deformation of a metal at a temperature below the recrystallization temperature.

(d) *Hot Forging:* The forging of a metal, usually in the temperature range of 1100–1150°C (2000–2100°F).

(e) *Annealing:* Furnace cooling from a temperature just above A_3.

(f) *Normalizing:* Air cooling from a temperature just above A_3.

(g) *Hardenability:* With respect to the isothermal transformation diagrams, Fig. 4-2, hardenability relates to the time available on cooling austenite to avoid transformation to ferrite-pearlite, and transform instead to bainite or martensite. The larger the time available, the greater the hardenability. One of the principal reasons for alloying a steel is to increase its hardenability, and this is clearly seen in comparing Figs. 4-2*a* and 4-2*b*. The hardenability of certain steels, such as the maraging steels, is sufficiently high that heavy sections transform to martensite on cooling from the austenitic range. An increase in the austenite grain size will also increase hardenability, as this reduces the number of sites at which transformation is initiated.

(h) *Tempering:* Heating a quenched alloy to a temperature below the transformation range to produce desired changes in properties. The resistance to fracture of an alloy can be strongly dependent upon heat treating and processing parameters. For example, HY 80 steel, an alloy use for submarine hulls, is significantly tougher in the quenched and tempered condition than in the normalized condition, due to differences in grain size and distribution of precipitates.

(i) *Spheroidizing:* A heat treatment carried out just below A_1 during which lamellar cementite becomes globular.

(j) *Stress Relieving:* Holding at a temperature near 650°C (1200°F), usually for several hours, to reduce residual stress. In the case of weldments, the weld microstructure will be tempered.

(k) *Carburizing:* The process of diffusing carbon into the surface layers of austenitic steel, usually carried out at temperatures of 925–980°C (1700–1800°F) in a time of the order of an hour.

(l) *Nitriding:* The process of diffusing nitrogen into the surface layers of ferritic steel, usually carried out at temperatures of 480–540°C (900–1000°F) in a time of 8–24 hours.

(m) *Banding:* A heterogeneous microstucture with segregates aligned in filaments or planes parallel to the direction of working. An example is shown in Fig. 4-3.

(n) *Preheating in Welding:* Heating the region to be welded just prior to welding in order to reduce cooling rates after welding to prevent the formation of martensite. The recommended preheat temperature varies with the carbon content of the steel. For a 0.2 C steel it would be 95°C (200°F). For a 0.7 C steel it would be 315°C (600°F).

Some problems associated with hot working and heat treating and processing are:

(a) *Hot Shortness:* This is a type of brittleness that develops in copper-containing steels in the hot working range where the solubility of copper in γ-iron is as much

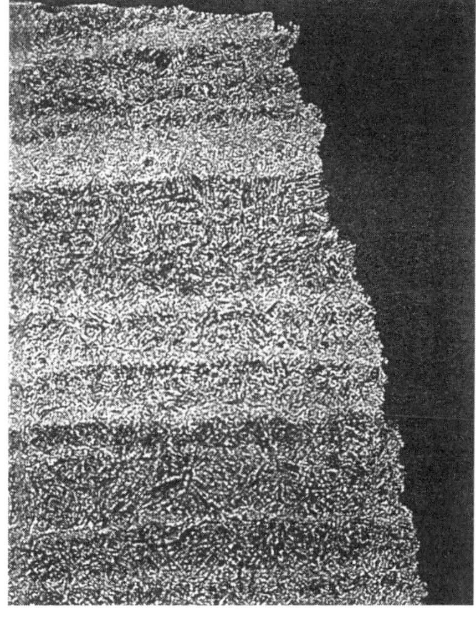

(*a*) (*b*)

Fig. 4-3. An example of banding in a medium-carbon steel. (*a*) Magnification: 50×, (*b*) Magnification: 400×. (Courtesy of S. Crosby.)

as 9%. The cracking associated with hot shortness occurs from the surface inwards. In the presence of oxygen, three layers of iron oxides form. The outer oxide is hematite, Fe_2O_3, and is reddish-brown to black in color. The intermediate layer is magnetite, Fe_3O_4, and is black in color, and the inner layer is wustite, FeO, which is also black in color. At elevated temperatures, a metallic phase (95% Cu, 5% Fe) appears in the FeO. This phase remains dispersed in the FeO at relatively low temperatures near 900°C, but is concentrated as a nearly continuous layer at temperatures of 1050–1200°C along the metal/oxide interface. Since the melting point of copper is 1084°C, detrimental penetration of the austenitic grain boundaries by the liquid or nearly liquid copper-rich phase occurs rapidly. Because of concerns about hot shortness, the amount of automotive scrap is limited in the making of quality steel due to the high copper content of this type of scrap.

(b) *Temper Embrittlement:* Temper embrittlement occurs at a higher temperature than does blue brittleness, and is due to the segregation of elements such as antimony and phosphorous to prior austenitic grain boundaries. In the heat treating of heavy sections, either during temperature rise or temperature fall, there is a danger of temper embrittlement because the relatively slow rate of temperature change may provide sufficient time for deleterious forms of segregation to occur. A similar type of embrittlement can occur during long-time service at elevated temperatures. For example (4), the flues of a furnace were constructed of a steel that contained 0.09% phosphorus, a relatively high value, and after 30,000 hours of service at 430°C, the steel was found to be severely embrittled. The room temperature Charpy energy of the steel had decreased from an initial value of 80 J to 4 J. The addition of molybdenum can offset the embrittling effects of phosphorus, but molybdenum was not an alloying element in this steel, and the embrittlement was attributed to the high phosphorus content.

(c) *Blue Brittleness:* This type of embrittlement is associated with strain aging and the precipitation of carbides at temperatures near 350°C. Since this reaction is promoted by cyclic or unidirectional plastic deformation, the working of steels in this range is not recommended. Figure 4-4 shows, for several low-carbon steels, the extent of the drop in toughness that occurs in this range. This type of embrittlement is referred to as the "blue brittleness," because of the color of the oxide formed on a steel.

(d) *Quench Cracking:* The objective in quenching a steel is to prevent transformation at temperatures above the martensite range. Susceptibility to quench cracking increases with the severity of the quench and with carbon content. The influence of carbon is primarily due to its effect on the M_s temperature. The M_s temperature decreases from about 425°C (800°F) at 0.2% C to 150°C (300°F) at 1.0% C, and as the M_s temperature decreases, the volume expansion associated with the transformation increases. If the hardenability is high enough, when the interior transforms, the already transformed, relatively brittle surface will be placed in tension and may crack. Further, the potential for cracking is increased by the presence of stress raisers. In order to avoid delayed cracking after the quench, a tempering

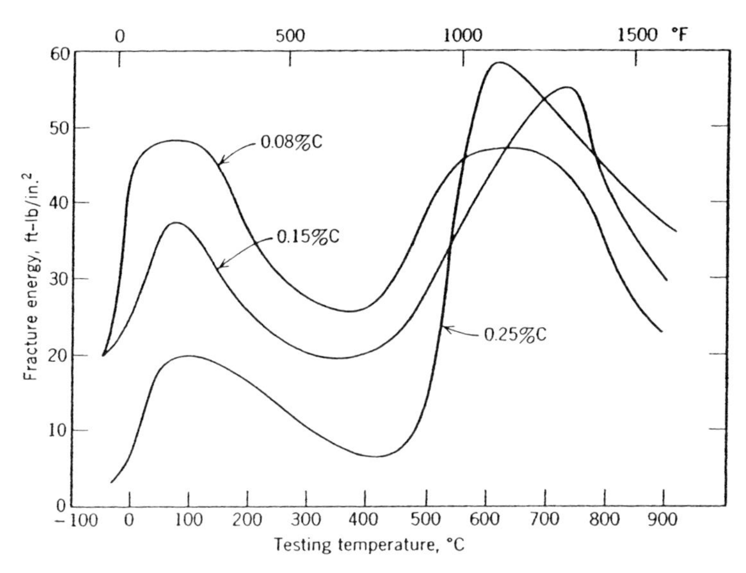

Fig. 4-4. Impact energy curves for low-carbon steels in which blue brittleness occurs around 300°C. (From Reinhold, 11.)

operation should immediately follow the quench. Allowing fully quenched pieces to stand overnight before tempering increases the probability of crack formation.

B. Aluminum Alloys

There are two classes of aluminum alloys, heat-treatable and nonheat-treatable. The former category attains strength properties by a process of quenching and aging, while the latter category is strengthened by cold work. The equilibrium diagram for the Al-Cu system, a typical heat-treatable alloy, is shown in Fig. 4-5 (5). Commercial alloys of this type also contain additional alloying elements such as silicon, iron, manganese, magnesium, lithium, and zinc. For those alloys in which copper is the principal alloying element, the amount of copper is about 4%. From Fig. 4-5, it is seen that this amount of copper is in solid solution at 500°C. However, on cooling from this temperature range, a phase boundary is crossed and precipitation occurs, either naturally at room temperature or artificially at a slightly elevated temperature. During aging the first precipitates to form are copper-rich metastable regions known as Guinier Preston (GP) zones. There are two types, the first to form is known as GP I, and upon further aging, the GP I zones evolve into GP II zones. Both of these zones are fully coherent with the parent fcc lattice, meaning that the crystallographic planes of the zones are continuous with the parent lattice in three dimensions, although the lattice parameters may differ slightly. As a consequence, at a sufficiently high shear stress, dislocations are able to pass from the parent matrix and shear

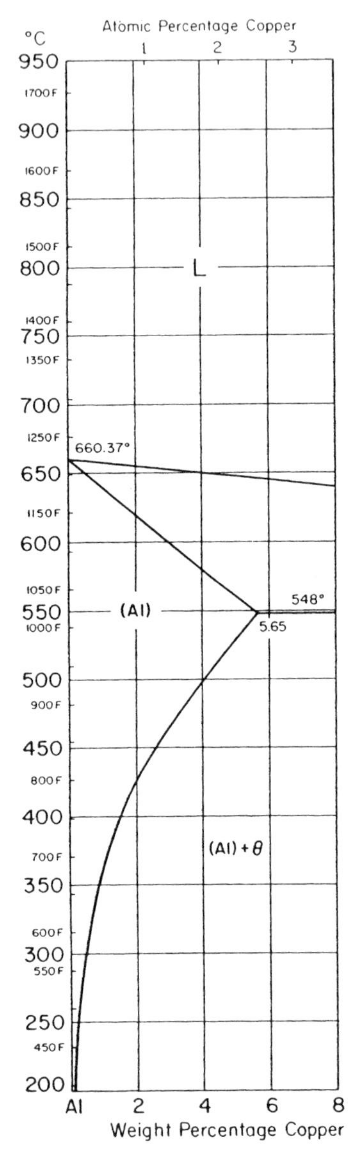

Fig. 4-5. Al-Cu phase diagram. (*L* = liquid) (Reprinted with permission of ASM International, 5.)

through the coherent zones. With further aging, the semicoherent phase θ' appears, and with yet further aging, the noncoherent equilibrium phase θ ($CuAl_2$) with its own crystal structure appears. Maximum strength is reached when both GP II and the θ' phases are present. When the precipitate is θ, the strength is less than in the peak-aged condition, and the system is considered to be in an overaged condition.

In contrast to the ferrite-pearlite system, the grain size is not a factor determining the strength of aluminum alloys; rather it is the spacing of the precipitates that governs the shear stress τ required for plastic deformation. The equation is

$$\tau = \frac{\mu b}{l}, \tag{4-3}$$

where μ is the shear modulus, b is the Burger's vector, and l is the spacing between particles. At a spacing of several microns, typical of the overaged condition, the ratio of b to l is about 10^{-4}. For a shear modulus of 3.4×10^6 psi, the value of τ would be only 340 psi, a negligible amount. For particle strengthening to be effective, the value l must be much smaller.

C. Titanium Alloys

The density of titanium, 4.507 gm/cc, is midway between that of steel, 7.87 gm/cc, and of aluminum, 2.699 gm/cc. It has a greater Young's modulus, 116 GN m^{-2} (16.8×10^6 psi), than aluminum, 69 GN m^{-2}(10×10^6 psi), and has useful long-term elevated properties up to 540°C (1000°F), as well as good corrosion resistance. All of these characteristics make titanium and its alloys attractive for aerospace applications, and they are therefore used up to moderate temperatures in jet engines and as major structural components of aircraft as well. The three principal alloy types are known as α alloys, α-β alloys, and β alloys.

In pure titanium, there is an allotropic transformation at 885°C from the low-temperature hexagonal α phase to the high-temperature bcc β phase. The alloying elements aluminum, tin, carbon, oxygen, and nitrogen stabilize the α phase, and Ti-5Al-2.5Sn is an example of an important α phase alloy. Oxygen, carbon, and nitrogen are usually considered as impurities in titanium, but small quantities of oxygen are used as strengthening agents in most alloys, although extra-low interstitial (ELI) alloys are available if desired. The elements vanadium, zirconium, and molybdenum stabilize the β phase and cause it to exist as the stable phase at temperatures well below 885°C. In fact, the Ti-4.5Sn-6Zr-11.5Mo alloy is β stable at room temperature, and is therefore known as a β alloy.

Figure 4-6 is a pseudobinary equilibrium diagram for the (Ti-6Al)-V system (6). In this system, the aluminum and vanadium atoms are in solid solution above the β-transus. The addition of α-stabilizer aluminum to titanium increases the range of the α field, whereas vanadium expands the β field. Both of these elements are present in the most common α-β alloy, Ti-6Al-4V in weight %, or Ti-10Al -4V in atomic %. In common with steels, a variety of microstructures can be developed in titanium alloys depending upon composition and cooling rate. Ti-6Al-4V can be strengthened by quenching from above or below the β-transus followed by aging at moderately elevated temperature. However, at temperatures above the β-transus, the β grains grow rapidly, and therefore heat treating above the β-transus may be undesirable. The alloy is therefore often heat-treated at a temperature some 15°C below the β-transus to reduce the amount of α phase. It is then rapidly cooled to trans-

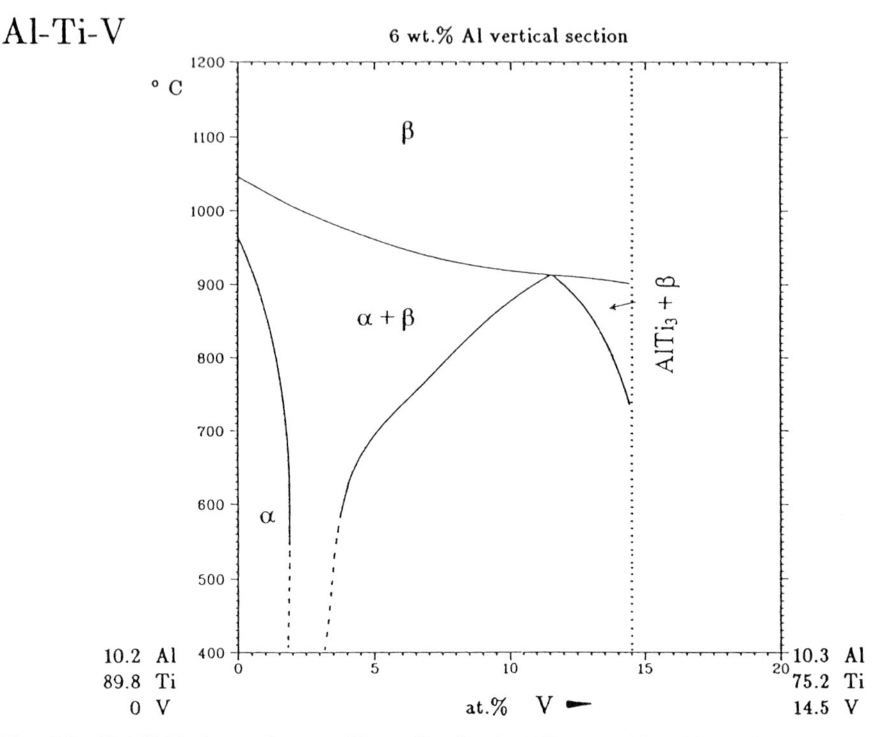

Fig. 4-6. (Ti-6Al)-V phase diagram. (From Handbook of Ternary Alloy Phase Diagrams, 6. Reprinted with permission of ASM International.)

form some of the β phase to martensite and then aged at a temperature of about 700°C. This type of heat treatment is referred to as a solution treated and aged (STA) heat treatment. If the aging treatment is carried out at a somewhat higher temperature, the material is said to be in a solution treated and overaged (STOA) condition. During these aging treatments, fine volumes of α form in the β. An aging heat treatment at temperatures below 510°C can result in a β-to-ω transformation. The ω phase embrittles titanium alloys, but aging for sufficient time, for example, eight hours serves to eliminate the ω phase and to restore ductility.

An undesired nitrogen-stabilized phase, referred to as hard α, can be present in Ti-6Al-4V and act as a site for relatively easy crack initiation. To avoid the presence of this phase, triple vacuum-remelt procedures are used. Another undesired phase is known as α case. This phase can form as a surface oxide of some thickness if the heat-treating atmosphere contains oxygen. It is undesired because of its hard, brittle nature.

D. Nickel-Base Superalloys

Nickel-base superalloys are widely used in the hot sections of gas turbine engines. They may be in the form of forged or cast polycrystalline alloys, cast directionally solidified alloys, or cast single crystals. The use of directional solidification and single crystals is aimed at minimizing or eliminating the deleterious effects of grain boundaries in the creep range.

The unique microstructure of nickel-base superalloys permits the alloys to be used for short periods of time at remarkably high fractions of their melting points, for example, $0.8T_M$. The microstructure is based upon a two-phase system known as $\gamma - \gamma'$, where γ is the fcc nickel-rich matrix phase, and γ' is a precipitate phase. The γ' phase is an ordered fcc phase based on the composition Ni_3Al. The unit cell of the γ' phase consists of aluminum atoms at the cube corners and nickel atoms at the centers of the cube faces. This ordered phase is coherent with the parent matrix and is often of cuboidal form, as indicated in Fig. 4-7. The lattice parameter of γ' can be either slightly larger or slightly smaller that of that of the parent matrix, depending upon the particular alloy and temperature, a factor contributing to the stability of the phase at elevated temperatures. The coherency strains are also considered to contribute to the strength of these alloys. Since the γ' phase is ordered, the dislocations gliding through it do so in pairs. As the first dislocation glides through, the ordered arrangement of atoms along the glide plane is disturbed. The second dislocation is needed to restore order. The relatively high resistance to plastic deformation of $\gamma - \gamma'$ alloys has been attributed to the difficulty that the first dislocation experiences in entering the γ' phase from the γ phase. The climb of dislocations to bypass the γ' phase is also a possibility at elevated temperatures.

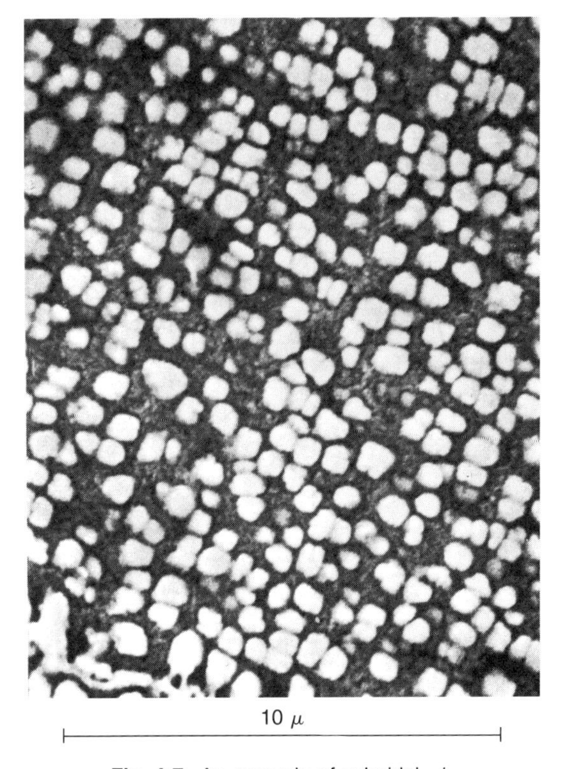

Fig. 4-7. An example of cuboidal γ'.

The phase diagram for the nickel-aluminum system is shown in Fig. 4-8 (7), and it is seen that the γ' phase (86.7 wt % Ni, 13.3 wt % Al) is ordered to the peritectic temperature, 1395°C (2543°F). However, commercial two-phase alloys contain less than 13.3 wt % aluminum and also contain titanium as well as a number of additional alloying elements, and therefore they are a mixture of the phases γ and γ'.

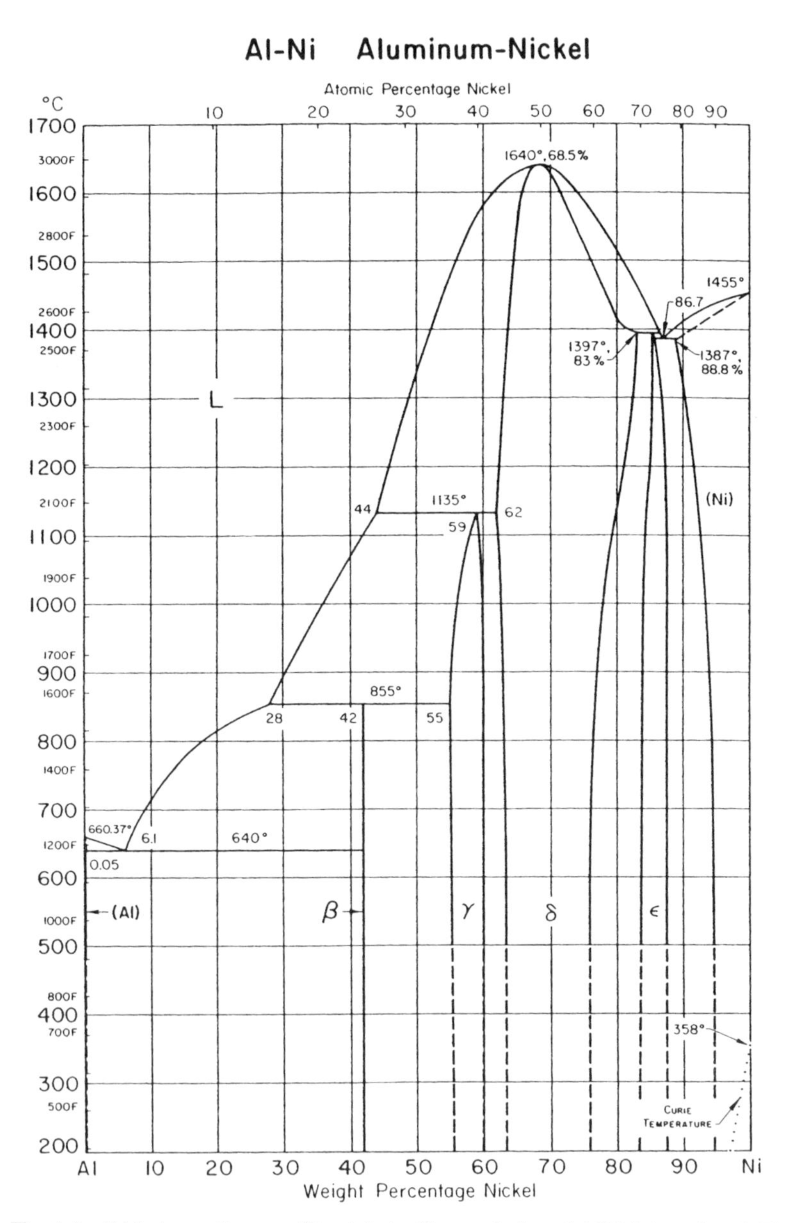

Fig. 4-8. Al-Ni phase diagram. (Reprinted with permission of ASM International, 7.)

As shown in the following table, the composition of an alloy depends upon whether the alloy is to be used as either a polycrystal or a single crystal. The single crystal contains fewer alloying elements because many of the elements contained in the polycrystal are there to strengthen the grain boundaries, which of course are not present in single crystals. Single crystal blades are inspected for casting porosity at the time of manufacture. However, if a turbine blade containing undetected porosity should enter service, there is a possibility that the pores will act as nuclei for fatigue crack formation. Another cause for concern is foreign object damage since the damaged area may recrystallize in polycrystalline form and be more prone to crack initiation than an undamaged single crystal would be. As a result, the creep and fatigue properties can be significantly degraded.

COMPOSITION OF MAR-M200

	Cr	W	Co	Al	Ti	Nb	C	B	Zr	Ni
Single Crystal	9.0	12.0	10.0	5.0	2.0	1.0	—	—	—	Bal.
Polycrystal	9.0	12.0	10.0	5.0	2.0	1.0	0.15	0.15	0.05	Bal.

Both polycrystal and single-crystal nickel-base superalloys exhibit an elevated temperature region in which the alloying elements are in solid solution. The γ' solvus marks the lower equilibrium boundary of this region. The temperature of the γ' solvus depends upon the sum of the aluminum and titanium weight percentages. At a sum of 4 wt %, the solvus is at 1900°F (1038°C), and at a sum of 8 wt % it is at 2100°F (1150°C). Upon aging below the γ' solvus, precipitation occurs. A typical aging treatment involves aging at a temperature 20°C below the solvus for several hours to develop primary precipitates, followed by further aging for up to 20 hours at a temperature some 200°C less than the solvus temperature to bring about secondary precipitate formation. The secondary precipitates are smaller than the primary, and are said to improve the creep strength. Figure 4-9 shows an example of the effect of aluminum plus titanium content on the 100-hour creep rupture strength of several nickel-base superalloys at 1600°F (870°C).

If the temperature of a blade should rise in service due to poor convective cooling or to improper operating conditions, it is possible for the γ' solvus to be exceeded. The γ' phase would then go back into solution and reprecipitate later on slow cooling as fine γ'. An example of this is shown in Fig. 4-10. Therefore, some information on the thermal history of a blade may be obtained through examination of microstructure.

Up to about 850°C (1560°F), the γ' particles retain their cuboidal shape and during deformation are cut by dislocations. At higher temperatures, the γ' precipitates tend to transform to platelike particles known as rafts, which for most superalloys under uniaxial stress lie perpendicular to the tensile stress axis or parallel to the compressive stress axis (8). Usually these rafts accelerate the creep rate and thus reduce the creep strength. The rate of fatigue crack propagation is also enhanced when the rafts lie perpendicular to the stress axis. Conversely, the rate is decreased when the rafts are parallel to the axis of cyclic stressing.

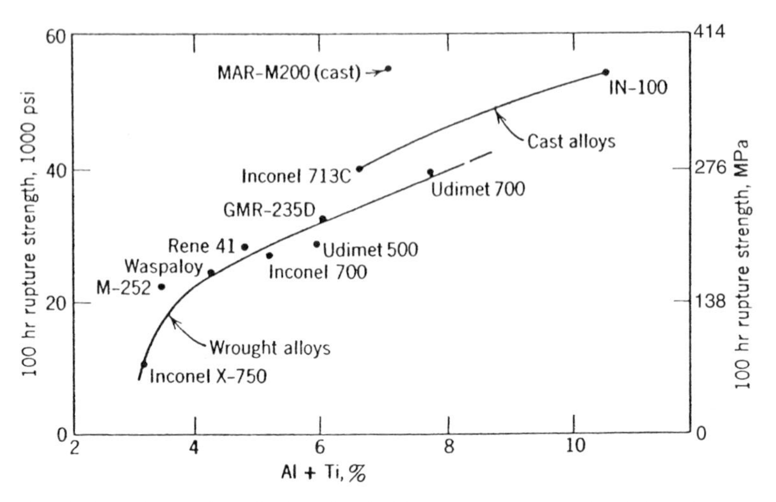

Fig. 4-9. Effect of the (Ti + Al) content on the 100-hour rupture strength of nickel-base superalloys at 870°C. (Reproduced by permission of ASM International, 8.)

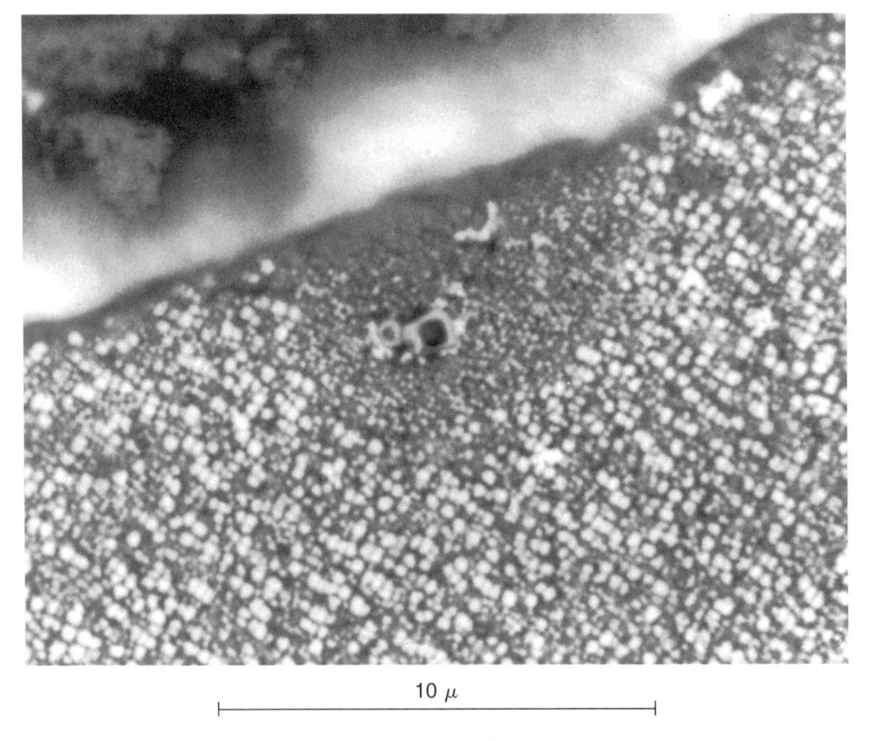

Fig. 4-10. Reprecipitated γ'.

III. COATINGS (9)

The turbine blades and vanes in jet aircraft engines and in stationary power generating units are often coated for protective reasons. After solution heat treating, the coating application process is carried out at a high temperature below the γ' solvus, and therefore primary γ' appears as the coating is being applied. After the coating is applied, secondary precipitation may be induced at a lower temperature. As a result, there will be a bimodal distribution of γ' sizes, as shown in Fig. 4-11, with the larger size being associated with precipitation during the coating process, and finer precipitate sizes being associated with aging at a lower temperature.

The coatings that are used to protect components in the hot section of gas turbine engines are of two types: (a) oxidation-resistant coatings and (b) thermal-barrier coatings (TBCs), which are used to reduce the operating temperatures of blades and vanes by lowering the heat flux across the component wall. Oxidation-resistant coatings are used on the first- and second-stage superalloy blades and vanes of most advanced gas turbines, as well as on internal cooling passages of the hotter parts of blades. There are two standard coating groups: diffusion coatings, including aluminides and modified aluminides; and overlay coatings of the MCrAlY type, where M can be nickel, cobalt, or a combination of the two. The coating systems are designed to form stable, continuous, and slow-growing protective alumina scales. Diffusion coating is a surface modification process wherein the coating elements are deposited on the surface, and then the coating species is diffused into the

Fig. 4-11. Bimodal distribution of γ'.

substrate to form a protective layer, which is typically 10–100 μm in depth. The phase providing the aluminum for the protective oxide scale is β-NiAl. Incorporating platinum in the aluminide diffusion coating improves the oxidation and hot corrosion and sulfidation resistance. MCrAlY overlay coatings are often deposited by plasma spraying, and contrary to diffusion coatings, they do not consume the substrate material. This is an advantage in repair, since stripping a diffusion coating can lead to appreciable loss of the substrate metal. In general, the mechanical properties of overlay coatings are better than those of aluminide coatings.

Thermal barrier coatings consist of either an MCrAlY or a platinum-modified aluminide diffusion coating, which is known as the bondcoat, together with a ceramic top coat. In electron beam physical vapor deposition (EB-PVD) coatings, a continuous, thin (1-μm), thermally grown oxide (TGO) layer is developed prior to the deposition of the ceramic. This layer imparts good adhesion to the TBC and protects the substrate from oxidation and corrosion. A ceramic outer layer, the topcoat of 6–8 wt % yttria-stabilized zirconia (YSZ) may then be deposited by EB-PVD. This top coat grows in the form of columnar grains on the bond coat. The grains are strongly bonded to the substrate, but weakly bonded to each other, providing good thermal strain tolerance. The typical thickness of a TBC on a turbine blade is 125–200 μm.

Air plasma sprayed (APS) coatings are also used. These coatings are deposited in layers, and their microstucture does not have a columnar structure, but consists instead of individual platelets formed from droplets impinging on the surface during the spraying process. For APS coatings, the TGO between the top coat and the bond coat develops under service conditions.

In service, the lifetime of components protected by an oxidation-resistant coating is often limited by the premature cracking of the protective layer, due to thermal-mechanical fatigue or thermal shock, exposing the substrate alloy to incipient cracking and to attack by the oxidizing atmosphere. The tendency for cracking under thermal-mechanical loading is a function of the difference between the coefficients of thermal expansion (CTE) of the coating and the substrate as well as of the thickness of the coating, for thicker coatings fail at a lesser strain than do thinner coatings. With TBCs on the other hand, the interface between the coating and substrate is often the weak link. Cracking of this interface can result in large-scale delamination due to the ceramic top coat spalling off from the underlying bond coat. In the failure process, the difference in CTE is again an important factor, with spallation of the top coat often being triggered by small buckles that form at defects in the interface.

IV. SUMMARY

This chapter has briefly reviewed some of the principal characteristics of widely used alloys and coatings. In a failure analysis, it is at times important that the microstructure of the material examined be in the condition specified. If not, the cause of the discrepancy may have to be established, and this may require a more detailed knowledge of material processing than has been provided herein. The many volumes of the *ASM Handbook* series are vauable sources of such information.

REFERENCES

(1) Smithells Metals Reference Book, 6th ed., ed. by E. A. Brandes, Buttersworth, London, 1983, 11–143.

(2) Isothermal Transformation Diagrams, US Steel, Pittsburgh, PA, 1963.

(3) P. Payson and C. H. Savage, Trans. ASM, vol. 33, 1944, p. 261.

(4) D. R. H. Jones, in Fracture Mechanics: Applications and Challenges, ed. by M. Fuentes et al., ESIS Pub. 26, Elsevier, Oxford, 2000, pp. 29–46.

(5) ASM Metals Handbook, vol. 8, ASM, Materials Park, OH, 1973, p. 259.

(6) Handbook of Ternary Alloy Phase Diagrams, ed. by P. Villars, A. Prince, and H. Okamoto, vol. 4, ASM, Materials Park, OH, 1995, p. 4389.

(7) ASM Metals Handbook, vol. 8, ASM, Materials Park, OH, 1973, p. 262.

(8) H. C. Cross, Metals Progr., vol. 87, 1965, p. 67.

(9) H. Mughrabi, in Fracture Mechanics, Applications and Challenges, ed. by M. Fuentes et al., ESIS Pub. 26, Elsevier, Oxford, 2000, pp. 13–28.

(10) J. Bressers, S. Peteves, and M. Steen, in Fracture Mechanics, Applications and Challenges, ed. by M. Fuentes et al., ESIS Pub. 26, Elsevier, Oxford, 2000, pp. 115–134.

(11) O. Reinhold, Ferrum, vol. 13, 1916, p. 97.

5

Examination and Reporting Procedures

I. INTRODUCTION

Important aspects of failure analyses are the safekeeping and recording of evidence, the determination of dimensional characteristics, alloy identification, the study of fracture surfaces (fractography), and the determination of residual stresses by X-ray analysis. Some of the tools used in the examination of materials are quite simple, such as a small permanent magnet used to determine whether or not a metal is ferromagnetic. On the other hand, some tools are quite sophisticated, such as a scanning electron microscope used in the determination of fractographic detail. This chapter describes some of the more common methods used in the examination of materials and fracture surfaces, both in the field and in the laboratory. In addition, the topics of report preparation and testifying are also discussed.

II. TOOLS FOR EXAMINATIONS IN THE FIELD

The equipment needed for an on-site investigation will vary depending upon the particular circumstances. Some of the more useful items are as follows:

(a) Video camera
(b) Polaroid camera
(c) 35 mm camera or digital camera equipped with close-up lens
(d) Permanent magnet to determine if part is ferromagnetic (steel) or not
(e) Micrometers of various sizes
(f) Rulers

(g) Tape measure
(h) Log book
(i) Writing equipment.
(j) Polishing, etching, and replicating equipment
(k) Dye penetrant kit
(l) Magnifying lens
(m) Tape recorder

III. PREPARATION OF FRACTURE SURFACES FOR EXAMINATION

One of the complications associated with the examination of a fracture surface is that corrosion subsequent to fracture may have obscured some of the fractographic features. This can be a particular problem with carbon steel and cast iron components, since they rust very quickly in humid atmospheres. Fortunately, aluminum alloys and stainless steels are much less prone to corrosion. In some litigations, steel components have been stored outdoors for years and have become heavily coated with rust prior to examination. If the rust coating on a fracture surface is a light one, it can be removed from the fracture surface by brushing on a commercial rust remover such as naval jelly and wiping it off after a few minutes. Nevertheless, sharp fractographic features are likely to be rounded. Where the rust coating is heavy, important fractographic features may be absent upon removal of the rust layer.

Ultrasonic cleaning is used to rid a fracture surface of debris, dirt, and oil coatings. Ultrasonic cleaning uses sound waves passed at very high frequency through liquid cleaners, which can be alkaline, acid, or even organic solvents. The passage of ultrasonic waves creates tiny gas bubbles, which offer vigorous scrubbing action on the parts being cleaned. This action results in an efficient cleaning process, and is used as a final cleaner only, after loosely adherent debris has been removed.

IV. VISUAL EXAMINATION

Visual examination is one of the most important steps in fractographic analysis, but there are some commonsense precautions to be observed. For example, in the examination of the two halves of a failed component, the broken halves should not be fitted together to see how they fit because fractographic information contained on the surfaces may be altered. The fracture surfaces should not be touched because the salt contained in the moisture on fingertips may lead to corrosion. If possible, apply a removable coating of oil or plastic compound to the surfaces to prevent any further corrosion.

The instruments used in visual examinations include dental mirrors, borescopes, and flexible fiber-optic scopes for use in examining internal surfaces that are not accessible to direct viewing. Steel rules, tapes, calipers, and weld gauges are among the tools used to check on dimensional characteristics of failed components. A variety of cameras, including Polaroid and video cameras, are used to record the macroscopic features of failures. When making observations or when taking photographs, it is good practice to maintain a detailed log listing the observations and the sequence of photographs together with the subject matter of each photograph. A tape recorder is also useful in this regard.

Much information about the nature of a fracture can be obtained through visual examination. The human eye has a good depth of field, so that even though a fracture may be rough, it will appear to be in focus to the viewer. Differences in color and evidence of corrosion or wear can be detected, the extent of plastic deformation or necking can be observed, and evidence of fatigue is often visible to the naked eye. The main limitation of a visual examination is the inability to resolve fine detail. The Rayleigh criterion for the limit of resolution gives the linear separation z of two just resolvable point objects at a distance of 250 mm (the minimum distance of distinct vision) as

$$z = \frac{0.61\lambda_o}{NA}, \tag{5-1}$$

where λ_o is equal to 550 nm, the wavelength to which the eye is most sensitive, and NA is the numerical aperture of the eye. The NA is expressed as $n \sin u$, where n is the index of refraction, equal to 1.0 in air, and $\sin u$ is the sine of the half-angle of the cone of light admitted to the eye from a point at 250 mm from the eye. Since the half-diameter of the pupil is 1 mm, the value of the NA is 0.004. Therefore, the value of z is approximately 0.1 mm or 100 μm, which, as a point of reference, is about the diameter of the human hair and 100 times larger than the micron, a common length standard in discussing metallurgical microstructures.

V. CASE STUDY: FAILURE OF A STEERING COLUMN COMPONENT

In the course of an accident investigation of an automobile crash in which there was a fire and the two occupants perished, it was found that a carburized pinion gear tooth of a steel component of the steering mechanism had fractured. Visual examination of the fracture surface of the component revealed two distinctly different color patterns. At the edge of the carburized tooth where the fracture had initiated, the fracture surface was flat and deep blue in color. Further into the tooth, the fracture was in shear at an angle to the flat portion of the fracture. The color in the sheared region was a purplish-pink. Because these two different color patterns were present on the same fracture surface, it was asserted that the part had been exposed to an elevated temperature twice in its history. It was claimed that the component must have been cracked during manufacture prior to final heat treatment, during which some oxide had formed on the cracked surfaces. It was further claimed that this crack caused the component to fail, and thus caused the accident. The claim was that both the old fracture surface and the new fracture surface were oxidized during the ensuing fire, but the old fracture surface had a different appearance because of two exposures to elevated-temperature oxidizing conditions.

To check on these allegations, a pinion gear was broken from an identical component, and then exposed to 350°C for twenty minutes. It was found that this single exposure developed a color pattern similar to that observed on the failed pinion gear tooth. Additional tests with fatigue-cracked steel specimens were carried out to support these observations. The specimens were precracked in fatigue to develop a short fatigue crack and a corresponding flat fracture surface. The remaining ligament was then broken by overloading to develop a rough fracture surface. These specimens were then heated in air to a temperature of 350°C. It was found that the flat fatigue-cracked regions developed a blue oxide, whereas the rough fractures associated with the overload developed a pink-purple oxide. On the basis of such tests, it was shown that the fracture of the pinion gear tooth was the result of the crash rather than its cause.

VI. OPTICAL EXAMINATION

The magnification of a simple magnifying lens is given as

$$M = \frac{250}{f}, \tag{5-2}$$

where f is the focal length of the lens expressed in millimeters. For a focal length of 50 mm, a magnification of five times would be obtained. Small magnifying lenses are useful in inspecting fracture surfaces and provide greater detail than is available by visual inspection alone.

The binocular stereomicroscope can provide magnifications up to fifty times, and is a useful tool for the examination of fracture surfaces. This type of microscope has a good depth of field, and when equipped with a camera is useful for taking macrographs of fracture surfaces.

Optical microscopes are capable of obtaining much higher magnifications, up to about four hundred times. Higher magnifications do not increase detail but may allow a larger image to be viewed more easily. If an oil immersion lens is used, the numerical aperture NA can be increased to 1.60, and the limit of resolution z (see Eq. 5-1) is equal to 0.2 μm, of the order of one-half the wavelength of light, and five hundred times better than that of the eye. The main drawback is that, as the magnification increases, the depth of field decreases, and specimens have to be extremely flat to be in focus. While this latter requirement can be met with metallurgically mounted and polished specimens, the inherent roughness of most fracture surfaces limits the usefulness of optical microscopes at high magnification.

In the examination of fracture surfaces, the use of oblique illumination is helpful in bringing out detail by creating a shadowing effect. Normal illumination is useful in observing differences in color or texture, but it is not recommended for fractography because it does not produce the degree of topological contrast obtainable with oblique illumination. A trial-and-error process should be used to find the best inclination angle of the light source with respect to the fracture surface being examined. Rotation of the specimen will help to bring out the fine detail and aid in the identification of fractographic features.

VII. CASE STUDY: FAILURE OF A HELICOPTER TAIL ROTOR

A helicopter containing three passengers and a pilot was on a sight-seeing flight when the tail rotor blade separated from the helicopter. The purpose of the tail rotor blade is to counteract a torque induced by the main rotor blades, and with loss of the tail rotor blade the helicopter goes in to a spin. To counteract this spin, the pilot is trained to disengage the main rotor blades from the rotor and allow the main blades to windmill. Ordinarily, he can then bring the helicopter down safely. However, in this case the pilot did not do this. As a result, one passenger was hurled to his death from the spinning helicopter, and the other three occupants were injured in the ensuing crash. In the subsequent investigation, it was found that a component connecting the tail rotor to the engine had failed in fatigue, and this failure allowed the tail rotor to separate.

What caused the fatigue failure of the component? Was it due to poor design or some other factor? The tail rotor blade was recovered, and it was noted that the tip of the blades showed signs of contact damage. This type of damage can occur as the helicopter is maneuvered on or near the ground on takeoff or landing, and the FAA regulations allow for some minor scratches and indents by specifying the allowable depth of such defects.

To check on the depth of the observed scratches and indents, a 100 power light microscope, which was equipped with a calibrated barrel, was used. Such a microscope has a limited depth of field, and by focusing first on the end surface of the tail rotor blade and then on the bottom of a defect, a determination of its depth could be made. The results of such an examination revealed that the depths of a number of the scratches and indents were in excess of the allowables, and it was concluded that tail strikes had induced abnormally high fatigue loads, which led to the failure of the component. In this case, a small depth of focus turned out to be an asset rather than a detriment. It is also noted that some helicopter manufacturers mount the tail rotor within a containment to protect the blade tips from tip-strikes.

VIII. THE TRANSMISSION ELECTRON MICROSCOPE (TEM)

Prior to the advent of the scanning electron microscope (SEM), the TEM was used to study the topography of fracture surfaces. This was usually done with thin replicas of the surface. These replicas allowed the electron beam to pass through and create image contrast that depended upon thickness variations in the replica. To increase contrast, the replicas were shadowed by evaporating a metal in a metal evaporator. Today, the SEM is the preferred instrument for such studies. However, the scanning TEM is used in research studies of thin sections of a metal that contains a fracture surface to obtain both an image of the surface as well as an image of the underlying dislocation microstructure.

A schematic of the essential features of an electron microscope is shown in Fig. 5-1. There is a source of electrons, the electron gun, and a series of electromagnetic lenses that focus the electron beam on the specimen. The resolving power of an electron microscope is quite high because of the short wavelengths associated with the electron beam. From Eq. 5-1 it can be seen that the smaller λ the smaller

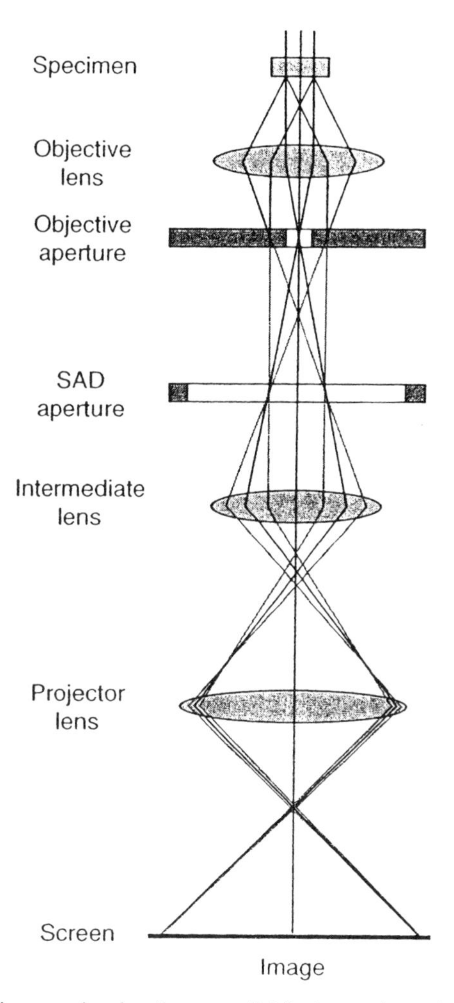

Fig. 5-1. Schematic diagram showing the essential features of an electron transmission microscope, (SAD = Selected Area Diffraction). (After Williams and Carter, 2, with permission of Plenum Press.)

is the separation of points that can just be resolved. An electron can be considered to have a wavelength that is related to its momentum mv by

$$\lambda = \frac{h}{mv}, \tag{5-3}$$

where h is Planck's constant, equal to 6.62×10^{-34} J-sec, m is the mass of an electron equal to 9.11×10^{-31} kg, and v is the velocity of the electron. The velocity of the electron can be expressed as

$$v = \sqrt{\frac{2eV}{m}}, \tag{5-4}$$

where V is the accelerating voltage, and e is the charge on the electron, equal to 1.602×10^{-19} coulombs. This leads to

$$\lambda = \frac{h}{\sqrt{2meV}} = \frac{1.224}{\sqrt{V}} \text{ nm}. \tag{5-5}$$

For an accelerating voltage of 100,000 volts, the value of λ is 0.0038 nm, orders of magnitude smaller than the wavelength of visible light. The theoretical resolution in a TEM image approaches the wavelength of the incident electrons, although this resolution is not attained due to such lens defects as spherical and chromatic aberration and aperture diffraction. Typical resolutions obtained in modern TEMs are approximately 0.2 nm.

IX. THE SCANNING ELECTRON MICROSCOPE (SEM)

The basic elements of an SEM, one of most valuable tools in failure analysis, are shown in Fig. 5-2. As in the TEM, there is an electron source and a series of lenses that focus the electron beam on the surface being examined. Good resolution, often about 10–20 nm, and a large depth of field are features of the SEM. The SEM is usually equipped with an auxiliary energy dispersive X-ray spectroscopy (EDS) unit, which can provide quantitative information on the chemical composition by analyzing the characteristic energies of the X-rays emitted through a depth of about 2 μm below the specimen surface. These characteristic X-rays are emitted when a primary electron displaces an inner electron from its orbit in an element and an outer electron replaces it. This information is displayed in chart form as a plot of X-ray intensity versus wavelength, and is usually sufficient for purposes of alloy identification. To detect light elements such as carbon, nitrogen, and oxygen, special "windowless" detectors, or detectors sensitive to wavelength rather than energy, are used. The X-ray information can also be displayed as a map indicating the distribution of the element within a grain.

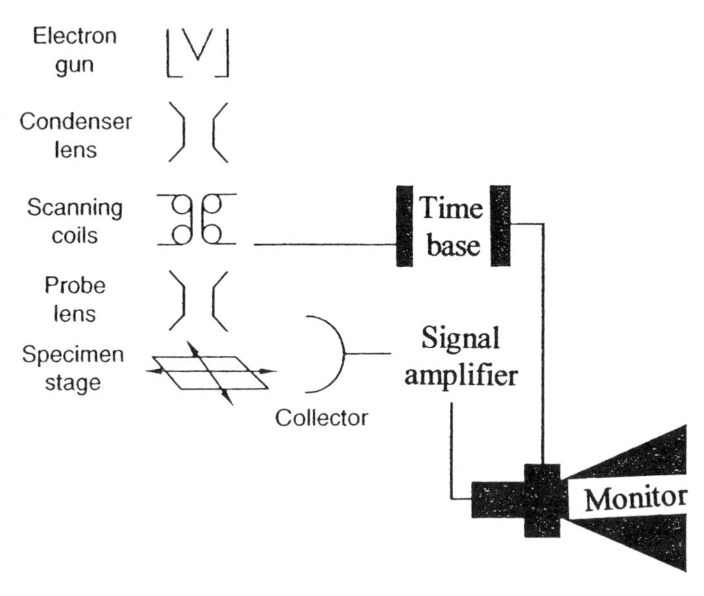

Fig. 5-2. Basic elements of a scanning electron microscope. In the scanning electron microscope, a fine probe of high-energy electrons is focused on to the sample surface and then scanned across the surface in a television raster. A signal generated by the interaction of the probe with the sample is collected, amplified and displayed on a monitor with the same time base as the raster used to scan the sample. (From Brandon and Kaplan, 1. Reprinted by permission of John Wiley.)

In the examination of fracture surfaces, the magnifications usually range from the order of 100 up to 5000 times. One limitation of the SEM is that there is an upper limit to the size of specimen that can be examined, which commonly is about 15 cm × 5 cm × 5 cm, although some specialized instruments have provision for much larger samples. If the available space is too small for the specimen being examined, then a replica technique may be employed. This technique is also useful in obtaining information about fracture surfaces in the field.

In the SEM, an electron beam is scanned (rastered) over the specimen's surface, and images are formed either by the backscattering of the primary electrons or by the emission of secondary electrons. The backscattered electrons are collected by a solid-state device located just below the objective lens pole piece of an SEM, see Fig. 5-3, and are useful in compositional analysis because brighter regions in the

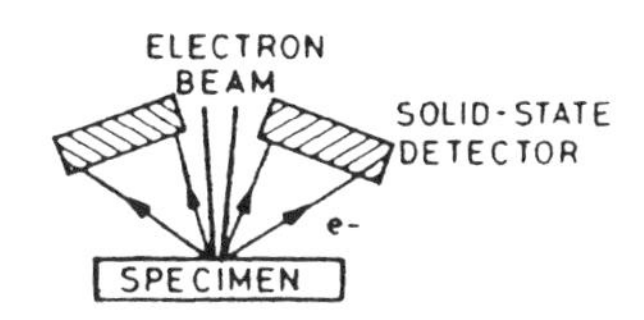

Fig. 5-3. Collector located below the pole piece for backscattered electrons.

image are an indication of material of higher average atomic number. These electrons can also be detected by a charge-coupled device in determining the orientation of individual grains in a polycrystal.

The secondary electrons are electrons that have been displaced from the specimen due to interaction with either primary or backscattered electrons. These electrons are important in bringing out the topography of a fracture surface. The energy associated with the primary electrons is usually in the range of 10–25 keV, whereas the energy of the secondary electrons is much less, for example, 10–50 eV. The emitted secondary electrons are collected by a Everhard-Thornley (ET) detector, which consists of a scintillator biased at +200 volts and a photoemission multiplier, Fig. 5-4. The image created by the collected secondary electrons is displayed on a cathode ray tube (CRT) and can be collected digitally or photographed. Image contrast is due to electron collection differences arising from the tilt of portions of the surface with respect to the electron detector. A region tilted toward the electron detector will give rise to an enhanced signal, while the signal will be reduced if the surface is tilted away from the detector. In addition, contrast is due to emission differences due to variations in the beam-surface tilt angle. Asperities on heavily oxidized materials as well as nonconductors such as ceramics tend to develop charges during SEM observation that detract from the quality of the image. The use of a lower operating voltage or the evaporation of a metallic coating onto the surface will help to reduce the extent of charging.

The ET detector collects both secondary electrons (SE) and backscattered electron (BS) signals, but there are differences in the characteristics of these signals

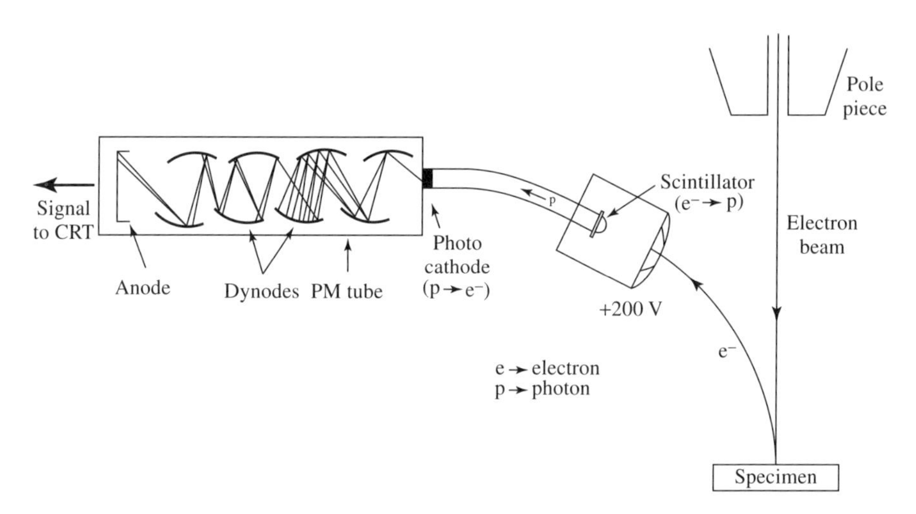

Fig. 5-4. An Everhard-Thornley detector. Electrons from the specimen surface are drawn into the scintillator, generating visible light, which travels via fiber optics to the photocathode. There the light is reconverted to electrons. The electrical signal is multiplied by several electrodes in the photo-multiplier (PM) tube and transmitted to the cathode ray tube (CRT). (After Williams and Carter, 2.)

that affect the topological contrast in the SEM. The SE signal is relatively weak and is symmetrically distributed in space. The ET detector attracts this signal by means of a electric field, so the signal loses directionality, and the resulting contrast is low and the image looks even (within a close range of grays) but with fine details revealed. On the other hand, the highly energetic BS signals are very directional. As a result, the ET detector collects not only SE but also those BS electrons that are in its "line of vision." Those edges and surfaces that are inclined toward the detector will produce a bright contrast, and the overall image (SE plus BS) will show bright lines in the edges and tilted surfaces (toward the detector) and gray and even contrast where the surface is flat and normal to the beam. This whole situation may produce a contrast that is not what the observer expects based on optical examinations, and very often it can be the opposite to the contrast obtained by light microscopy. For example, a pore may appear to be a particle. Therefore, the microscopist needs to be aware of this situation when interpreting SEM images.

The yield of both SE and BS electrons increases with atomic number, and this fact gives rise to atomic number contrast, which is used in compositional analysis. However, if the observer is trying to obtain compositional images by atomic number contrast, since an inclined surface or a sharp edge will produce a stronger BS signal it might be mistaken as a region of higher atomic number (Z), since a heavy Z element also produces more signal.

The resolution of the SEM is determined primarily by the magnification, the spacing of the lines on the CRT screen, and the probe size. The spacing of lines on the CRT, δ, is usually about 0.1 mm, as is the size of the pixels, that is, close to the resolution of the eye. In order for the image to be resolved on the screen, the probe size on the specimen multiplied by the magnification must equal this limit of resolution, that is, the spacing of the lines. At a magnification of 1000×, the optimum probe size would be $0.1/1000 = 10^{-4}$ mm $=$ 100 nm. With secondary electrons and a small probe size, the resolution is usually of the order of 10–20 nm. For a 10-nm resolution, the minimum magnification needed to bring 10 nm up to the 100 μm that the eye can resolve on the CRT screen would be about 10,000 times. At a magnification of 1000 times, a resolution of only 100 nm would be needed.

The depth of field D in an SEM can be expressed as

$$D = 0.2 \frac{WD}{R \times M} \text{ mm} \tag{5-6}$$

where WD is the working distance (focal distance below the final aperture, approximately 5–25 mm), R is the radius of the objective aperture of the microscope, 50–200 μm, and M is the magnification (less than 10,000). Equation 5-6 indicates that the depth of field can be increased by using a long working distance and a small aperture. (For high resolution, a short working distance is required.) For $M = 1000$,

$R = 100\ \mu m$, and WD 25 mm, the value of D is 50 μm, a much greater value than can be obtained in the light microscope. For example, for an oil immersion lens with a numerical aperture of 1.60, a working distance of 1 mm, and a magnification of 400 times, the depth of field would be only 0.3 μm.

In the examination of rough fracture surfaces in the SEM, the use of the stereo-pair technique is helpful in obtaining a better sense of the three-dimensional nature of fractographic features. Our normal depth perception is due to the viewing of an object by both eyes, each from a slightly different angle than the other. In the stereo-pair technique, two micrographs are obtained, one with the specimen rotated 5° clockwise to the beam direction, and the second at 5° counterclockwise to the beam direction, as indicated in Fig. 5-5. To observe the three-dimensional image, the two micrographs are then viewed simultaneously in a viewer that superposes the micrographs and allows the left eye to see one micrograph and the right eye the other.

X. REPLICAS

The replication of a limited region of a surface is a useful nondestructive method for the examination of large components. Repeated replication can be used to provide a historical record of an area of interest over a period of time, as in fatigue testing. Fatigue cracks as small as 10 μm in length have been detected by this method.

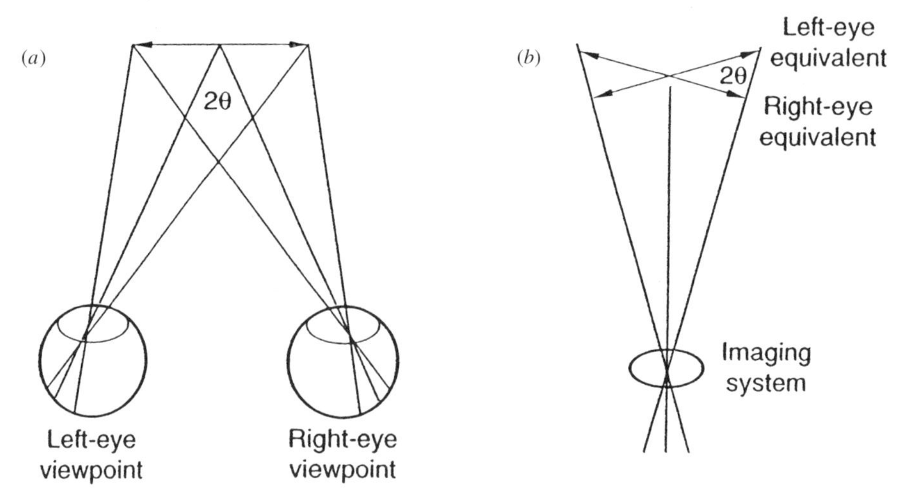

Fig. 5-5. Stereo pairs. The twin images observed by the left and right eyes (*a*) are equivalent to the image pair recorded before and after tilting a specimen about a known angle (*b*). (After Brandon and Kaplan, 1. Reprinted by permission of John Wiley.)

In making a replica of a fracture surface, either a single-stage or a two-stage replica is used. The first step in either process is to clean the surface to be replicated to remove rust, loose debris, and any oily deposits. A solvent such as acetone is often suitable for this purpose.

To make a replica once a clean surface is available, one side of a strip of cellulose acetate film is moistened with acetone to soften it, and then the soft side of the film is pressed lightly against the fracture surface and allowed to dry. This process takes approximately five minutes. The film is then carefully peeled from the fracture surface. In some cases, the area being replicated may be polished and etched prior to the application of the acetate film in order to bring out microstructural features. Once the replica is obtained, it can then be observed in a light microscope, although in order to enhance the contrast of features, the film can be shadowed in a metal evaporator unit with a metal such as gold or chromium. The replica can be examined either with a light microscope or by scanning electron microscopy. However, the single-stage replica is not well suited for transmission electron microscopy, because the film does not stand up well in a high-energy electron beam. To get around this difficulty, the two-stage replica process has been developed. In this procedure, a single-stage replica is prepared as described above. The replica is then placed in a metal evaporator, replica side up, and shadowed at 45°with a metal, such as germanium, to enhance contrast in the electron beam. A thin layer of carbon is then deposited at normal incidence on the plastic replica. The original cellulose acetate film is then dissolved away in acetone, and the resultant shadowed carbon film is ready for examination in the transmission electron microscope. Under the electron beam, it is much more robust than the single-stage plastic replica because the carbon conducts heat away much more effectively than does the cellulose acetate film. The production and evaluation of metallographic replicas obtained in the field is covered in ASTM Designation E 1351.

XI. SPECTROGRAPHIC AND OTHER TYPES OF CHEMICAL ANALYSIS

If more precise analyses of chemical compositions are desired than provided by the EDS technique, then some form of spectroscopic analysis is usually used. Spectroscopy is the study of emission and absorption spectra, and is widely employed in quantitative chemical analyses. In a prism spectrograph, a small sample of an alloy is vaporized, and as the electrons of each element present fall to an orbit of lower energy, electromagnetic radiation in the visible and ultraviolet ranges of a wavelengths characteristic of that element is emitted. The index of refraction of a substance is defined as the ratio of the speed of light in vacuum to that in the substance. This index is also dependent upon wavelength, and is larger the shorter the wavelength. This results in a greater deviation for violet light and a lesser deviation for red light. In spectroscopy the emitted radiation passes through a prism that diffracts and separates the various wavelengths that are present, and the resultant spectrum is photographed. The determination of the diffracted angle, as well as the intensity of the spectral lines, is then used to determine the nature and amount of each element present in the alloy. This technique works well for sodium and higher elements in the periodic table. Therefore, for the determination of the amount of such light elements such as carbon, nitrogen, or oxygen, other methods are used. For example, one commercial unit determines the carbon content of a steel by combining the carbon with oxygen to form CO. The amount of CO produced is then used to determine the carbon content.

The etching characteristics of the various aluminum alloys are sensitive to composition, and this fact can be used to determine if a particular sample is of one type or another, for example, Al-Zn-Mg or Al-Cu.

The relative amount and distribution of sulfur in steel can be of concern. A sulfur print provides a macrographic method for obtaining the distribution of sulfide inclusions. To obtain a sulfide print, a sheet of photographic bromide paper is immersed for three to four minutes in a dilute solution of sulfuric acid. The emulsion side of the sheet is then placed in contact with a polished steel surface for one or two minutes under moderate pressure. The sheet is then removed from the surface, rinsed, and fixed by placing it in a photographic fixing solution for about 15 minutes. A sulfur print will display the location and extent of sulfur inclusions on the prepared surface of the steel. ASTM Designation E 1180 provides detailed instructions for preparing sulfur prints for macrostructural examination.

A macroscopic examination of a polished steel surface can yield information about the type, number, and distribution of inclusions in general. ASTM charts are used to characterize the nature of the distribution of inclusions. Modern methods of quantitative metallography are also used in the analysis of such distributions.

XII. CASE STUDY: FAILURE OF A ZINC DIE CASTING

Die castings made of zinc alloys are relatively inexpensive and are used in a variety of applications. In one case, a zinc die casting was used in an electrically-heated apparatus that circulated 180°F (82°C) water to keep an automobile engine warm overnight under extremely cold climatic conditions. One day, the car owner happened to notice that the zinc die casting was leaking, and made the mistake of jostling the die casting. It fractured and scalding hot water was spewed on him.

One of the well-known problems with a zinc die casting is that, if lead is present in the alloy, it can lead to weakening due to segregation of the lead to the grain boundaries. In fact, the phase diagram for the Pb-Zn system, Fig. 5-6, shows that lead has no solid solubility in zinc. Therefore, during casting, as the alloy is cooled from the molten state, any lead present will segregate to the grain boundaries. Concern about lead is reflected in chemical specifications for zinc alloys, which limit the amount of lead to less than 0.005 wt %. Unfortunately, lead is often found to-

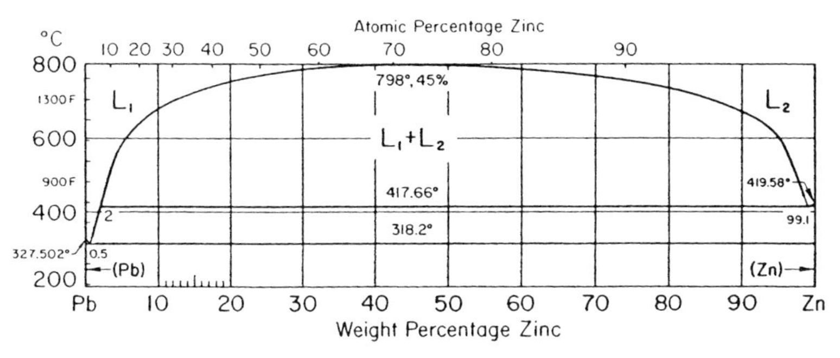

Fig. 5-6. Equilibrium phase diagram for the Pb-Zn system, 3. (Reprinted by permission of ASM International.)

gether with zinc ores, and it is therefore important that they be separated in the refining process.

A spectrographic analysis was carried out to determine the lead content of the zinc die casting. The results showed that there was 0.009 wt % lead present, well above the allowable level.

The melting point of lead is 318°C, and that of zinc is 418°C. At an operating temperature of 90°C, the lead in the grain boundaries was unable to withstand the stresses developed in the circulating water system, and grain boundary cracks developed, which led to the final failure.

XIII. SPECIALIZED ANALYTICAL TECHNIQUES

(a) Small spot electron spectroscopy (EPS) is a surface analysis technique that provides composition and chemical bonding analysis for the first 2 nm of a sample beneath the surface. EPS can detect all elements except hydrogen and helium. In failure analysis, it is useful in determining corrosion products and the chemical degradation of surfaces.

(b) Fourier transform infrared spectroscopy (FTIR) is used to determine the chemical composition and the nature of bonding of organic, polymeric, and many inorganic materials.

(c) Scanning Auger microanalysis (SAM) or Auger electron spectroscopy (AES) is a technique for the determination of composition within 2–3 nm of the surface. It can detect all elements except hydrogen and helium. The sample is scanned with a focused electron beam that causes Auger electrons of low energy, 20–2500 eV, to be emitted from the surface. The energies of the emitted Auger electrons are measured to provide an elemental analysis of the top few layers of the surface.

(d) Electron probe microanalysis (EPMA) uses a finely focused beam of electrons to generate X-rays to analyze the composition of a volume as small as a cubic micron. For example, the composition of nitride layers deposited within a grinding crack can be analyzed by this technique.

(e) In electron energy loss spectroscopy (EELS), electrons are transmitted though a thin film of metal. As the electrons interact with atoms in the specimen, they lose energy, and the amount of energy lost is a characteristic of the atoms with which the electrons have interacted. Spectrographic analysis of the energy distribution of the electrons can be used for chemical analysis of the specimen. This technique is well suited to the detection of elements of atomic number 10 or less.

(f) Secondary ion mass spectrometry (SIMS) is used to provide high sensitivity spatial information about the elements present in top few atomic surface layers with 100-nm resolution, and has the unique ability of detecting hydrogen. In SIMS, a beam of focused ions is directed at the surface in high vacuum. These primary ions cause sputtering of the surface atoms and molecules. The sputtered ions are termed secondary ions, and are mass analyzed using a mass spectrometer. A focused gallium ion gun is employed, together with a magnetic sector mass analyzer. A secondary ion image of the surface can be generated to provide a spatially resolved analysis of the surface. Secondary electrons are also emitted, and an ET electron detector can also be used to obtain secondary electron images.

(g) Scanning atomic force microscopy (AFM) permits the study of surface topography of an area 30 μm $\times$ 30 μm with atomic-scale resolution. A probe that has one atom at its tip is suspended by a thin cantilever beam. The probe is scanned across a surface, and the supporting beam flexes in response to topographical changes. The beam deflections can be measured optically or with a capacitor. The method has been used to periodically monitor changes in slip-band topography during fatigue cycling, for example.

XIV. STRESS MEASUREMENT BY X-RAYS (4)

When stress is applied to a body in the elastic range, elastic strains are induced that manifest themselves at the atomic level as changes in the normal separation distance between atomic planes. X-ray diffraction techniques are used to measure these changes at the surface of a body. These changes are converted into strains, and a modification of Hooke's law is then used to convert the strains into the corresponding stresses. Residual stresses that may be introduced during manufacture or in service are often studied by this technique, since they may play a role in promoting fatigue or stress corrosion cracking failures. The following analysis indicates the nature of the procedures used in the analysis of stresses by X-ray techniques.

Consider a polycrystalline specimen loaded in tension. To measure the longitudinal strain by X-rays would require diffraction from planes perpendicular to the axis of the bar. Since this is usually impossible, the Poisson's contraction is measured instead. Similar considerations apply to a plane surface loaded under biaxial stress conditions.

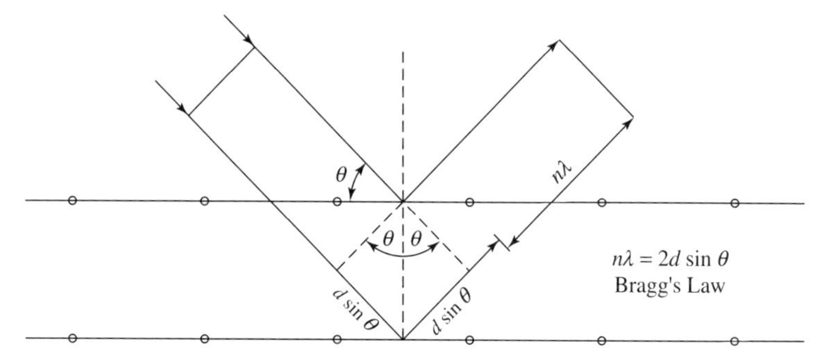

Fig. 5-7. X-ray diffraction and Bragg's law.

Bragg's law (see Fig. 5-7) is used in the analysis. This law is

$$n\lambda = 2d \sin \theta, \tag{5-7}$$

where n is the number of wavelengths, λ is the wavelength, d is the spacing between planes, and θ is the angle of reflection, that is, the angle between the tangent to the plane and the reflected beam. When the lattice is strained by an amount $\Delta d/d$, there will be a change in sin θ. It is desired to make the change in θ as large as possible in order to increase the sensitivity of the method. By taking the differential of Eq. 5-7, this condition can be written as

$$d \cos \theta \, \Delta\theta + \Delta d \sin \theta = 0, \qquad \text{or } \Delta\theta = -\frac{\Delta d}{d} \tan \theta. \tag{5-8}$$

Therefore, for a given change in d, the larger the value of θ, the larger will be the change in θ, the quantity being measured. This means that the angle of incidence should be as close to 90° as can be arranged.

The following indicates the principles involved in making an X-ray stress analysis. Consider a free surface with a biaxial stress present in the plane of the surface. The strain in the direction normal to the surface is a principal strain, and can be written as

$$\varepsilon_z = -\frac{\nu}{E}(\sigma_1 + \sigma_2), \tag{5-9}$$

where σ_1 and σ_2 are principal stresses in the surface plane. This equation can be rewritten as

$$\frac{\Delta d_z}{d_{oz}} = -\frac{\nu}{E}(\sigma_1 + \sigma_2), \tag{5-10}$$

and from Eq. 5-8,

$$\Delta\theta = \frac{\nu}{E}(\sigma_1 + \sigma_2)\tan\theta. \tag{5-11}$$

In general, only the sum of the principal stresses can be obtained from a diffraction pattern obtained at normal incidence. With information about the loading conditions or state of residual stress, it may be possible to determine the values of σ_1 and σ_2. For example, the compressive residual stresses in a shot-peened surface are equal in all directions in the surface.

If instead of obtaining the sum of principal stresses it is desired to measure the stress acting in a given direction in the surface, then a different procedure has to be followed. Figure 5-8 shows the angular relations between the stress to be measured σ_ϕ, the principal stresses σ_1 and σ_2, and the arbitrary x-, y-, and z-axes. In this procedure, the diffraction peak is measured at the surface normal and at one or more angles ψ, which are in the plane of the normal and the desired surface direction. Here only one value of ψ will be considered for purposes of illustration. The normal-incidence pattern measures the strain approximately normal to the surface, and the inclined-incidence pattern measures the strain approximately parallel to OA. These measured strains are therefore approximately equal to ε_3 and ε_ψ, respectively, where ε_ψ is the strain in a direction at an angle ψ to the surface normal.

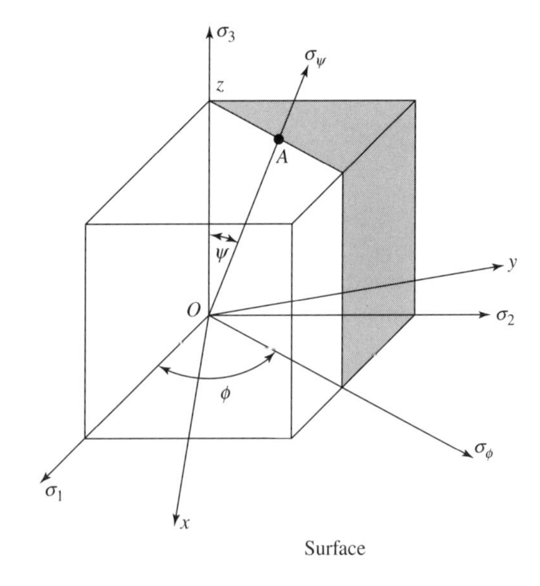

Fig. 5-8. Angular relationships in X-ray stress analysis.

It can be shown that the following relation gives the difference between these two strains:

$$\varepsilon_\psi - \varepsilon_3 = \frac{\sigma_\phi}{E}(1 + \nu)\sin^2 \psi. \tag{5-12}$$

But

$$\varepsilon_\psi = \frac{d_i - d_o}{d_o}, \tag{5-13}$$

and

$$\varepsilon_z = \frac{d_n - d_o}{d_o}, \tag{5-14}$$

where d_i is the spacing of the inclined reflecting $\{hkl\}$ planes, approximately normal to OA, d_n is the spacing of the $\{hkl\}$ planes reflecting at normal incidence under stress, and d_o is the spacing of the same $\{hkl\}$ planes in the absence of stress. Combining Eqs. 5-12, 5-13, and 5-14, we get

$$\frac{d_i - d_o}{d_o} - \frac{d_n - d_o}{d_o} = \frac{d_i - d_n}{d_o} = \frac{\sigma_\phi}{E}(1 + \nu)\sin^2 \psi. \tag{5-15}$$

Since d_o can be replaced by d_n with little error, Eq. 5-15 can be written as

$$\sigma_\phi = \frac{E}{(1 + \nu)\sin^2 \psi}\left(\frac{d_i - d_n}{d_n}\right), \tag{5-16}$$

and a knowledge of d_o is not required.

Modern X-ray stress analysis diffractometer units are programmed to make a series of readings at least five different ψ settings. The experimental and analytical procedures are similar to that described above, but provide a more accurate evaluation of σ_ϕ.

There are also other methods for the determination of residual stresses. For example, ASTM Designation E 837 deals with a method for determining residual stresses by the hole-drilling strain-gauge method. The method is semidestructive, and the residual stresses are determined near the surface of isotropic elastic materials. Strain gauges in the form of a three-element rosette are cemented to a selected area, and a hole is drilled at the center of the rosette to a depth 1.2 times greater than the hole diameter. The creation of the hole causes the residual stresses in the vicinity of the hole to relax, and the relieved strains are recorded. From these measurements, the initial state of residual stress can be calculated.

XV. CASE STUDY: RESIDUAL STRESS IN A TRAIN WHEEL

A derailment of a passenger train was caused by the failure of one of the forged steel wheels. A chemical analysis indicated that the wheel met the specifications for a Class A passenger wheel. Its carbon content was 0.56 wt %. In contrast, the carbon content of freight car wheels is more like 0.75 wt %. The wheel had been shot-peened by the manufacturer to develop a residual biaxial compressive stress of magnitude 25–30 ksi in both the rim and plate of the wheel. There was concern that the possible relaxation of the original residual stresses in service might have contributed to the failure. The X-ray method utilizing chromium $K\alpha$ was therefore undertaken to determine the state of residual stress in the failed wheel. It was found that the residual stresses were still within the manufacturer's original range, and therefore that relaxation of residual stress did not play a role in this failure. This determination was made based on the results of tensile tests, fracture toughness tests, fatigue tests, sulfur print analysis, inclusion studies, and metallographic determination of the microstructure, all of which indicated that the wheel itself was not defective in any way. The actual cause of failure was traced to the presence of a fatigue crack that had been initiated at the corner of the rim due to an overhanging brake shoe. Normally, a brake shoe is seated on the tread of the wheel away from the rim corner. During braking action, there is a increase in temperature at the tread-shoe interface, which can be high enough for the steel to be heated into the austenitic range. Upon subsequent cooling, the austenite can transform to untempered martensite, which appears as a white layer on etching. In addition, thermal checks (fine cracks) can develop. When the brake shoe is in its proper position, these thermal checks are in the central part of the tread and are worn away before they can grow to a critical size for fracture. However, when a crack is created due to an overhanging brake shoe at a rim corner, there is no subsequent wear at this location, and a crack can grow to critical size, as happened in this case. To guard against overhanging brake-shoe problems, visual inspections of wheels and the placement of the brake shoes are made at specified intervals. If discoloration of the rim corner due to overheating is noted, the wheel should then be more carefully inspected for the presence of cracking. However, in the case described above, the inspection procedure was not successful in detecting the problem in time, and the accident occurred.

XVI. THE TECHNICAL REPORT

An important aspect of any investigation is the preparation of the technical report. The following section provides an outline of the steps to be taken in involved in report preparation.

An Outline for Preparing the Technical Report

1. Introduction
   A. What is the problem?
      a. Manner and date of your first involvement with case at hand.
      b. Known circumstances relating to the failure.

c. Description of the components received, date received. (It is important to maintain the "chain of evidence.")
d. Documentation received (depositions, technical literature, manuals, etc.).
e. Purpose of investigation.

B. What is known about the failure?
a. More detailed, overall description of failed component. Include macrophotographs.
b. History of component; loads, environments, and so on.

C. What will you do?
a. Approach to be taken during analysis (hardness tests, chemical analysis, SEM work, site visits, and so on.

2. The Investigation
A. Observations and related photographs at site, if applicable.
B. Laboratory analyses (SEM, micrographs, chemical, hardness, etc.).
a. In a product liability case, it is generally preferred that the investigator take all photographs. If this is not feasible, the investigator should be present when the photographs are taken. This includes SEM work. The name of the person making the observations, if not the investigator, should be listed, along with the date and place of observation.
b. For chemical analyses carried out by a commercial laboratory, a description of the samples provided and the type of examination requested should be recorded.

C. Stress analyses
a. It is preferred that the investigator carry out the stress analysis. If not, the person and organization that did the actual analysis should be identified.

3. Current Status of Evidence
A. Was it returned or is it still in your possession?

4. Conclusions: On the basis of the above observations it is concluded that:
A. The type of failure was ("undoubtedly" or "most likely") _____.
B. The cause of the failure was ("undoubtedly" or "most likely") due to _____.

XVII. RECORD KEEPING AND TESTIMONY

The importance of keeping good records cannot be overemphasized, particularly in legal matters. The following section reviews some of the steps that should be taken in preparing for trial testimony in a product liability case.

A. Maintain a "Chain of Custody"

(a) Be sure to keep a log of the date a piece of evidence was received and from whom. The party giving the part to you should get a receipt, a copy of which should be in your file. Photograph the part to show its "as received' condition.

(b) When the part is returned, enter the date and recipient's name in your file, and obtain a receipt. It is good practice to photograph the part again to show its condition at the time of transfer.

(c) While a part is in your custody, record types of examinations made and associated dates.

B. Photos and Other Records

(a) When taking a series of photographs, use a voice recorder to identify each photo by number, and also indicate the subject matter.

(b) When prints are received, label each print with subject matter identification and date taken. Keep prints with case file.

(c) If copies of prints are supplied to other parties, record the name of recipient and date given.

(d) Copies of EDS readouts should be labeled with date and subject matter and then placed in the case log book.

(e) SEM photos should be dated and identified, and then stored with case file.

C. Examination in Your Laboratory (or Service Laboratory)

(a) Initially, use only nondestructive types of examination (NDE). If destructive examination is required, be sure to get permission in writing from your attorney as to the specific form of examination. He or she may have to contact the opposing side for their permission as well. For example, the removal of a rust layer to better observe a fracture surface should only be done with the attorney's permission. Similarly, a hardness indentation may be considered by some to be destructive, and therefore permission should be obtained before doing such testing. Obviously, the sectioning of a part requires prior approval.

(b) Record in the log book the date and types of examinations made.

(c) List all data obtained: photographs, hardness values, chemical compositions, SEM work, and so on.

D. Storage

Keep any parts received in a safe and secure place while in your custody, preferably in a locked cabinet or secure room. The parts may be in your custody for a year or more at times. Protect the parts from corrosion; for example, coat the fracture surfaces of steel specimens with a light oil.

E. Depositions

If you give a deposition before a trial, be sure to answer all questions as directly and simply as possible. If you feel that an answer should be further clarified, ask for a break and discuss the matter with your attorney. He or she will have an opportunity to ask you some questions for clarification after the opposing attorney has completed his or her portion of the questioning. Be prepared. Any uncertainties or inaccuracies or inconsistencies may be brought up later at the trial.

F. Pretrial Preparation

Most trial lawyers are not technically trained, and in a technical case they may not have grasped all of the details of the technical aspects of the matter completely. Their lack of understanding may lead them to ask questions that do not bring out all the points you as an expert would like to make to convince the jury of the correctness of your views. One way of avoiding difficulties would be to meet with the attorney prior to the trial and to provide a list of questions that will develop your testimony in a logical and complete fashion. The following is a suggested list of questions.

(a) What is your name?

(b) What is your professional address?

(c) Please describe your education.

(d) Please describe your employment history and consulting experience, particularly as it relates to the present case.

(e) Please describe your publications in terms of numbers and general content.

(f) Have you been previously qualified to be an expert witness at a trial? Approximately how many times? Over what period of time?

(g) Have you done any research or consulting work on topics related to the present case?

(h) How and when did you become involved in this case?

(i) What documents did you receive?

(j) What meetings of a technical nature did you attend in connection with this case? (These meetings may have been with a group of consultants on your side of case, or with manufacturer's representatives. There may have also been plant and site visits.)

(k) What components did you receive for examination?

(l) What types of examinations did you carry out? (As you answer this question during the trial, your photos, chemical analyses, similar occurrences, etc. may be placed in evidence.)

(m) Did you examine the whole unit rather than just a component?

(n) Did you examine any similar units?

(o) What were your overall evidentiary, factual findings? (For example, "the part failed in fatigue.")

(p) Did you discuss your findings with anyone else who may have helped in reaching your conclusions (professionals, technicians, mechanics, etc.)?

(q) What were your conclusions?

(r) How did you reach these conclusions? (In a failure analysis, there is often a sequence of events leading up to the final failure. You may have to offer your opinion as to what was the most likely sequence.)

These questions can obviously be modified to fit a particular case. But if you and your attorney agree beforehand on an appropriate list and sequence of questions, and your attorney follows them at the trial, you will be better prepared and more confident on the witness stand.

In addition,

(a) Before the trial, review the depositions of opposing side in the case, if any.

(b) Prepare a list of questions for your attorney to ask during cross-examination.

G. Trial Testimony

(a) Be well groomed; dress in a professional style (a shirt, tie, and business suit for a man, suitable equivalent for a woman).

(b) After you have responded to the first seven questions from the list in Section F, the opposing attorney may challenge your expertise as it applies to this particular case, and the judge will make a final decision. Normally, this is not a problem for an expert in the type failure analysis that relates to the matter of the trial, and the answers to the above questions should be sufficient to qualify you. Obviously, if you have never worked in the field of electronics, you would not want to put yourself forward as an expert in that field. Once you are qualified as an expert witness, you are then permitted to give your opinion concerning technical aspects of the matters at hand.

(c) During the trial, the judge may make some comments relating to your testimony. Listen carefully and don't interrupt. Some judges may seem to be rather abrupt and impatient. Take a deep breath and try not to get flustered. Answer questions calmly, accurately, and directly.

H. Cross-Examination

Answer the questions simply and directly. If you don't know an answer, say so. Do not attempt to argue; hopefully your attorney will know if there is any need for further questioning during the redirect portion of the trial to give you the opportunity to clarify an answer given during the cross-examination. For example, during the cross-examination you may have said that it is possible for a part to have failed in a certain way (although you don't think it likely). During the redirect examination, your attorney may ask the question that will give you a chance to state that the probability of such an event is quite small in your opinion, given the general circumstances of the failure at hand.

I. A Final Point

If, during the trial, you are present for the testimony of the opposing side, prepare a list of comments and possible questions concerning this testimony for your attorney to use on cross-examination of the opposing witnesses.

XVIII. SUMMARY

This chapter has provided an introduction to many of the most common procedures used in carrying out a failure investigation. These procedures range from fairly simple to highly sophisticated. The investigator needs to be aware of these resources, and to make use of the appropriate ones in a given analysis. The importance of: maintaining proper custody of evidence; and record keeping, report writing, and presentation of results can not be overstressed.

REFERENCES

(1) D. Brandon and W. D. Kaplan, Microstructural Characterization of Materials, Wiley, New York, 1999.
(2) D. B. Williams and C. B. Carter, Transmission Electron Microscopy, Plenum, New York, 1999.
(3) ASM Metals Handbook, 8th ed., vol. 8, ASM, Materials Park, OH, 1973, p. 330.
(4) B. D. Cullity, Elements of X-Ray Diffraction, Addison-Wesley, Reading, MA, 1956.

6

Brittle and Ductile Fractures

I. INTRODUCTION

In a polycrystalline material, there are only two types of fracture, either transgranular or intergranular. Under tensile loading, the transgranular mode can be subdivided into brittle and ductile types of failure. Failures that occur under cyclic loading are usually transgranular, whereas long-term creep failures occur in an intergranular mode. Other types of failure, such as stress corrosion cracking, can take place in either a transgranular mode or an intergranular mode.

This chapter deals with brittle fracture and ductile fractures, and the transition from brittle to ductile fracture that occurs in steels as the temperature is increased from a low to a higher level.

II. BRITTLE FRACTURE

Transgranular brittle fracture in steel is characterized by a separation process known as cleavage, which occurs along the {100} crystallographic planes of the ferrite. If one examines a piece of steel that has failed in a brittle manner, either by eye or with a low power microscope, bright, specular reflections can be seen. These planar, bright facets are the {100} cleavage planes. On a microscopic level, these facets usually contain tear lines, visible at 100×, Fig. 6-1. These are created at a grain boundary as the crack front passes from one grain to another, since the (100) plane of one grain is in general not aligned with that of its neighbor.

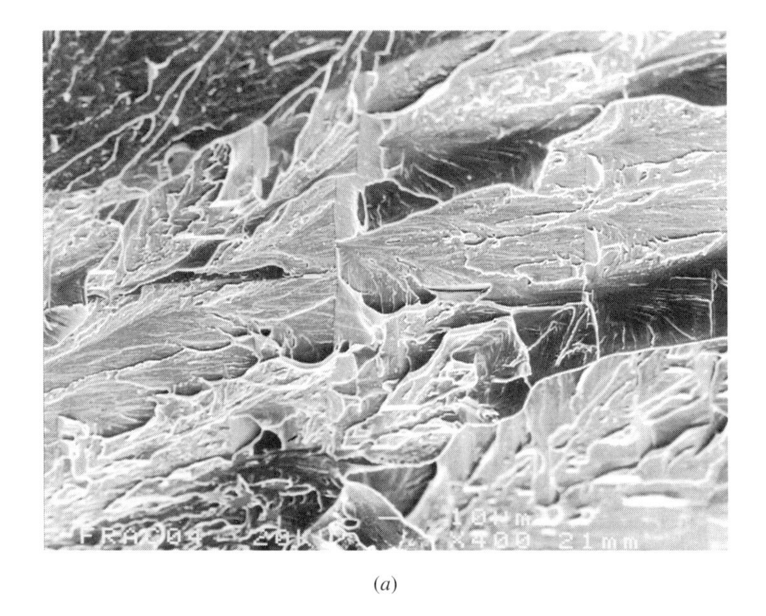

(*a*)

(*b*)

Fig. 6-1. Examples of brittle fracture. (*a*) Cleavage facets. (*b*) River patterns. (Courtesy of Dr. J. Gonzalez.)

These tear lines are steps within a grain that join parallel (100) planes, and their creation adds to the energy required for fracture. As a crack front advances further into a grain, there will be a tendency for the number of tear lines to merge and disappear, thus creating what is known as a river pattern, that is, a merging of the tributaries with the main stream. These river patterns are a useful aid in the determination of the local direction of crack propagation. Since tear lines tend to be perpendicular to the crack front, the merging of the tear lines indicates that the local crack front curvature was convex with respect to the direction of crack propagation. Since the creation of tear lines requires an expenditure of energy, they exert a drag on crack propagation, and therefore can affect the local shape of the crack front.

On a macroscopic level, the brittle fracture of a steel plate is characterized by so-called chevron markings. These markings are useful in failure analysis as the tips of the chevrons point toward the origin of the fracture. An example is shown in Fig. 6-2. These chevron markings are macroscopic tear lines that run perpendicular to the curved crack front and diverge as the crack extends. Crack front curvature is due to a slower rate of crack propagation in the plane-stress surface region than in the plane-strain interior region.

Brittle cracks in steel travel at a high velocity, that is, 0.1–0.2 times the speed of an elastic wave in steel. This means that the strain rates along the crack front are extremely high. As will be discussed, high strain rates, as well as low temperatures, both contribute to the tendency for brittleness in steel. Brittleness also means that there is a lack of ductile deformation. Note that the fracture in Fig. 6-2 is quite flat and there is no lateral contraction, as would be the case if a process akin to tensile necking had taken place.

Intergranular fractures are also brittle. Such fractures occur because of a weakness of the grain boundaries, which is often due to the segregation of impurity elements to the grain boundaries during processing and heat treatment. The term "rock candy" is used to characterize the appearance of intergranular fractures, Fig. 6-3. At low magnification, regions of intergranular fracture will appear to be bright because of the planar rock-candy topography, but the crystallographic planarity associated with cleavage fracture is absent.

Fig. 6-2. An example of chevron markings on the fracture surface of a 25-mm thick steel plate involved in the collapse of an Allegheny County, PA, oil tank in 1988. Origin of fracture is to the left. (Courtesy of Dr. I. Le May.)

Fig. 6-3. Intergranular fracture in steel. (Courtesy of Dr. J. Gonzalez.)

In contrast to these two types of brittle fracture, a ductile fracture is noncrystallographic and takes place by plastic shear deformation. Because of the absence of crystallographic facets, it is duller in appearance than a brittle fracture.

III. SOME EXAMPLES OF BRITTLE FRACTURE IN STEEL

Some interesting examples of well-known brittle fractures in steel are listed below.

(a) In January of 1919, a steel tank in Boston that contained 2,300,000 gallons of molasses suddenly collapsed. In the ensuing flood of molasses, 12 persons along with several horses either drowned or died of the injuries sustained (1).

(b) In the 1930s, several truss bridges failed in Europe at low temperatures. The failures were subsequently found to have initiated at weld defects. Charpy impact tests showed that the bridge steel was brittle at room temperature (1).

(c) During World War II, of 4694 merchant ships studied from February 1942 to March 1946, 970 had developed cracks that required repairs. Between 1942 and 1952, more than 200 ships sustained fractures classified as serious, and at least nine T-2 tankers and seven Liberty ships broke completely in two as a result of brittle fractures. These events led to investigations that resulted in a rapid decrease in the number of these failures. Whereas there had been 130 failures per month in March 1944, there were only five per month in March 1946, although the total number of ships had nearly doubled. The improvements that were made included changes in design (including elimination of square corners at hatchways, etc.), changes in fabrication procedures, retrofits, as well as impact requirements on the materials of construction (1, 2).

(d) At 5 pm on December 15, 1967, a suspension bridge spanning the Ohio River to connect Point Pleasant, WV, with Kanauga, OH, suddenly collapsed. Forty-six people were killed in the accident. The cause of the failure was traced to two adjacent small semicircular flaws of the order of a few millimeters in radius, Fig. 6-4, in one of the eyebars that were linked together to make up the suspension system. These eyebars were 30 feet or so in length, arranged in pairs so that four bars met at a pin connector, Fig. 6-5. However, the structure was not redundant, for if one member failed, the resulting load on its companion member would be sufficient to bring about its failure as well. One of the flaws was 1/8-inch in radius and the other was 1/16-inch in radius. These flaws had initiated at corrosion pits and had developed due to a combination of corrosion fatigue and stress-corrosion cracking, processes that were not known to occur in bridges when the bridge had been designed in 1927. Unfortunately, these flaws developed in a portion of the structure not accessible to inspection. The steel had a yield strength of 80 ksi, but because it was lacking in fracture toughness, the small flaws were able to initiate brittle fracture, which led to the collapse (3).

(e) On January 2, 1988, a large aboveground fuel storage tank located in Allegheny County, PA, suddenly collapsed, releasing 3.9 million gallons of diesel fuel. The crest of this wave washed over nearby earthen dikes, and much of the oil found its way into the Monongahela and Ohio Rivers. The steel shell of the failed tank was twisted and contorted, and massive steel columns and other internal support

Fig. 6-4. Two flaws, upper left, 3 mm and 1.5 mm in radius, that existed prior to fracture of the critical eyebar of the Point Pleasant Bridge. (From NTSB, 3.)

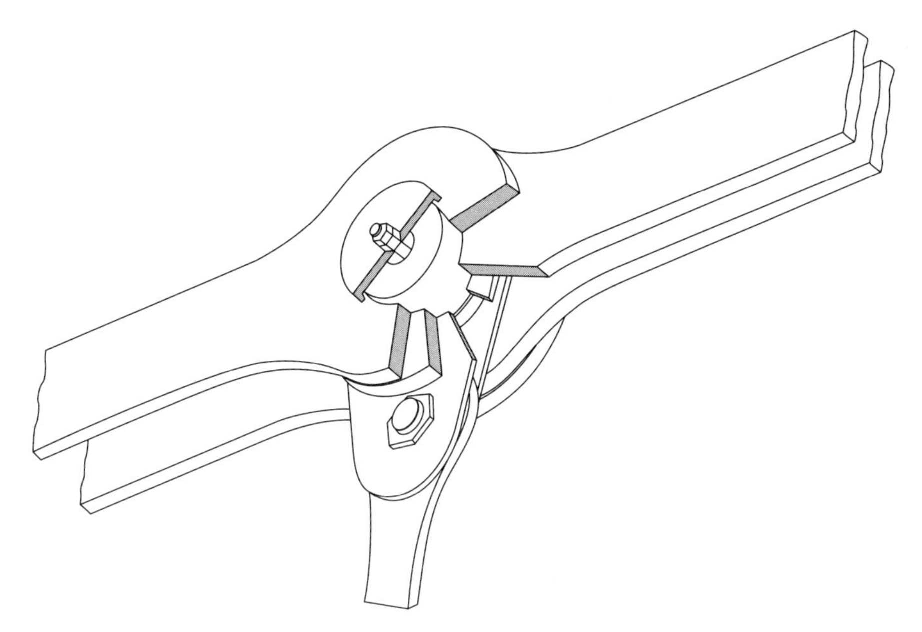

Fig. 6-5. A typical eyebar chain joint of the Point Pleasant Bridge. (From NTSB, 3.)

members were bent and thrown dozens of feet from their original locations. The shell itself was displaced about 120 feet to the east of its original location. This tank had originally been located at another site and had been taken apart and reassembled at the Pennsylvania site, and failed on the first filling. Construction deficiencies attributable to reassembly, however, did not materially contribute to the collapse of the tank. Rather, the immediate causes of the collapse were found to be:

(1) A decades-old flaw due to a burn from a cutting or welding torch located near the top edge of a steel plate in the first level (in burning, steel is permanently damaged by incipient melting or intergranular oxidation)
(2) Ambient temperatures cold enough that the steel in use was prone to brittle failure
(3) Static stress from filling the tank to its permissible capacity

In addition, the tank had not been subjected to a complete hydrostatic test after reassembly.

IV. DUCTILE-BRITTLE BEHAVIOR OF STEEL

A. The Charpy Test

Steels, and many other materials as well, fail in a brittle manner below a certain temperature range, and fail in a ductile manner above this range. Aluminum alloys

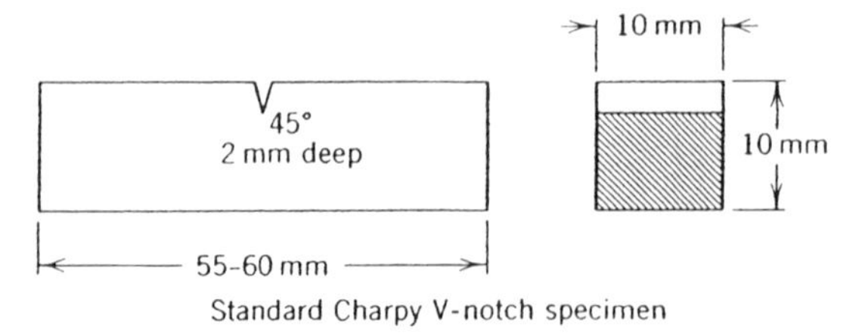

Fig. 6-6. The Charpy V-notch specimen.

and copper alloys, on the other hand, remain ductile even at extremely low temperature. The Charpy V-notch (CVN) impact test, ASTM Designation E 23, is often used in the study of the ductile-to-brittle transition. A common specimen for this test is shown in Fig. 6-6. The specimen is a square bar, 10 mm × 10 mm in cross section and 55 mm in length, with a notch at midlength transverse to the bar. The notch is 2 mm deep, with a 45° included angle, and a tip radius of 0.25 mm. In some cases the specimen is precracked by cyclic loading prior to impact testing.

In a test, the bar is removed from its cooling or heating medium (if any) and placed horizontally on the anvils of a pendulum-impact machine, with the V-notch vertical. Within five seconds after positioning a cooled or heated specimen, the pendulum is released from a predetermined height, and its potential energy is transformed into kinetic energy. A striker on the pendulum impacts the specimen on the side opposite to the notch, generally breaking the specimen. The height to which the pendulum rises after breaking the specimen is then compared to its initial height, and the difference is used to determine the energy absorbed in the impact process. The striker may be equipped with strain gauges to provide a force-time readout during the fracture process (instrumented Charpy test). If a fatigue precracked specimen is used, this information can be used to determine the dynamic fracture toughness of the material.

The Charpy test can be used in conjunction with, and at times to substitute for, more expensive fracture mechanics type tests. Consider the following case (4). A steel producer has a contract to supply plate steel for an offshore platform. The plate material needs to meet mechanical properties that are quite rigorous for safety and end product reliability reasons. Before full-scale production of the order can begin, the steel supplier needs to demonstrate to the buyer that the material is capable of meeting such criteria. To accomplish this, the supplier qualifies the material for the project. The process begins by making the steel grade and then testing a portion of the plate to determine if all requirements are met. Steel mill equipment imposes limits on plate size; therefore individual units need to be welded together in the field to produce lengths that can reach to great depths. Small sections of the sample plate are welded together and fracture mechanics tests are conducted to determine the crack tip opening displacement (CTOD) toughness in the heat affected zone (HAZ) and along the fusion line where the weld metal meets the base metal. Then a steel supplier might correlate the CTOD results with CVN 50% ductile-to-brittle transition temperature (DBTT). By agreement, this correlation can allow the steel supplier to use the Charpy test instead of the more expensive and time consuming CTOD testing. Empirical correlations between CVN test results and K_{Ic} and K_{Id} (the fracture toughness obtained at a high strain rate) values have been obtained by Rolfe and Barsom (1) for particular grades of steels as a function of temperature. It is noted that design criteria for bridge steels have been based upon such correlative procedures.

An example of the influence of temperature and carbon content on the amount of energy absorbed in a Charpy test is shown in Fig. 6-7. The midpoint of the transition temperature range for a given steel is a measure of what is called the ductile-to-brittle transition temperature (DBTT). Other measures of the transition temperature may be related to the amount of lateral expansion on the compression side of the specimen, or to the amount of shear fracture present on the fracture surface (picture window). In other cases, the transition temperature may be specified as a function of the level of energy absorbed, for example, 41 J (30 ft-lb).

The failure of the welded ships at temperatures of the order of 30°F during World War II focused attention on the factors that led to brittle behavior. Whether or not a steel will behave in a brittle or a ductile manner depends upon a number of factors, principal among them being composition, strength level, thickness, temperature, and strain rate. Higher levels of carbon, phosphorous, molybdenum, and arsenic in steel increase the transition temperature, whereas nickel, silicon, manganese, and copper decrease the transition temperature. If loading conditions are such that dislocations within the steel move and multiply to accommodate the imposed strain rate before fracture is initiated, then the steel will behave in a ductile manner; otherwise the steel will be brittle. In steels, dislocation motion at low temperatures is strongly influenced by interaction with the lattice. This interaction is known as the Peierls force, and it is much more pronounced in body-centered cubic (bcc) material such as ferritic steels and in hexagonal close-packed material (hcp) such as zinc than it is in the face-centered cubic (fcc) materials such as aluminum and copper.

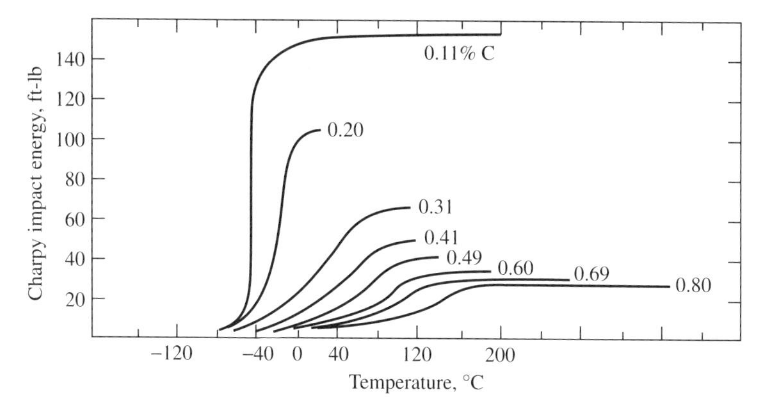

Fig. 6-7. The influence of the carbon content of steels on the Charpy V-notch energy as a function of temperature. (From Burns and Pickering, 14. Reprinted by permission.)

Steels are referred to as strain-rate-sensitive materials in that their low-temperature strengths are dependent upon strain rate. In contrast, the strength properties of copper and aluminum are much less dependent upon strain rate. This strain-rate sensitivity has consequences, not only with respect to fracture, but to other areas, such as high-rate forming, as well. A simple expression can be derived for the stress required to move a dislocation at a given strain rate $\dot{\varepsilon}$ and temperature T. It is assumed that the process is thermally activated and follows Arrehenius-type behavior, and that the activation energy Q is reduced by the applied stress acting within an activation volume v. The resultant equation is

$$\dot{\varepsilon} = Ae^{(Q-v\sigma)/RT}, \tag{6-1}$$

where A is called a frequency factor, R is the universal gas constant (8.314 J $\mathrm{mol^{-1}K^{-1}}$), and T is the absolute temperature.

Thus

$$\ln \frac{\dot{e}}{A} = -\frac{Q - v\sigma}{RT}, \tag{6-2}$$

which can be expressed as

$$\sigma \frac{v}{R} = \frac{Q}{R} - T \ln \frac{A}{\dot{\varepsilon}}. \tag{6-3}$$

Equation 6-3 indicates that the stress to move a dislocation at a given strain rate decreases with increase in temperature, and increases with increase in strain rate. The parameter $T \ln A/\dot{\varepsilon}$ is known as the rate-temperature parameter, and it can be used to correlate yield strength behavior for steels as a function of temperature and strain rate, as shown in Fig. 6-8, with A equal to 10^8/sec. Note that the strain-rate senstivity decreases with increase in yield strength.

It is noted that some limited amount of plastic deformation usually takes place in steel even for nominally completely brittle fracture, as indicated by the fact that the fracture stress in tension of smooth bars tested in liquid nitrogen is equal to the yield strength measured in compression (5). This is interpreted to mean that plastic deformation is necessary to trigger brittle fracture. When this plastic deformation occurs, locally high stresses develop due to the blockage of dislocations at barriers (dislocation pileups). In carbon steels, these stresses then initiate the local fracture of grain boundary carbides, which act as the barriers to the dislocations.

One way to view the brittle fracture process in steel is to consider that the steel has a fixed local fracture strength, independent of temperature and strain rate. Should the stress in a component reach this value, brittle fracture will occur. In Chapter 2, we saw that under plane-strain conditions, the yield stress ahead of a U-shaped notch could be increased due to constraint by a factor of three over the yield strength when

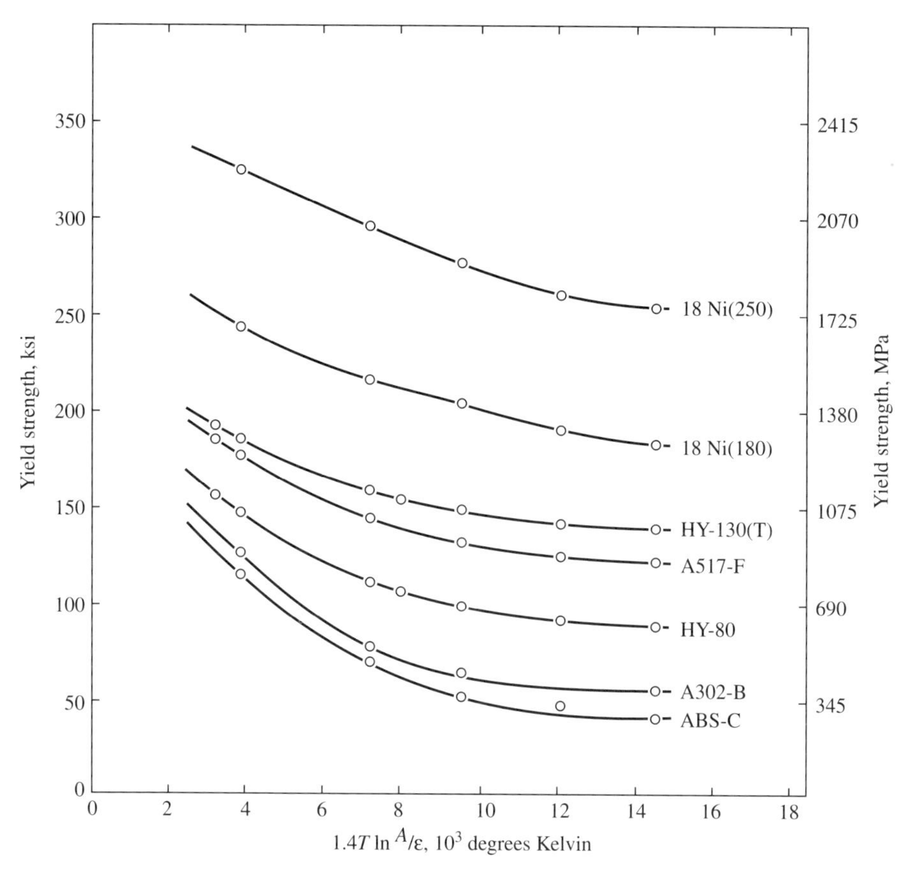

Fig. 6-8. Yield strength of several steels as a function of the parameter $T \ln A/\dot{\varepsilon}$ with $A = 10^8$/sec and the temperature T in degrees Kelvin. (After Rolfe and Barsom, 1. Based on Bennett and Sinclair, 13.)

the von Mises criterion was used. In the case of the V-notched Charpy bar, this factor is somewhat less, 2.5, since the flanks of the V-notch are at 45° to each other rather than parallel. If we assume that brittle fracture at low temperatures in steels occurs when the local yield stress σ_Y equals the fracture stress σ_f, then brittle fracture of a Charpy bar will occur at a stress 1/2.5 that of the fracture stress required for brittle fracture. That is, the stress at point *C* in Fig. 6-9 is 1/2.5 that of the stress at point *B*. This has the effect of moving the temperature at which the specimen is brittle to a higher value as compared to the brittle temperature for an unnotched specimen, Fig. 6-9. Similarly, an increase in strain rate will increase the transition temperature, since it increases the yield strength in a strain-rate-sensitive material such as carbon steel. The Charpy notched-impact test, therefore, provides a conservative measure of the transition temperature, since it involves both notch constraint and a high strain rate.

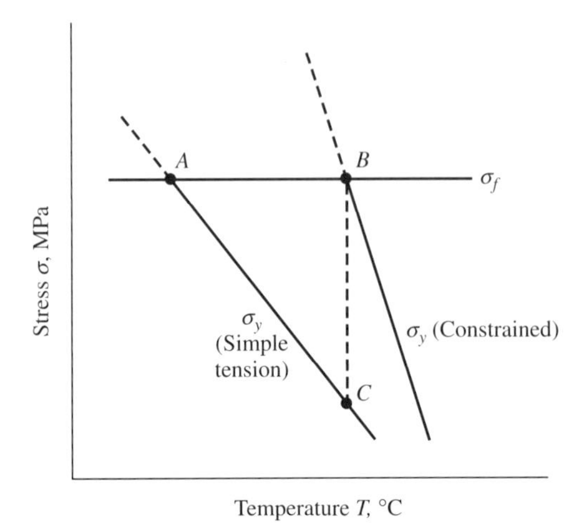

Fig. 6-9. Effect of a notch on the brittle transition temperature.

Grain boundaries per se block plastic deformation from progressing from one grain to the next because of differences in orientation of the slip systems from grain to grain. As a result, the grain size d can influence the yield strength of annealed polycrystalline alloys in which precipitation is absent. As a simplification, we consider a grain of square in cross section, as shown in Fig. 6-10. Under the action of an applied shear stress, the shear strain γ is given as

$$\gamma = \frac{\tau}{G} = \frac{\Delta}{d}. \tag{6-4}$$

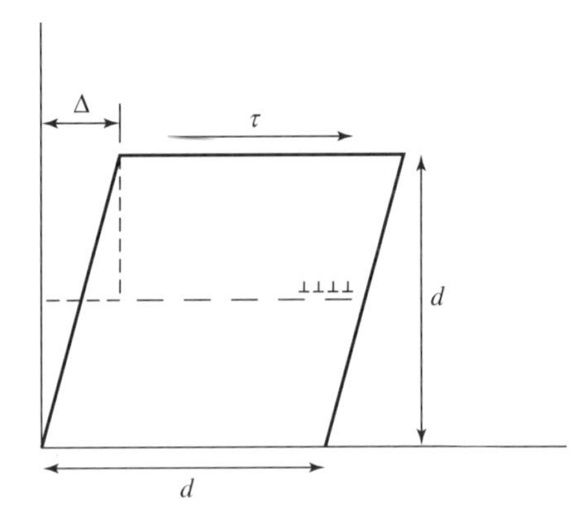

Fig. 6-10. Square grains under shear before and after the relaxation of elastic strain due to the motion of dislocations.

When plastic deformation occurs in this grain, the elastic displacement Δ is relaxed and replaced by a plastic displacement, that is,

$$\Delta = \left(\frac{\tau}{G}d\right)_{el} = (nb)_{pl}, \tag{6-5}$$

where n is the number of dislocations and b is their Burger's vector. These dislocations are blocked and pile up at the grain boundary, and as a result, a stress concentration develops whose magnitude τ_{max} is given by the simple expression

$$\tau_{max} = n\tau. \tag{6-6}$$

The quantity n, appearing in Eq. 6-6, can therefore be replaced by τ_{max}/τ, which leads to

$$\tau_Y = (Gb\tau_{max\ c})^{1/2}\, d^{-1/2}, \tag{6-7}$$

where τ_Y is the yield stress in shear, and $\tau_{max\ c}$ is the local stress at the tip of the dislocation pileup required to initiate slip in the next grain. This equation is rewritten in terms of the tensile yield stress as

$$\sigma_Y = \sigma_i + k_y\, d^{-1/2}, \tag{6-8}$$

where $k_Y = 2(Gb\tau_{max\ c})^{1/2}$ and σ_i is the lattice friction, that is, the stress required to move a dislocation in a single crystal. This equation is known as the Hall-Petch relation.

An interesting and useful outcome of this derivation is that, in single-phase materials or carbon steels having a ferritic-pearlitic microstructure in which plastic deformation occurs primarily in the soft ferritic phase, a decrease in grain size is seen to result in an increase in the yield strength. Further, if a certain level of $\tau_{max\ c}$ has to be reached before fracture of a grain-boundary carbide can occur, then a decrease in grain size requires an increase in τ_Y. Therefore, the yield strength and the fracture resistance are both raised by a decrease in grain size. This is a singular result, for generally any other modification that increases the yield strength of a given alloy results in a decrease in fracture resistance.

As the test temperature increases above the highest temperature at which a steel is completely brittle, a brittle-to-ductile transition occurs. The results of Charpy impact tests are shown in schematically in Fig. 6-11. The term "upper shelf" refers to the region in Fig. 6-11 that is above the transition region, and the term "lower shelf" refers to that portion of the diagram below the transition region. Whereas high strain rates below the transition temperature promote brittle behavior by lessening the time available for the motion and multiplication of dislocations, a high strain rate above the transition temperature may increase the resistance to fracture, that is, the fracture toughness, since the fracture toughness depends upon both strength level and

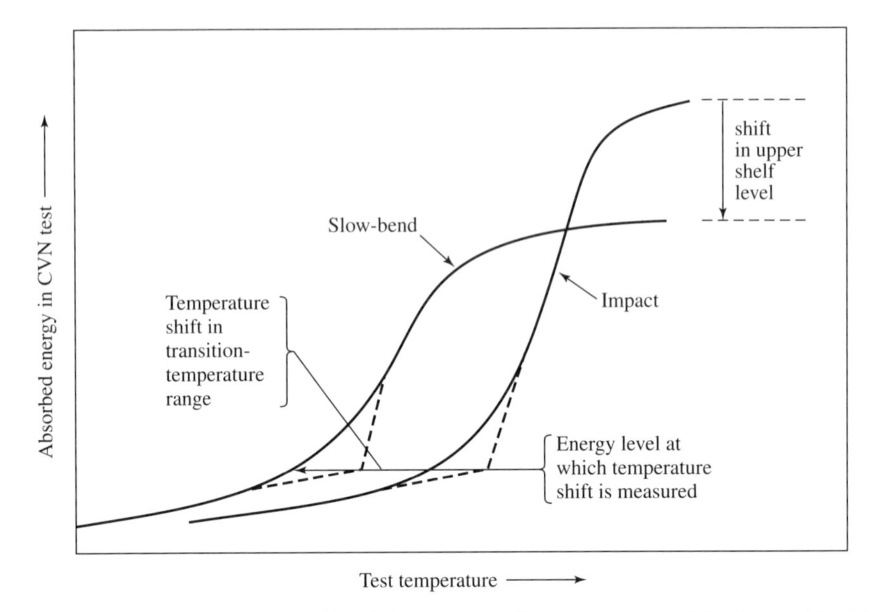

Fig. 6-11. A schematic representation of the upward shift in temperature of the Charpy V-notch transition temperature and the upper shelf energy level due to an increase in strain rate. (After Barsom and Rolfe, 1.)

ductility (extension to fracture). Therefore, a high strain rate in the ductile range that increases the yield strength may actually improve toughness, as indicated in Fig. 6-11.

V. CASE STUDY: THE NUCLEAR PRESSURE VESSEL DESIGN CODE

A. Prevention of Brittle Failure (4)

In a nuclear power plant, it is essential that the nuclear reactor be operated in a manner that ensures that the pressure vessel integrity is maintained under both normal and transient operating conditions. This is accomplished by postulating limiting flaws and using linear-elastic fracture mechanics (LEFM) to calculate the allowable coolant temperature (T) and pressure (P) during heat-up, cooldown and leak/hydro testing (P-T curves). The P-T limits are revised periodically throughout the life of a plant to account for neutron damage to the pressure vessel. Over a period of time, neutron bombardment in the reactor beltline region shifts the ductile-brittle transition as measured by the 41-J (30 ft-lb) energy level, to a higher temperature. The amount of the shift in the ductile-to-brittle transition is then used to shift the ASME reference stress intensity factor (K_{IR}) curve upward in temperature to account for the effects of embrittlement due to neutron bombardment, as shown in Fig. 6-12.

Since it is impractical to test large fracture toughness specimens throughout the life of a nuclear power plant, surveillance programs use Charpy and tensile speci-

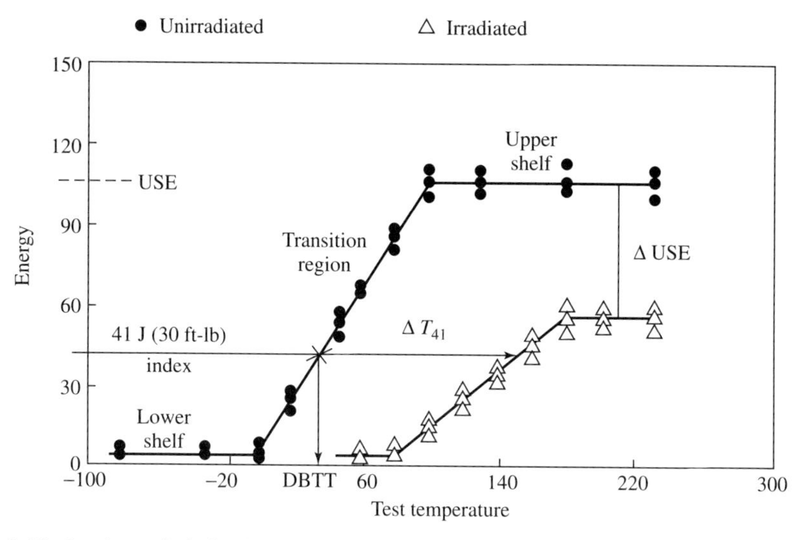

Fig. 6-12. A schematic indication of the effect of neutron irradiation on Charpy impact behavior. (USE = upper-shelf energy.) (From Manahan et al., 4. Copyright ASTM. Reprinted by permission.)

mens to track the neutron-induced embrittlement. The nuclear industry uses the 41-J index to define a ductile-brittle transition temperature (DBTT). Neutron irradiation shifts the transition region to higher temperatures (ΔT_{41}) and the Nuclear Regulatory Commission (NRC) sets screening limits on the maximum shift that can occur during the life of the plant. If the screening limits are exceeded, then the plant must be shut down or a thermal anneal must be conducted to restore the material properties.

The ability of a nuclear pressure vessel to withstand ductile fracture is judged by the upper shelf energy (USE). In older plants built before fracture toughness testing was widely used, Charpy testing was used to qualify individual heats of material. The ASME code and the Code of Federal Regulations prescribe minimum plate properties that must be satisfied prior to service (e.g., at least 102 J of energy on the upper shelf prior to service). The NRC requires an in-depth fracture mechanics assessment if the Charpy USE is expected to drop below 68 J during the operating life of the plant. If Charpy data can be used to extend the life of a plant by one year beyond the initial design life, a plant owner could realize revenues as large as \$150,000,000. Further, the cost avoidance from a vessel-related fracture is expected to be in the billion-dollar range. To date, the NRC has shut down one U.S. plant (Yankee Rowe) as a result of Charpy data trends (one concern was that the surveillance specimens were not from the same heat as the plates used for the pressure vessel). This plant's pressure vessel was constructed from a one-of-a-kind steel (contained a small but higher level of copper than now allowed, which promoted neutron damage at 550°F) and is not representative of the U.S. reactor fleet.

The specified toughness requirements are obtained using Charpy V-notch test specimens coupled with the determination as described below of the nil-ductility transition temperature (NDTT). The NDTT is the highest temperature at which fracture occurs without macroscopic plastic deformation. The approach involves a reference temperature, designated RT_{NDT}, and the reference fracture toughness curve K_{IR}.

The value of RT_{NDT} is obtained by measuring the drop weight NDTT of steels 5/8 inch and thicker. To determine the NDTT temperature (ASTM Designation E 208) for a plate of one-inch thickness, a specimen 14×3.5 inches is used. A single-pass, crack-starter weld, 2.5 inches long, is deposited in the center of the plate running longitudinally, and then a notch is made in the weld. The plate is mounted weld side down in a drop weight testing rig and is impacted by a drop weight whose kinetic energy equals 816 J (600 ft-lb) at the temperature of interest. Several specimens are needed to determine the NDTT. Above the NDTT, plastic deformation occurs and the specimen is restrained from breaking. At the NDTT the specimen breaks without plastic deformation.

After the NDTT is determined, a set of three Charpy V-notch specimens is tested at a temperature 33°C (60°F) higher than the NDTT to measure the toughness, T_{CV}, to insure an increase of toughness with temperature. A Charpy energy of 68 J (50 ft-lb) and a lateral expansion of 0.89 mm (35 mil) are required. The NDTT becomes the RT_{NDT} if the Charpy results meet or exceed these requirements. If not, additional tests are performed at increasing temperatures (in steps of 10°F) until the requirements are satisfied. The RT_{NDT} then is designated to be this temperature minus 33°C. (See Fig. 6-13).

The K_{IR} curve is then referenced to the RT_{NDT}. The data used to construct the lower bounding function K_{IR} curve used dynamic fracture toughness at the lower end as well as crack arrest toughness results to fix the upper end, and is used only for steels having a yield strength of 345 MPa (50 ksi) or less. The expression for the K_{IR} curve is

$$K_{IR} \text{ (in ksi } \sqrt{\text{inches}}) = 26.777 + 1.223 \exp \{0.014493[T(°\text{F}) - (RT_{NDT} - 160]\}, \tag{6-9a}$$

$$K_{IR} \text{ (in MPa } \sqrt{\text{m}}) = 24.36 + 1.129 \exp \{0.2029[T(°\text{C}) - (RT_{NDT} - 91.4)]\}. \tag{6-9b}$$

VI. CASE STUDY: EXAMINATION OF SAMPLES FROM THE ROYAL MAIL SHIP (RMS) TITANIC (6)

In April 1912, the RMS Titanic struck an iceberg while steaming at 20+ knots and sank off Newfoundland in less than three hours with a loss of over 1500 people. The wreckage was not located until 1985, and in 1991, a piece of the hull material was recovered and Charpy impact tests were carried out on test specimens machined from the sample. It was found that the steel fractured in a 100% brittle fashion at ice brine temperatures. These results led to the speculation that the impact with the

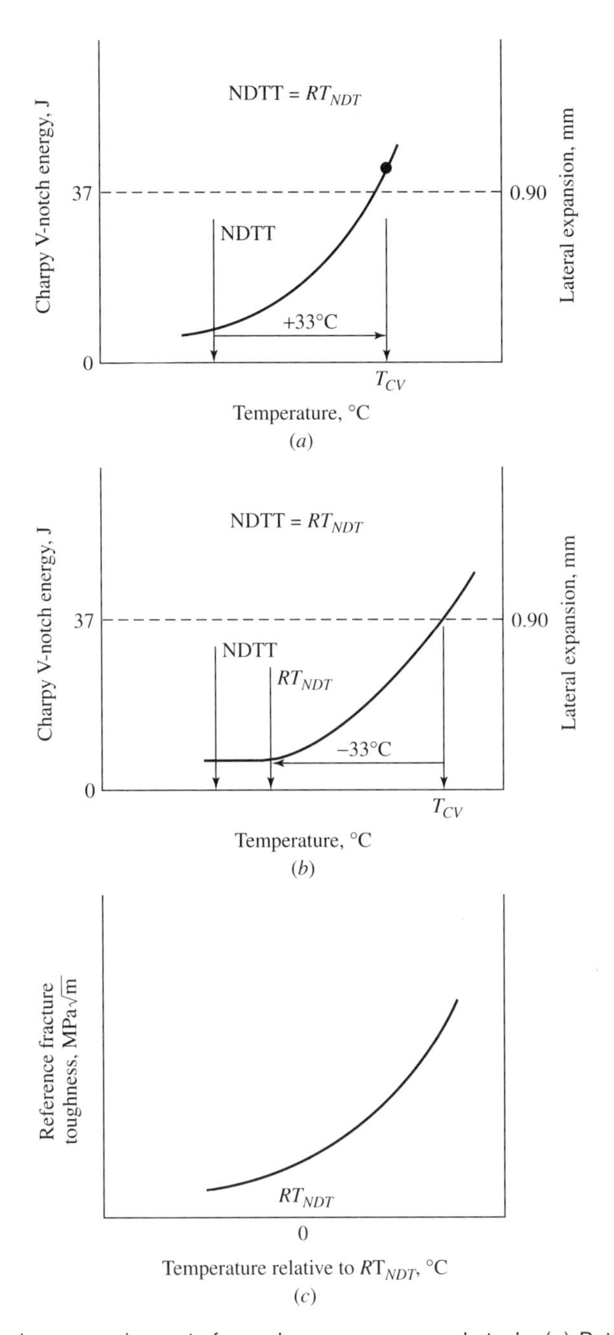

Fig. 6-13. Toughness requirements for nuclear pressure vessel steels. (*a*) Determined using the reference temperature, RT_{NDT} approach, where $NDTT = RT_{NDT}$. (*b*) Determined using the reference temperature approach where $NDTT \neq RT_{NDT}$. (*c*) Determined using the reference fracture toughness, K_{IR}, curve. See text for details. (4).

iceberg had shattered the brittle hull plates, thereby allowing rapid flooding of the ship. In 1996, a section of the hull plating and several hull and bulkhead rivets were recovered, and metallurgical and mechanical analyses were performed on them. It was again found that the pieces of the hull steel possessed a ductile-to-brittle transition temperature that was very high with respect to the service temperature, making the material brittle at ice-water temperatures, Fig. 6-14. This brittleness was attributed to both chemical and microstructural factors. The following tables provide information on the chemical and mechanical properties of the hull steel.

A. Chemistry

The following table gives the composition of the Titanic hull steel in weight % and compares this composition with that of a comparable modern steel.

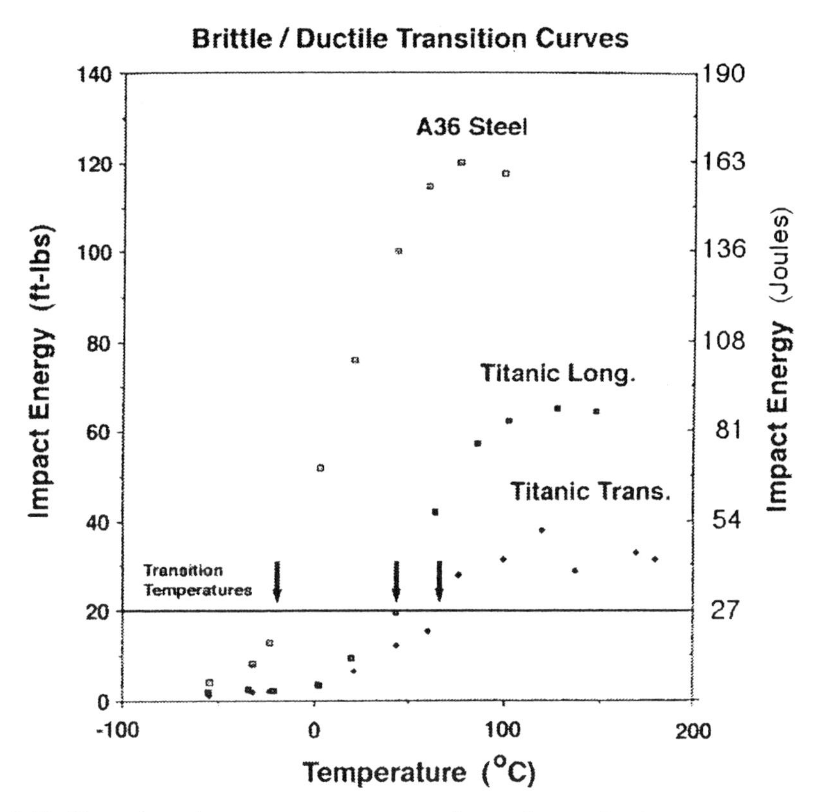

Fig. 6-14. Charpy impact energy versus temperature for specimens taken longitudinal and transverse to the rolling direction of the Titanic hull steel. Results for a modern steel, A36, are shown for comparison. The transition temperature is arbitrarily defined to be at the 27-J level. (From Foecke, NIST, 6.)

Element	Titanic	AISI 1018
Carbon	0.21	0.18–0.23
Sulfur	0.065	0.05 max
Manganese	0.48	0.6–1.0
Phosphorous	0.027	0.04 max
Silicon	0.021	——
Copper	0.025	——
Nitrogen	0.004	0.0026
Oxygen	0.013	——
Rare earths	——	——
Mn/S ratio	7.4 1	12 1–20 1
Mn/C ratio	2.3 1	3 1–7 1

Both sulfur and phosphorous decrease the upper shelf toughness. It is noted that both the sulfur and phosphorous levels in the Titanic's ship plates are much higher than is available in modern steels, where sulfur levels of 0.002% and phosphorous levels of less than 0.01% are common. The steel was low in manganese, which could lead to embrittlement if there were not enough manganese (atomic weight 54.93) to combine with sulfur (atomic weight 32.06) in the form of MnS particles, where the ratio of manganese to sulphur is 1.71 1. However, in the case of the Titanic's hull steel, there was sufficient manganese to tie up the sulfur in the form of MnS. Manganese is also a solid-solution toughening agent, and the relatively low amount of manganese in the hull steel as well as a low manganese-to-carbon ratio may have had a detrimental effect on impact toughness.

B. Tensile Properties

Yield strength	40 ksi
Tensile strength	62 ksi
Percent elongation	30 (50-mm gauge length)

These results showed that the hull steel met the specifications for the mechanical properties.

C. Microstructure

Figure 6-15 shows the microstructure of the Titanic's hull steel and compares it to a modern steel of similar composition. The microstructure consists of large, coarse pearlite colonies (roughly 0.2-μm lamella thickness) and large ferrite grains (ASTM grain-size number 4–5, 100–130 μm in diameter.) The ASTM grain-size number n is related to N^*, the number of grains per square inch at a magnification of 100×, by the relationships

$$N^* = 2^{n-1}, \tag{6-10}$$

(*a*)

10 μ

(*b*)

Fig. 6-15. (*a*) The microstructure of a longitudinal section of modern 1018 steel. (*b*) In comparison, that of the Titanic hull steel. The grain size, the pearlite lamellar spacing, and the MnS particle sizes are all larger in the Titanic hull steel. (From Foecke, NIST, 6.)

and

$$n = 3.32 \log 2N^*. \tag{6-11}$$

The microstucture shows a large amount of banding in the rolling direction. MnS and oxide particles are present throughout the material, and are quite large, occasionally exceeding 100 μm in length. The MnS particles had been deformed into lenticular shapes instead of being in the form of a series of elongated stringers, an indication of a low rolling temperature (6). The large grain size and the coarse pearlite indicate the plate was hot rolled and air cooled, consistent with production in a low-speed rolling mill as was the norm in the early 1900s. A large grain size is detrimental in the brittle range, because the intensity of stress where dislocations pile up against microstructural barriers such as pearlite or a carbides increases with increase in grain size. On the other hand, a small grain size not only improves toughness by reducing the size of pileups, but also raises the yield strength. The yield strength can be expressed as a function of grain size by means of the Hall-Petch relation, Eq. 6-8.

Modern steels processed in high-speed rolling operations have much smaller grain sizes and much finer pearlite as compared to the Titanic's hull steel. In the hull steel, both the large ferrite grain size and the coarse lamellar structure of the pearlite contributed to a higher transition temperature and lower upper shelf toughness. The high strain rate associated with impact with the iceberg at 20+ knots would also have contributed to brittle behavior because of the strain-rate sensitivity of the steel.

D. Fractography

The fracture surfaces of Charpy specimens fractured at 0°C were 95% cleavage and 5% ductile. No evidence for intergranular fracture was found. 10% of the cleavage fractured area was seen to originate at fractured MnS particles.

E. Corrosion

Corrosion was not a factor in the sinking. However, at the 12,000-foot depth at which the wreck rests, there is sufficient oxygen to allow the corrosion process to proceed, and "rusticules," large stalactites of rust, have been observed. It is estimated that the ship, originally consisting of 60,000 tons of steel, is losing a ton per day to corrosion. At that rate, by 2076 AD the hull will be completely gone.

F. Rivets

In 1997, a panel concluded that the principal damage to the hull was due to parting of the plates, and not a due to a 90-m gash in the brittle plate steel, as had been proposed, thereby implying that failure of the rivets had played a role. Only two hull rivets were examined in detail (out of 3,000,000 total), and the mechanical properties of these rivets figured in further speculation as to the sequence of events leading to the rapid flooding of the ship. Wrought iron usually consists of relatively pure iron, which contains 2–3% by volume iron silicate slag. However, metallographic analysis of the rivet shown in Fig. 6-16 revealed that it contained 9.3% slag, more than three times the expected level. Some of this slag was in the form of stringers more than 200 μm in length. These stringers are not bound to the matrix, and at low temperatures can fracture and nucleate cracks in the iron. At the time of construction, the wrought iron rivets had been hydraulically driven through the hull plates and were flattened on the inside. Residual tensile stresses develop in the rivets on cooling, and these stresses reduced the resistance of the rivets to fracture.

The mechanical properties of wrought iron are highly anisotropic. In the direction of the stringers, wrought iron is as strong as mild steel. However, it is much weaker in the direction perpendicular to the stringers, and the ductility measured perpendicular to the stringers can be an order of magnitude less than in the direction of the stringers. The two recovered rivets lacked their interior heads, an indication that they were popped off on impact due to a component of tensile stress act-

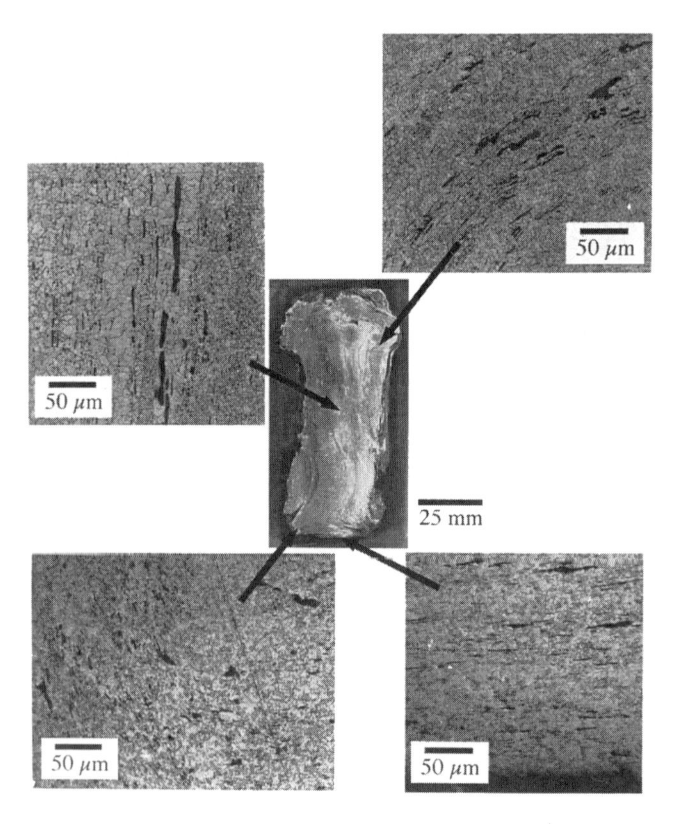

Fig. 6-16. The orientation of silicate slag at various locations within a cross section of a Titanic hull rivet. At the bottom right, near the fracture surface created when the inner rivet head popped off, the stringers are oriented perpendicular to the tensile axis due to deformation when the inner head was formed. (From Foecke, NIST, 6.)

ing across the stringers in the rivet heads. Therefore, pending further information, the investigators concluded that fracture of the rivets and the resulting parting of seams, rather than brittle fracture of the hull plates, may have been the critical factor involved in the sinking of the Titanic.

VII. DUCTILE FRACTURE

The process of ductile rupture on both the macroscale and the microscale differs from brittle rupture in that extensive plastic deformation is involved. The term "ductile failure" includes not only the separation process but, in certain forming operations, the process of necking as well. In this chapter, the topics of necking, which involves the use of the concepts of true stress and true strain, and ductile separation are treated. The microstructural characteristics of ductile fracture are also discussed.

VIII. DUCTILE TENSILE FAILURE, NECKING

In sheet forming, metals are extensively deformed. One of the forming limits imposed upon the extent of deformation in operations such as the forming of external automobile body sheet is that necking should not occur during the forming operation. Understanding of the causes of necking is therefore needed in dealing with this potential problem in manufacturing.

A. Condition for Necking of a Bar

On a load-extension plot, Fig. 6-17, the point at which necking begins is at P_{max}. The rate of increase of load-carrying capacity per unit of area is given by

$$\frac{dF}{A\,d\varepsilon} = \frac{d\sigma}{d\varepsilon}, \tag{6-12}$$

where σ is the true stress and ε is the true strain. The rate of decrease of load carrying capacity per unit of area due to loss of cross-sectional area is given by

$$\frac{dF}{A\,d\varepsilon} = -\frac{\sigma\,dA}{A\,d\varepsilon}. \tag{6-13}$$

We next make use of the fact that, for purely plastic deformation, the volume is constant, and we neglect any elastic dilatation. Therefore, $d(V) = 0$, and since $V = Al$, where l is the length of the element, we can write

$$d(V) = d(Al) = A\,dl + l\,dA = 0, \tag{6-14}$$

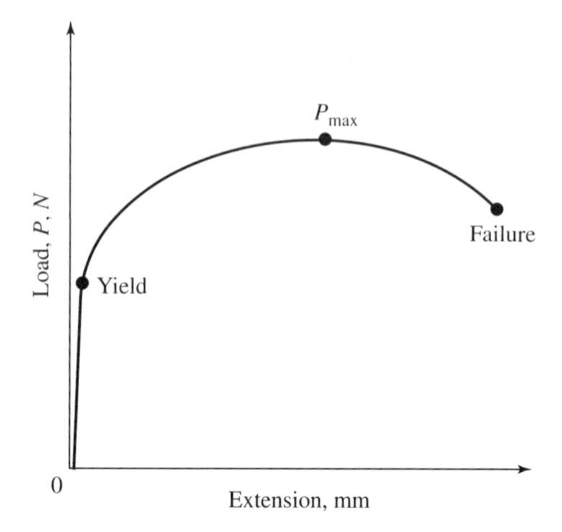

Fig. 6-17. Load (*P*) versus extension (Δ) plot used in the analysis of necking.

or

$$\frac{dA}{A} = -\frac{dl}{l} = -d\varepsilon, \tag{6-15}$$

and Eq 6-12 can be rewritten as

$$\frac{dF}{A\ d\varepsilon} = \sigma. \tag{6-16}$$

At P_{max}, the rate of increase in load-carrying capacity due to strain hardening is no longer able to compensate for the rate of area decrease, and an instability develops, the condition for which is

$$\frac{d\sigma}{d\varepsilon} = \sigma. \tag{6-17}$$

This can also be expressed in the following way: At P_{max}, $\Delta P = 0$, and since $P = \sigma A$, where σ is the true stress and A is the current cross-sectional area, it follows that

$$\sigma\ dA + A\ d\sigma = 0, \qquad \text{or } \frac{d\sigma}{\sigma} = -\frac{dA}{A}. \tag{6-18}$$

Hence

$$\frac{d\sigma}{\sigma} = d\varepsilon, \qquad \text{or } \frac{d\sigma}{d\varepsilon} = \sigma, \tag{6-19}$$

which is the same condition for necking of a bar as given in Eq. 6-17 above.

The true stress-strain curve can be approximated by the following expression due to Ludwik:

$$\sigma = \sigma_o + k\varepsilon^n, \tag{6-20}$$

where n is the strain-hardening exponent. At large strains as assumed in the present case, σ_o can be neglected. Hence,

$$\frac{d\sigma}{d\varepsilon} = nk\varepsilon^{n-1}. \tag{6-21}$$

If this expression for $d\sigma/d\varepsilon$ is substituted in the condition for necking, it is found that the strain at necking is

$$\varepsilon_n = n. \tag{6-22}$$

Hence, the greater the strain-hardening exponent, the greater is the strain to necking.

It is noted that another stress-strain relationship that is often used in the analysis of plastic deformation is known as the Ramberg-Osgood equation, that is,

$$\varepsilon = \frac{\sigma}{E} + k_{RO}\left(\frac{\sigma}{E}\right)^m, \tag{6-23}$$

where k_{RO} and m are material constants.

B. Strain Localization

In 4340 steel treated to have a high yield-strength-to-tensile-strength ratio, that is, 94%, when plastic deformation occurs, particularly if hydrogen is in the steel, a deleterious form of nonuniform localized deformation can occur due to plastic instability. Irregular grooves form in plastically deformed regions, as shown in Fig. 6-18, and in one case, the crash of a helicopter occurred because of the development of such markings in the main rotor spindle lugs. The subsequent fatigue crack growth from these grooves, which acted as stress raisers, resulted in the crash (7). The damaging plastic deformation was thought to have occurred under the severe loading conditions experienced during practice emergency landings.

C. Axisymmetric Stress in Necking

When a neck forms in a round bar, material at the minimum cross section of the of the neck attempts to contract more than the material just above and below the minimum cross section. As a result, the material above and below the minimum cross section constrains the material at minimum cross section from freely contracting, and a triaxial state of hydrostatic stress is developed within the neck. This hydrostatic stress does not contribute plastic deformation since no shear stresses are as-

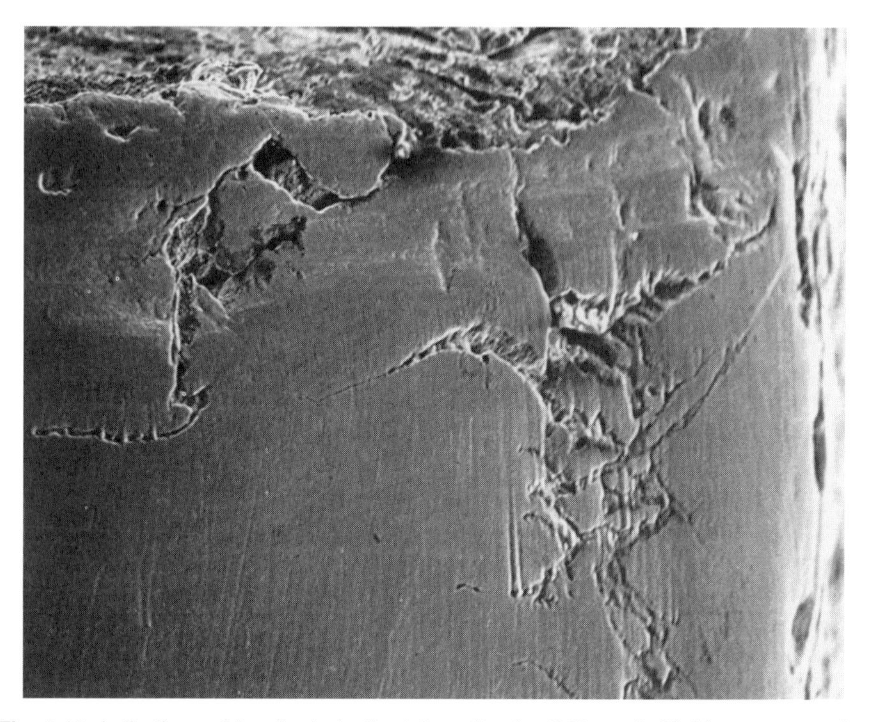

Fig. 6-18. Indications of localized plastic deformation in 4340 steel. (Yield-stress-to-tensile-strength ratio 0.94). (Reprinted from Materials Characterization, vol. 26, A. J. McEvily and I. Le May, Hydrogen assisted cracking, pages 253–268, Copyright 1991, with permission of Elsevier Science.)

sociated with a state of hydrostatic stress, but the hydrostatic stress does increase the true tensile stress P/A required for plastic flow. By increasing the peak tensile stress, the hydrostatic stress does promote fracture of particles and interfaces, and thereby affects the fracture process.

In the minimum cross section of a neck in a round tensile bar, the hydrostatic components of stress are given by

$$\sigma_{\theta\theta} = \sigma_{rr} = \sigma_{zz} - \overline{\sigma}. \tag{6-24}$$

Bridgman (8) derived the following relation to indicate the amount by which the local tensile stress acting at each point in the minimum cross section of a necked tensile bar had to be increased over the flow stress to overcome the effects of triaxiality:

$$\frac{\sigma_{zz}}{\overline{\sigma}} = 1 + \ln\left(\frac{a}{2R} + 1 - \frac{r^2}{2aR}\right), \qquad \sigma_{\theta\theta} = \overline{\sigma}\ln\left(\frac{a}{2R} + 1 - \frac{r^2}{2aR}\right), \tag{6-25}$$

where $2a$ is the diameter of the necked region, r is distance measured from the centerline in the plane of minimum cross section, and R is the radius of the neck.

Note that $\sigma_{\theta\theta}$ is zero at the surface and rises to a maximum at the centerline of the bar.

The above equation is used to determine how much the applied tensile stress must be raised as a function of the bar size and the radius of the neck to obtain plastic flow. The hydrostatic stress must be subtracted from the total stress to find the equivalent flow stress. Figure 6-19 (9) gives the ratio of the equivalent stress $\overline{\sigma}$ to total nominal true stress P/A in a necked tensile specimen. The calculation for the average applied stress is obtained by integrating $\sigma_{zz}\, 2\pi r\, dr$ over the cross-sectional area to get the total force and then dividing by the area to get the average stress. The nonhydrostatic component is simply $\overline{\sigma}$.

D. Necking in a Thin Strip Under Tension

In a thin strip under tension, two types of necks can develop. One is referred to as diffuse, the other as localized. The condition for diffuse necking is the same as for bar, and this type of neck precedes the formation of the localized neck. A localized neck is a form of plane-strain deformation, the strain along the direction of the neck being zero. It is of interest that the localized neck in a thin sheet strip under tension does not develop in the transverse direction, but rather at a pronounced angle to the

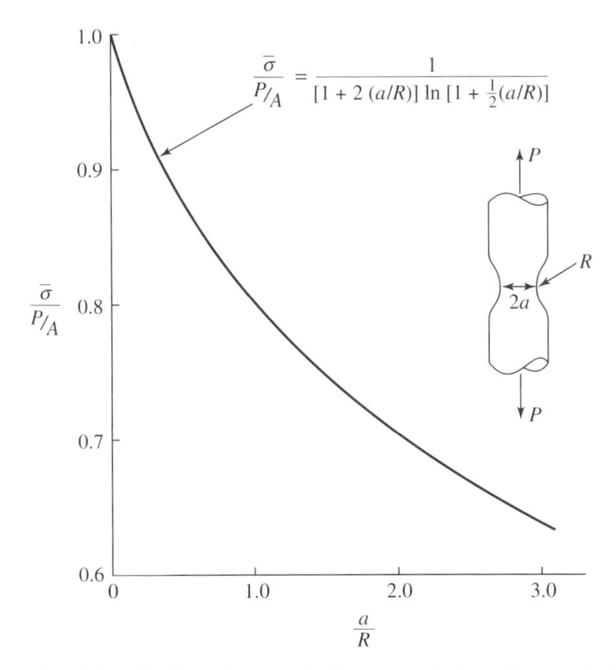

Fig. 6-19. The ratio of the effective stress, $\overline{\sigma}$, to the nominal stress in a necked tensile specimen, where $2a$ is the diameter of the bar at the neck, and R is the radius of the neck (8, 9).

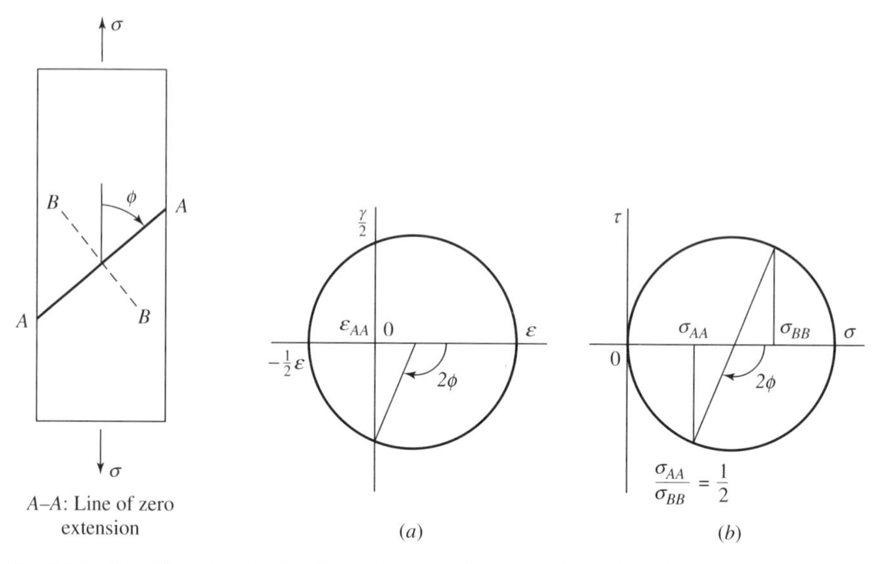

Fig. 6-20. Conditions for localized necking in a sheet. (*a*) Mohr circle for strain. (*b*) Mohr circle for stress.

transverse direction. If a localized neck were to develop in the transverse direction, a constraint on plastic deformation would develop in the transverse direction, and according to the von Mises criterion, further deformation would require that the tensile stress be increased by $2/\sqrt{3}$. A higher axial stress would be required to cause the local deformation in the neck than is required to cause general deformation throughout the strip. Therefore, localization will not occur in the transverse direction. A direction of the neck in which localized deformation is possible is needed. From the Mohr circle for strain, Fig. 6-20*a*, it is seen that the strain is zero in a direction 54.7° from the axial direction. It also follows, from the Mohr circle for stress, Fig. 6-20*b*, that the stresses parallel and normal to the direction of zero strain are in the ratio 1:2, which is consistent with the stress state required for plane strain plastic deformation. Analysis (10) indicates that the localized neck develops when the following condition is reached:

$$\frac{d\sigma}{d\varepsilon} = \frac{\sigma}{2}. \tag{6-26}$$

Use of the Ludwik relation indicates that, at localized necking, $\varepsilon_n = 2n$. Therefore, it is to be expected that the localized neck will develop at twice the strain required for the formation of the diffuse neck. This is important in some sheet-forming operations where diffuse necking of the sheet is acceptable, but localized necking is not, especially since fracture rapidly follows the development of a localized neck.

An example of a forming limit diagram for a low-carbon steel is given in Fig. 6-21. To use this diagram, the ratio of the minor to the major engineering strain is

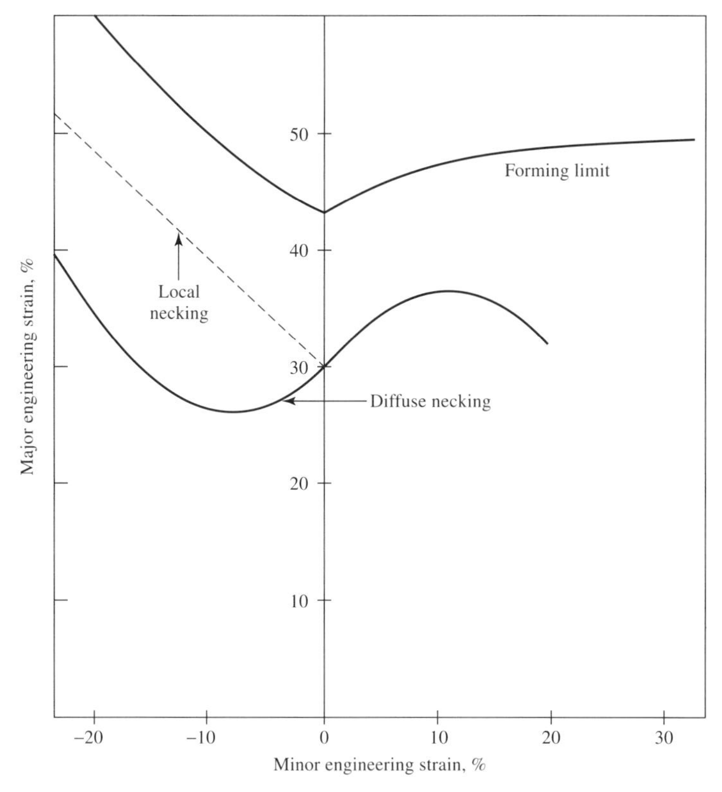

Fig. 6-21. An exampe of a forming limit diagram (FLD) for sheet metal. (Adapted from Hosford and Caddell, 10.)

determined for a given forming operation. This ratio determines the slope of a line drawn from the origin. Where this line intersects either the diffuse necking, localized necking, or forming limit (fracture) lines on the diagram, the corresponding forming limit is reached, depending upon the particular considerations associated with a given forming operation. Note that, for biaxial tension straining, localized necking does not occur, since for this loading condition there is no direction in which the strain is zero.

IX. FRACTOGRAPHIC FEATURES ASSOCIATED WITH DUCTILE RUPTURE

As the result of the triaxial state of stress developed in a neck in alloys containing carbides, oxides, and so on, the particles may break or the interface between particle and matrix may separate. By either process, transgranular, noncrystallographic voids are formed within the central portion of the neck. With further deformation these voids grow and link up, thereby reducing the effective cross-sectional area and bringing about final separation. The final overall appearance of the broken tensile bar is usually in the form of cup and cone, where the central, flat portion of the fracture developed by the void linking process, and the outer cone developed by a shearing-off process at the surface where the triaxial stresses were low.

On the microscopic level, these linked-up voids give rise to distinctive fractographic features known as dimples. In the flat part of the fracture, these dimples are circular in shape, whereas in the cone portion of the fracture rupture where shearing is dominant, the dimples are elongated. Figures 6-22 and 6-23 provide examples of both types of features. These features are usually of a size that they are best examined in a SEM. They are not usually found using optical microscopy.

Each of these dimples is nucleated at a particle. These fractographic features are always found on the tensile rupture fracture surfaces of face-centered cubic (fcc) alloys, for example, aluminum, copper, and nickel alloys. They are also found on the fracture surfaces of steels that have failed in tension above their ductile-to-brittle transition temperature.

In a copper specimen of normal commercial quality, oxides are dispersed in the copper, and in a tensile test it is at these oxides that the voids are nucleated, dimples form, and a cup-cone type of fracture results. It is of interest that, if copper of exceptionally high purity is tested in tension, the final separation process is quite different. Since there are no particles to nucleate voids upon the onset of necking,

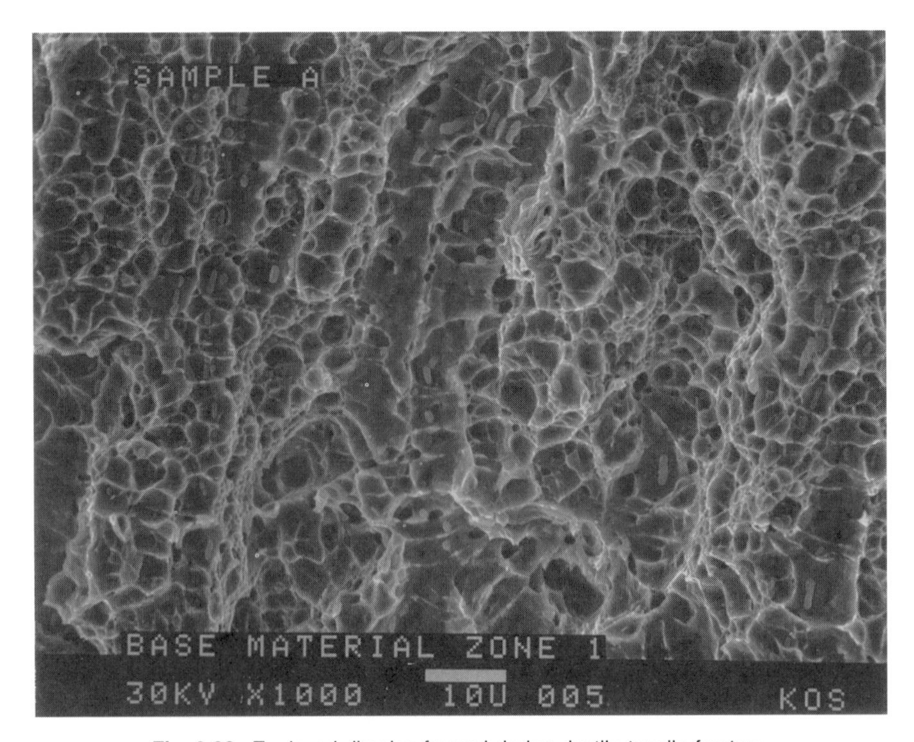

Fig. 6-22. Equiaxed dimples formed during ductile tensile fracture.

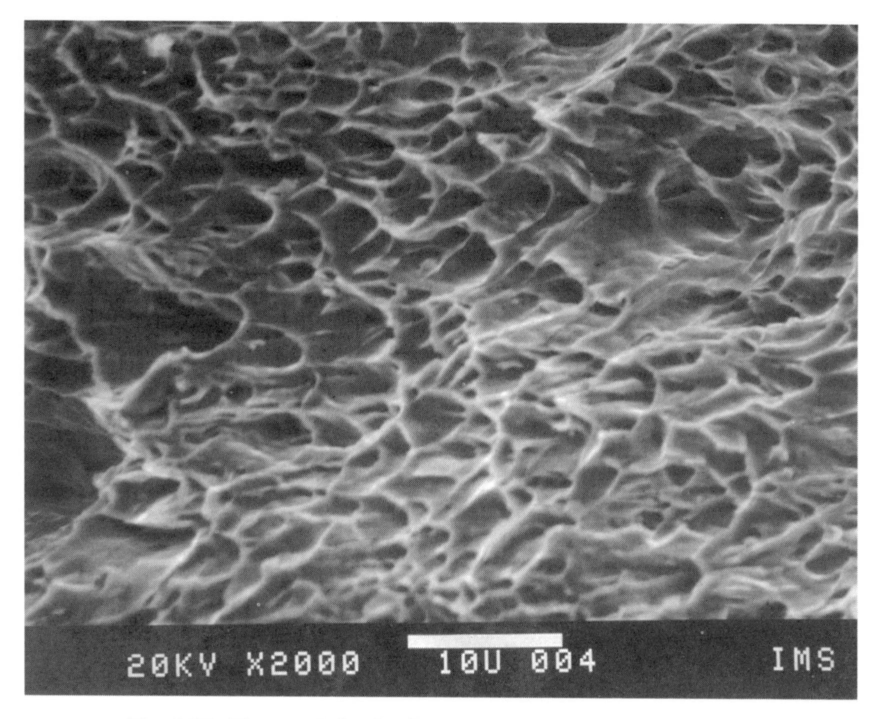

Fig. 6-23. Elongated dimples formed during ductile shear fracture.

the specimen will continue to elongate and neck down to a point much like taffy before finally separating. At sufficiently elevated temperatures, where the flow stress is much less, even copper of commercial purity can behave in this way, since the stresses are low, strain hardening is eliminated by dynamic recrystallization, and the triaxial stresses developed during necking are not sufficient to nucleate voids within the necked region.

The fracture toughness of a material is a function of strength, modulus, and elongation to failure. For a very low carbon steel alloy, a change from air melting to vacuum melting may have little effect on the strength and modulus, but the elimination of carbides and oxide particles can lead to an increase in ductility, as measured by the percent of elongation and the percent of reduction in area, which in turn can improve toughness. This increase in ductility comes about because of the absence of void-nucleating particles.

X. FAILURE IN TORSION

A round bar of a low-carbon steel can be twisted through several revolutions before it fails, and when it does so, the failure is in a plane of maximum shear stress at right angles to the axis of the bar, and there is little reduction in area and little, if any, elongation. On the other hand, when a round bar of brittle material such as gray cast iron is twisted to fracture, the strains involved are much less, and the failure is of a helical nature. The fracture plane is at 45° to the axis of the bar, perpendicular to the direction of principal tensile stress. In has been noted that when the hardness of a tool steel is above 720 Vickers Hardness (HV), it behaves more like cast iron and fractures in a helical manner (9).

The cyclic loading of a round bar of a ductile material subjected to fully reversed torsion in the low cycle regime will also result in a flat fracture that is perpendicular to the bar. However, the fracture surface of the same material that develops in the high cycle fatigue has been described as a "factory roof" because of the zig-zag nature of the cracking process as cracks grow under the influence of the principal tensile stress.

XI. CASE STUDY: FAILURE OF A HELICOPTER BOLT (12)

In failure analysis, in order to determine whether or not a component has failed under monotonic or cyclic loading, it is sometimes necessary to develop reference standards for each type of failure, especially when the fracture surface of a part that has failed in service contains features that may be ambiguous in their interpretation. Such is the case in the present case study, where in order to characterize the fractographic features of an AISI 9310 carburized steel, failures under monotonic and cyclic loading had to be obtained.

The component under investigation was a trunnion bolt, a critical part of a helicopter rotor mechanism. The helicopter had crashed in the ocean, and it was alleged that the bolt had failed due to fatigue, thereby causing the accident. To complicate matters, it was several months before the helicopter was recovered, and during this time some corrosion had occurred. The bolt was made of AISI 9310 steel, and had been carburized. Examination of the fracture surface revealed the presence of markings in the case that could be interpreted to be fatigue markings. In addition, an unusual secondary crack was present that ran circumferentially in the case. It was suggested that this secondary crack had developed due to stress corrosion cracking during the time the component was submerged in the ocean. Reference standards such as the ASM Metals Handbook volume on Fractography did not contain comparable fractographic examples, and it was therefore decided to carry out a set of tests to develop fractographic standards for this carburized steel. Since the fracture had occurred in bending at a reduced section of the bolt, a carburized specimen containing a semicircular notch was tested in four-point bending, as shown in Fig. 6-24. Two types of testing were carried out, monotonic bending to fracture and cyclic fatigue tests to fracture.

Figure 6-25 is a plot of the bending moment versus the deflection of a specimen in a monotonic test to failure. Discontinuities accompanied by loud cracking sounds were observed, which indicated progressive cracking in the case prior to final separation. Similar behavior is observed in tension tests of carburized steel where the case first cracks with an accompanying loud noise at a load below that required for fracture of the tougher core.

The macroscopic fracture appearance of this specimen is shown in Fig. 6-26. As with the failed helicopter bolt, a circumferential crack in the case can be seen, a di-

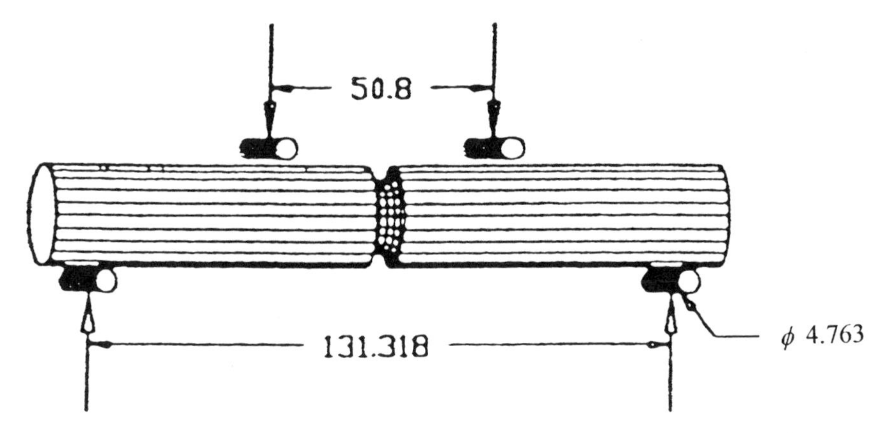

Fig. 6-24. Four-point bend test of a carburized, notched bar simulating the bolt, 12. (Reprinted from Materials Characterization, vol. 36, No. 4/5, A. J. McEvily et al., A Comparison of the Fractographic Features of a Carburized steel Fractured under Monotonic or Cyclic Loading, pp. 153–158, Copyright 1996, with permission of Elsevier Science.)

rect indication that this crack did not develop because of stress corrosion. It was surmised that the state of triaxial stress at the root of the notch, together with a radial tensile stress due to a difference in Poisson ratios between the case and the plastically deforming core, had developed to cause the circumferential crack.

Figure 6-27 shows the appearance of the case at higher magnification. The case contains markings that could be interpreted as being due to fatigue, whereas the fracture is known to be due to monotonic loading. The appearance of the core is shown in Fig. 6-28. Here, dimples representative of ductile overload failure are evident. In addition, manganese sulfide stringers can be seen.

For comparison, Figs. 6-29 and 6-30 show the fracture appearance after fatigue failure. The case is smooth, and markings known as fatigue striations can be observed

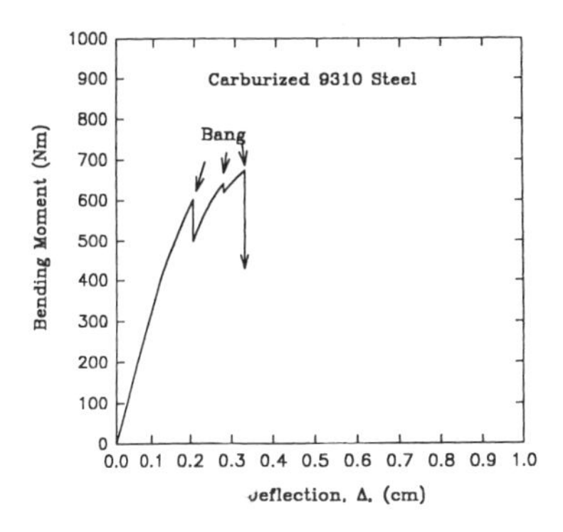

Fig. 6-25. Load-deflection plot of the carburized test specimen. (Reprinted from Materials Characterization, vol. 36, No. 4/5, A. J. McEvily et al., A Comparison of the Fractographic Features of a Carburized steel Fractured under Monotonic or Cyclic Loading, pp. 153–158, Copyright 1996, with permission of Elsevier Science.)

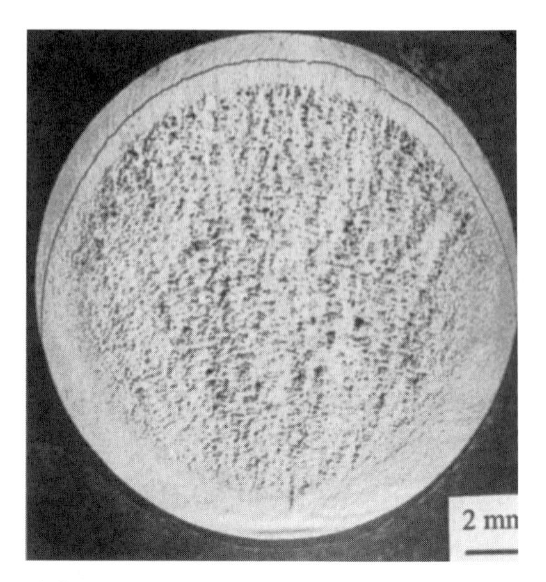

Fig. 6-26. Macroscopic fracture appearance after failure under monotonic loading. (Reprinted from Materials Characterization, vol. 36, No. 4/5, A. J. McEvily et al., A Comparison of the Fractographic Features of a Carburized steel Fractured under Monotonic or Cyclic Loading, pp. 153–158, Copyright 1996, with permission of Elsevier Science.)

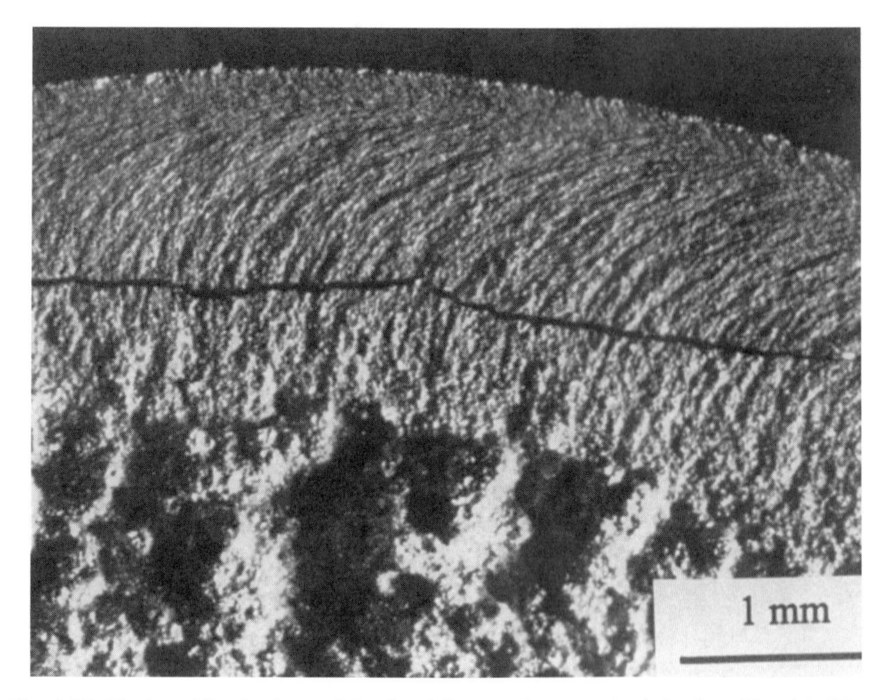

Fig. 6-27. Region of the fracture origin after failure under monotonic loading. (Reprinted from Materials Characterization, vol. 36, No. 4/5, A. J. McEvily et al., A Comparison of the Fractographic Features of a Carburized steel Fractured under Monotonic or Cyclic Loading, pp. 153–158, Copyright 1996, with permission of Elsevier Science.)

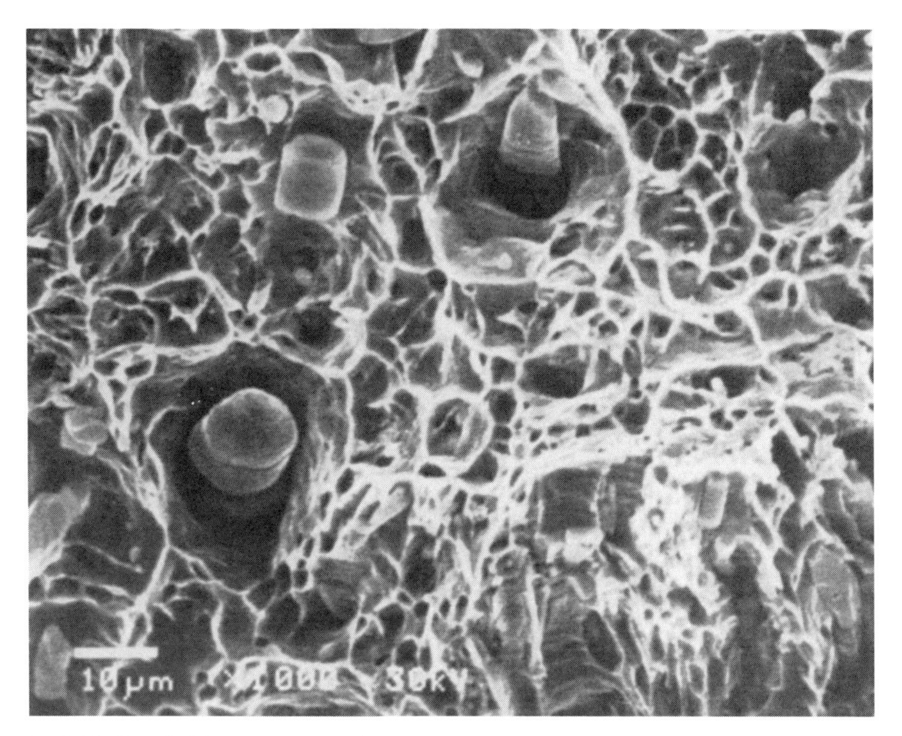

Fig. 6-28. Detail of fracture surface after monotonic loading showing dimples and manganese sulfide particles in core. (Reprinted from Materials Characterization, vol. 36, No. 4/5, A. J. McEvily et al., A Comparison of the Fractographic Features of a Carburized steel Fractured under Monotonic or Cyclic Loading, pp. 153–158, Copyright 1996, with permission of Elsevier Science.)

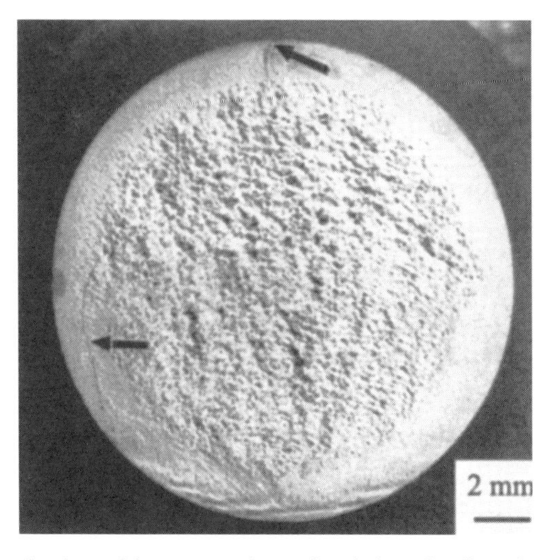

Fig. 6-29. Macroscopic view of fracture surface after fatigue loading. Arrows indicate fatigue crack origin and secondary cracking. (Reprinted from Materials Characterization, vol. 36, No. 4/5, A. J. McEvily et al., A Comparison of the Fractographic Features of a Carburized steel Fractured under Monotonic or Cyclic Loading, pp. 153–158, Copyright 1996, with permission of Elsevier Science.)

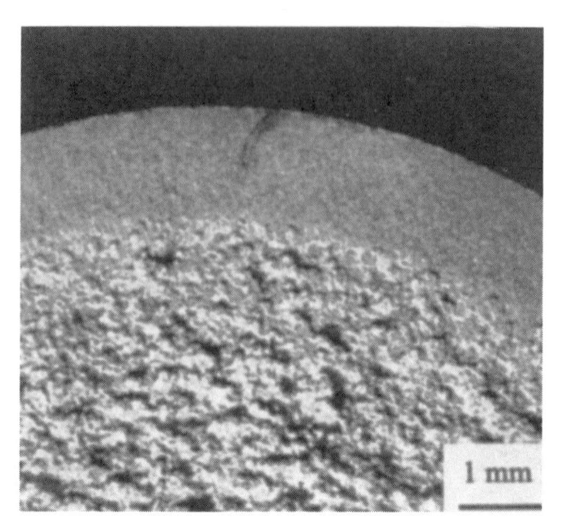

Fig. 6-30. A view of the area of fatigue crack initiation. (Reprinted from Materials Characterization, vol. 36, No. 4/5, A. J. McEvily et al., A Comparison of the Fractographic Features of a Carburized steel Fractured under Monotonic or Cyclic Loading, pp. 153–158, Copyright 1996, with permission of Elsevier Science.)

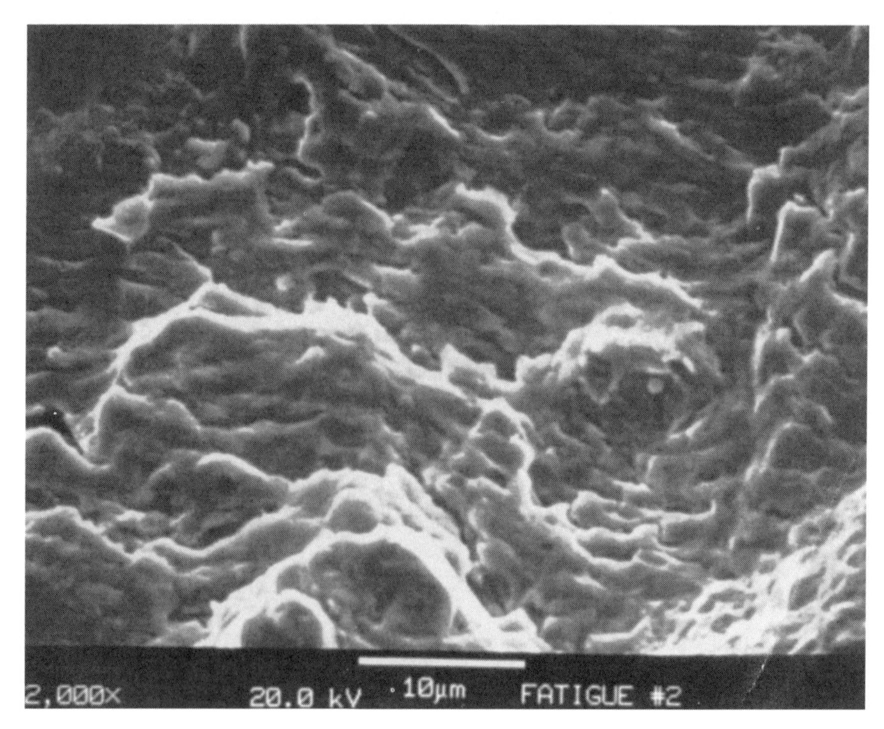

Fig. 6-31. A view of the core area showing fatigue striations. (Reprinted from Materials Characterization, vol. 36, No. 4/5, A. J. McEvily et al., A Comparison of the Fractographic Features of a Carburized steel Fractured under Monotonic or Cyclic Loading, pp. 153–158, Copyright 1996, with permission of Elsevier Science.)

in the core at higher magnification, Fig. 6-31. (Such markings are discussed more fully in Chapter 10.) Note that a secondary crack had developed in the case at the point where the fatigue crack had grown through most of the cross section and the remaining material failed by overload.

On the basis of such information, it was concluded that a fatigue fracture of the trunnion bolt had not occurred. Fracture of the bolt was a consequence of the crash rather than its cause.

XII. SUMMARY

This chapter has focused on the mechanical behavior and fractography of both brittle and ductile materials. A number of case studies have indicated the steps involved in carrying out an investigation to ascertain the nature of either brittle or ductile overload failure. In order to establish the type of failure, the failure analyst should be familiar with a wide variety of fractographic features.

REFERENCES

(1) S. T. Rolfe and J. M. Barsom, Fracture and Fatigue Control in Structures, Prentice-Hall, Englewood Cliffs, NJ, 1977.

(2) A. S. Tetelman and A. J. McEvily, Fracture of Structural Materials, Wiley, New York, 1967.

(3) Collapse of U.S. 35 Highway Bridge, Point Pleasant, West Virginia, December 15, 1967, NTSB Report NTSB-HAR-71-1, Washington, DC, 1971.

(4) M. P. Manahan, C. N. McCowan, T. A. Siewert, J. M. Holt, F. J. Marsh, and E. A. Ruth, ASTM Standardization News, Feb. 1999, pp. 30-35.

(5) J. R. Low, Jr., in Relation of Properties to Microstructure, ASM, Cleveland, 1953, p. 163.

(6) T. Foecke, Metallurgy of the RMS Titanic, National Institute of Standards and Technology Report NIST-IR 6118, Gaithersburg, MD, 1998.

(7) G. Wold and J. Skaar, 39th Annual Forum of the American Helicopter Soc., St Louis, MO, May 1983.

(8) P. W. Bridgman, Trans. ASM, vol. 32, 1944, p. 553.

(9) F. A. McClintock and A. S. Argon, Mechanical Behavior of Materials, Addison Wesley, Reading, MA, 1966.

(10) W. F. Hosford and R. M. Caddell, Metal Forming, 2nd ed., Prentice-Hall, Englewood, NJ, 1993.

(11) G. E. Dieter, Jr., Mechanical Metallurgy, 1st ed., McGraw-Hill, New York, 1961.

(12) A. J. McEvily, K. Pohl, and P. Mayr, A Comparison of the Fractographic Features of a Carburized Steel Fractured under Monotonic or Cyclic Loading, Proc. International Metallography Conf., MC95, Colmar, France, ASM Int., Materials Park, OH, 1996, pp. 317–322.

(13) P. E. Bennett and G. M. Sinclair, Parameter Representation of Low Temperature Yield Behavior of Body-Centered-Cubic Transition Metals, Trans. ASME, J. Basic Eng., Series D, vol. 88, 1966.

(14) K. W. Burns and F. B. Pickering, J. Iron Steel Inst., vol. 202, 1964, p. 899.

7

Thermal and Residual Stresses

I. INTRODUCTION

Thermal stresses arise in components that are subject to temperature gradients, and the thermal transients associated with engine operations are of particular concern. Residual stresses arise because of nonuniform plastic deformation, and they are important in that they affect resistance to fatigue and stress corrosion cracking, and also can cause distortion during machining and heat treating. This chapter briefly reviews how these types of stress are developed.

II. THERMAL STRESS, THERMAL STRAIN, AND THERMAL SHOCK

A. Thermal Stresses

Consider the bimetallic strip shown in Fig. 7-1. In order to avoid bending during heating, equal amounts of material 1 are placed on either side of material 2. The initial length of the strip is L_0, and the area of material 1 is designated as A_1, and that of material 2 as A_2. The corresponding coefficients of expansion are α_1 and α_2.

If the strips were free to expand due to a temperature change of ΔT, the lengths of each material would be

$$L_1 = L_0(1 + \alpha_1 \Delta T), \tag{7-1}$$

$$L_2 = L_0(1 + \alpha_2 \Delta T). \tag{7-2}$$

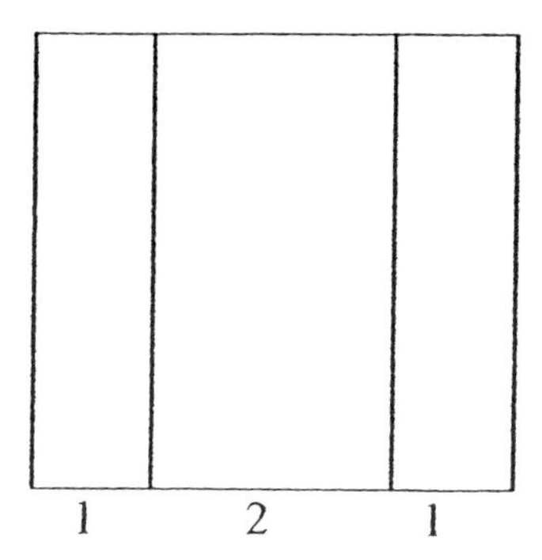

Fig. 7-1. A bimetallic strip.

But the final lengths must be equal, or $L_{1f} = L_{2f}$, or

$$L_0(1 + \alpha_1 \, \Delta T) + \frac{\sigma_1}{E_1} L_0 = L_0(1 + \alpha_2 \, \Delta T) + \frac{\sigma_2}{E_2} L_0. \tag{7-3}$$

Also, for equilibrium, $P_1 + P_2 = 0$, or $\sigma_1 A_1 + \sigma_2 A_2 = 0$.

There are thus two equations with two unknowns, and the stresses σ_1 and σ_2 can therefore be determined:

$$\sigma_2 = \frac{(\alpha_1 - \alpha_2)\Delta T E_1 E_2}{E_1 + E_2(A_2/A_1)} = \frac{(\alpha_1 - \alpha_2)\,\Delta T E_2}{1 + (E_2/E_1)/(A_2/A_1)}, \tag{7-4}$$

$$\sigma_1 = -\sigma_2 \frac{A_2}{A_1}. \tag{7-5}$$

As long as the stresses are in the elastic range, there will be no residual effects when the temperature returns to its initial value,and the above equations hold. However, if the stresses exceed the yield strength of the materials involved, a state of residual stress will develop when the temperature returns to its initial value. For example, if material 2 in the above example had a lower coefficient of expansion and a relatively low yield stress, it could undergo plastic extension as the temperature increased. Subsequently, when the temperature was lowered, material 2 would be left in a state of residual compression, and material 1 would be in residual tension.

B. Thermal-Mechanical Cyclic Strains

Transient thermal strains of a cyclic nature can arise in jet engine components such as disks, blades, and vanes. Figure 7-2 is a simplified thermal-mechanical cycle for blades in a gas turbine (1). Under cruise conditions, the blades are at 600°C. When engine power is increased, the surface temperature rises to 1100°C in 105 seconds, Fig. 7-2*a*. Engine power is then returned to the cruise condition, and the tempera-

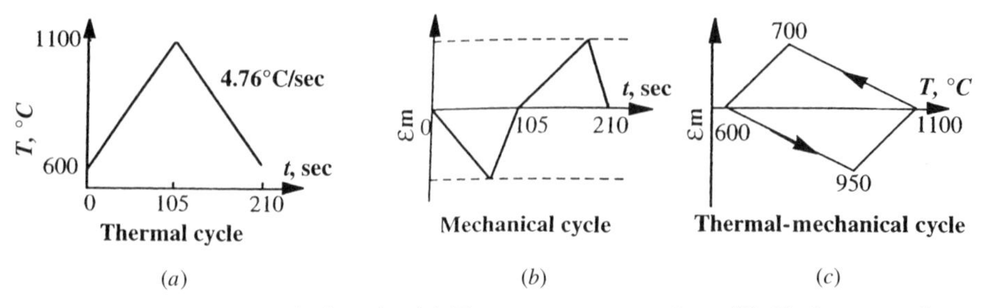

(*a*) (*b*) (*c*)

Fig. 7-2. Thermal-mechanical cycle. (*a*) Temperature versus time. (*b*) Strain versus time. (*c*) Thermal-mechanical hysteresis loop. (From Remy, 1. Reprinted from Low Cycle Fatigue and Elasto-Plastic Behaviour of Materials, edited by K.-T. Rie and P. D. Portella, pp. 119–130, Copyright 1998, with permission of Elsevier Science.)

ture falls to 600°C in another 105 seconds. The rate of heating and cooling is 4.76 °C/sec. The total strain is the sum of the strain due to thermal expansion plus the strain due to the thermal stresses developed. The mechanical strain versus time plot, shown in Fig. 7-2*b*, includes both elastic and plastic strains. On heating, the outer surface initially goes into compressive strain, but then the strain reduces to zero as the interior of the blade heats up. On cooling, the reverse procedure occurs. Figure 7-2*c* shows the counterclockwise diamond cycle that is the thermal-mechanical hysteresis loop for this transient thermal history. The corresponding stress versus mechanical strain hysteresis loops for single crystals in the [001] and [111] orientations are shown in Fig. 7-3. It is noted that the mean stress for the cycle is close to zero.

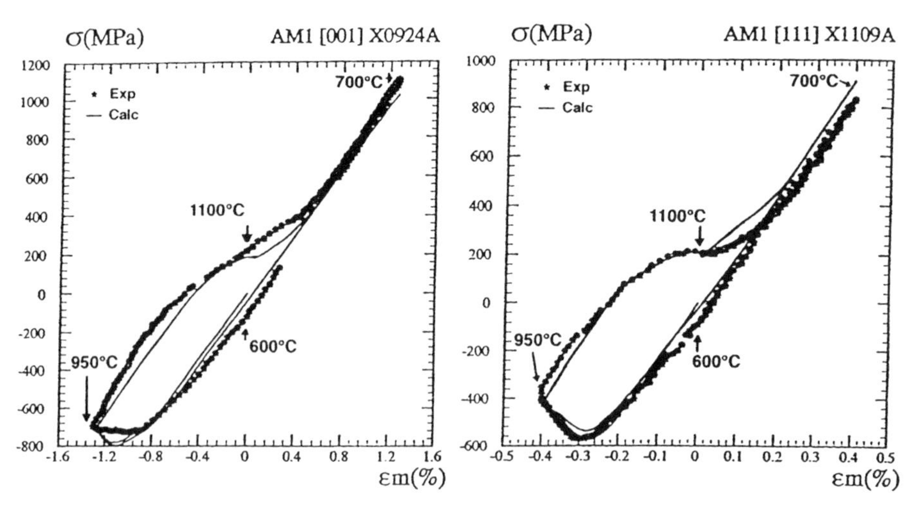

Fig. 7-3. Stress versus mechanical strain hysteresis loops for [001] and [111] thermal mechanical fatigue (TMF) single crystal specimens of AM1 superalloy using the cycle depicted in Fig. 7-2. Comparison between a viscoplastic model (solid line) and experiment (symbols). (After Remy, 1. Reprinted from Low Cycle Fatigue and Elastic-Plastic Behaviour of Materials, edited by K.-T. Rie and P. D. Portella, pp. 119–130. Copyright 1998, with permission of Elsevier Science.)

For the [001] orientation, plastic deformation (or viscoplastic deformation) is pronounced at a strain of -0.8. For the [111] orientation, plastic deformation begins at a strain of -0.25. Such thermal mechanical histories are complex, but they are also obviously important in assessing the fatigue life of such components.

C. Thermal Shock

Thermal shock denotes the rapid development of a steep temperature gradient and accompanying high stresses that can result in the fracture of brittle materials. It can occur either on heating or cooling. For example, the sudden shutdown of turbine engine can result in cracking of protective platinum-aluminide coatings due to the tensile stresses that develop as the surface cools and tries to shrink but is restrained by the interior.

An example of thermal shock that occurs during heating is as follows. Consider a glass bowl whose thermal conductivity is low.

1. A hot liquid is poured into the bowl.
2. The inside surface of the wall tries to expand because of the sudden rise in temperature.
3. A biaxial compressive stress develops on the inside surface because expansion of the is resisted by the surounding, still cool, wall material.
4. This compressive stress system sets up a balancing tensile stress system in the outer, still cool portion of the wall.
5. Fracture can initiate in the outer region if the magnitude of the tensile stress developed is sufficient to nucleate a crack at a weak point.

III. RESIDUAL STRESSES CAUSED BY NONUNIFORM PLASTIC DEFORMATION

Residual stresses arise because of a gradient in plastic deformation caused either by mechanical deformation or by a thermal gradient caused during cooling of a metal or alloy from a high temperature to a low temperature. The sign of the residual stress is always opposite in sign to the sign of the applied stress that gave rise to the residual stress. Residual stresses are important in fatigue and stress corrosion cracking where they can be either beneficial or detrimental. If residual stresses are present prior to heat treatment or machining, they can be detrimental because they can result in distortion (warping).

A. An Example of Mechanically Induced Residual Stresses: Springback after Bending into the Plastic Range (2)

In sheet bending, the width w is much greater than the thickness t, and width changes are negligible. Therefore, bending can be considered to be a plane-strain operation with $\varepsilon_y = 0$, $\varepsilon_z = -\varepsilon_x$. Let z be the distance measured from the mid-plane of the

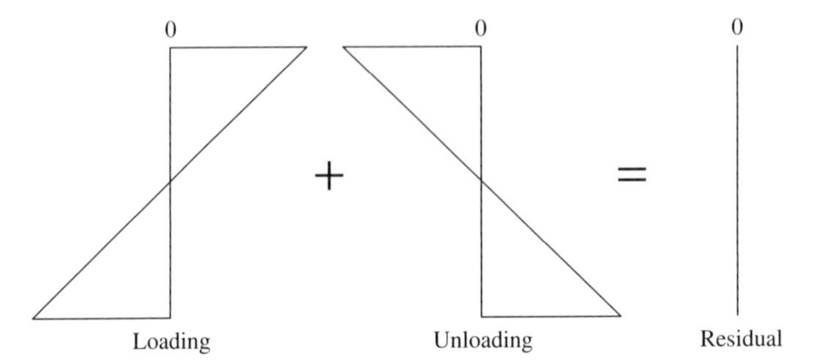

Fig. 7-4. Strip bending, elastic strains.

sheet in the thickness direction, and let r be the radius of curvature of the mid-plane. The value of ε_x varies linearly from $-t/2r$ at the inside of the bend ($z = -t/2$) to zero at the mid-plane ($z = 0$), to $+t/2r$ at the outside of the bend ($z = t/2$). Figure 7-4 shows the stress through the cross section. The principal of superposition is used to show that unloading can be considered to be the reverse of loading, so that for purely elastic bending there is no residual stress after unloading.

Next, assume that the material is elastic ideally plastic, that is, there is no strain hardening in the plastic range. If the tensile yield stress is Y, the flow stress in plane strain σ_0 will be $1.15Y$. Figure 7-5 shows the stress distribution throughout the sheet for fully plastic behavior. Except for an elastic core at mid-plane (which will be neglected), the entire section will be at a stress, $\sigma_x = \pm\sigma_0$.

To calculate the bending moment M needed to create this fully plastic bend, note that $dF_x = \sigma_x w\, dz$, and that $dM = z\, dF_x = z\sigma_x w\, dz$. Therefore, the fully plastic bending moment is

$$M = \int_{-t/2}^{+t/2} w\sigma_x z\, dz = 2\int_0^{t/2} w\sigma_x z\, dz = w\sigma_0 \frac{t^2}{4}. \tag{7-6}$$

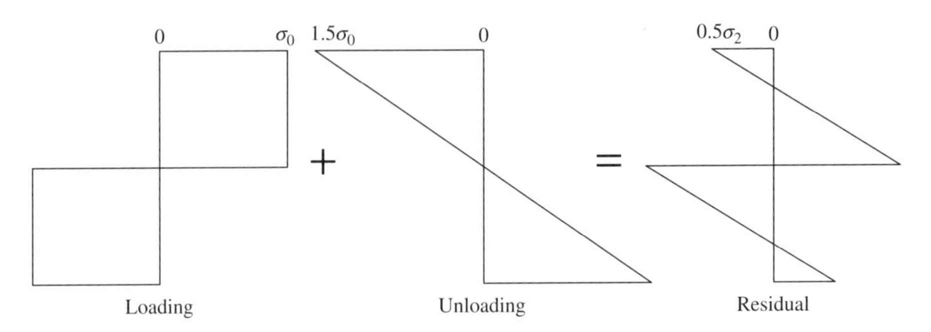

Fig. 7-5. Strip bending, plastic strains on loading, elastic on unloading.

(Note that the elastic bending moment that is needed to have σ_x just equal to σ_0 would be $w\sigma_0(t^2/6)$, so that the fully plastic bending moment is 50% higher.)

When the external moment is released, the internal moment must go to zero. As the material springs back elastically, the internal residual stress distribution must result in a zero bending moment. Since the unloading is elastic,

$$\Delta\sigma_x = \frac{E}{1-\nu^2}\,\Delta\varepsilon_x = E'\,\Delta\varepsilon_x. \tag{7-7}$$

The change in strain is given as $\Delta\varepsilon_x = z/r - z/r'$, where r' is the radius of curvature after springback. This causes a change in bending moment ΔM, where

$$\Delta M = 2w\int_0^{t/2}\Delta\sigma_x z\,dz = 2w\int_0^{t/2}E'\left(\frac{1}{r}-\frac{1}{r'}\right)z^2\,dz = \frac{wE't^3}{12}\left(\frac{1}{r}-\frac{1}{r'}\right)\cdot \tag{7-8}$$

$M - \Delta M = 0$ after springback, and therefore equating M and ΔM gives

$$\frac{w\sigma_0 t^2}{4} = \frac{wE't^3}{12}\left(\frac{1}{r}-\frac{1}{r'}\right), \tag{7-9}$$

or

$$\left(\frac{1}{r}-\frac{1}{r'}\right) = \frac{3\sigma_0}{tE'}\cdot \tag{7-10}$$

The resulting residual stress

$$\begin{aligned}\sigma_x' &= \sigma_x - \Delta\sigma_x = \sigma_0 - E'\,\Delta\varepsilon_x \\ &= \sigma_0 - E'z\left(\frac{1}{r}-\frac{1}{r'}\right) = \sigma_0 - E'z\left(\frac{3\sigma_0}{tE'}\right) = \sigma_0\left(1-\frac{3z}{t}\right)\cdot\end{aligned} \tag{7-11}$$

On the outside surface, $z = t/2$, and the residual stress σ_x' equals $-\sigma_0/2$. On the inside surface, the residual stress is $+\sigma_0/2$. The distribution of stresses is shown in Fig. 7-5.

Note that the sign of the residual stress is opposite to the sign of the stress that caused it. Also, plastic deformation is required, but the plastic deformation must be nonuniform. There is no residual stress associated with a tensile bar that has been uniformly stretched into the plastic region. On the other hand, residual stresses will develop at notches, and so on, if the material at the base of the notch is strained into the plastic range while the surroundings remain elastic.

An important state of residual stress is that formed at the tip of a growing fatigue crack due to an overload in plane specimens. During fatigue crack growth, if a 100% overload is applied, the fatigue crack will subsequently undergo a period of reduction in rate of crack growth as the crack penetrates the plastic zone created by the overload. This slowdown is related to an increased level of compressive residual stress brought about by the overload. This residual stress is largely a plane stress, surface regions, and is brought about by the lateral contraction of material at the surface that occurs in the overload plastic zone. In ductile materials, this lateral contraction leads to the formation of an obvious dimple immediately ahead of the crack tip. The compressive stress develops as the load is reduced from the overload level, and for now, because of the lateral contraction, there is more material in planes parallel to and just below the surface of the overload plastic zone than before the overload. As the crack grows through the overload zone, these residual stresses are released, giving rise to an increased level of crack closure in the wake of the crack tip (see Chapter 10), which results in a slowing down of the rate of fatigue crack propagation. The extent of the slowdown in crack rate is a function of the overload level and the thickness of the specimen, being much more pronounced in thin as compared to thick specimens. The reason for this difference is that, in thick specimens, plane-strain conditions prevail throughout most of the specimen's thickness, and the lateral contraction associated with plane stress is absent under plane-strain conditions. Hence, the level of enhanced residual compressive stress due to an overload is much less in thick specimens than in thin.

Surface compressive residual stresses are beneficial in fatigue and in stress corrosion cracking. For this reason, they are often deliberately introduced, by shot-peening, for example. Any cracks that form are retarded in their growth rates, much as in the case of an overload.

B. Case Study: Shaft of a Golf Club

At a golf driving range in Connecticut the shafts (tapered hollow tubes) of the clubs were sometimes bent or otherwise damaged in use and had to be replaced. A supply of replacement shafts was kept in a storage shed, which was at times damp and humid. The shafts were made of a low-alloy steel that had been chrome plated, which protected the exterior of the shaft from corrosion. Whereas a new shaft is sealed to the club head as well as at the grip end, and thereby protected from corrosion on the inside of the shaft, the replacement shafts were not sealed, and as a result the interiors of these replacement shafts underwent corrosion over a period of time.

A golfer was driving balls with a club whose shaft was a replacement. As he struck a ball, the shaft fractured, and the end he was holding struck him in the eye. Fortunately, the damage to his eye was not serious, and he recovered completely.

Upon examination of the shaft it was noted that the fracture origin was at a small dent in the shaft that preexisted the accident. The fracture had initiated at this dent on the inside of the shaft. The fracture origin, shown in Fig. 7-6, was more planar and brittle in appearance than was the fracture away from the origin, shown in Fig. 7-7, which exhibited the dimples characteristic of ductile fracture. It was concluded that the fracture was due to hydrogen embrittlement, which was associated

Fig. 7-6. Macroscopic view of the area of the fracture origin in a failed golf shaft. The rough area at bottom is the corroded inside surface of the shaft. (Reprinted from *Materials Characterization,* Vol. 26, A. J. McEvily and I. Le May, pp. 253–268, Copyright 1991, with permission of Elsevier Science.)

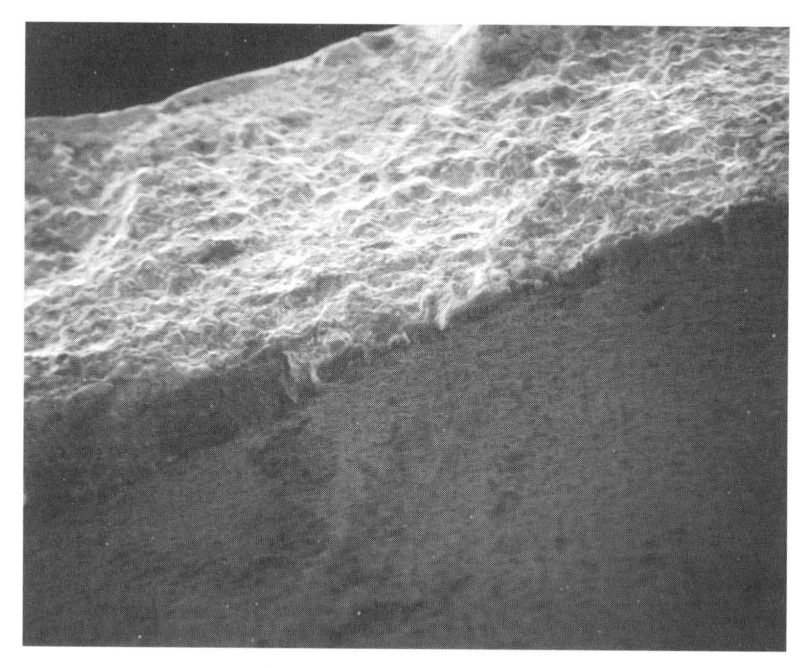

Fig. 7-7. Macroscopic view of fracture surface away from fracture origin in failed golf shaft. (Reprinted from *Materials Characterization,* Vol. 26, A. J. McEvily and I. Le May, pp. 253–268, copyright 1991, with permission of Elsevier Science.)

with corrosion and the seemingly innocuous dent which, because of springback, resulted in a residual tensile stress on the inside of the shaft at the periphery of the dent. When the dent was formed, material on the inside near the periphery of the dent went into compression, and material near the center of the dent went into tension. Upon springback, the signs of the stresses were reversed, and a residual tensile stress was left on the inside surface at the periphery of the dent.

IV. RESIDUAL STRESSES DUE TO QUENCHING

On cooling a metal part from an elevated temperature, residual stresses may be developed. For example, if a massive piece of copper is cooled rapidly, the surface layers will cool before the interior, thus setting up tensile strains and stresses in the surface that are counterbalanced by compressive stresses in the interior. At elevated temperatures, these tensile and compressive stresses are relaxed due to the low yield strength of the material. However, at a later point in the cooling process, the already cooled surface layer will be subjected to compression as the interior finally cools and shrinks. In general, the resultant compressive stress in the surface layers is beneficial, except when machining follows and distortion results from the nonuniform removal of the surface.

When steel is quenched to form martensite at a low temperature, the transformation from austenite to martensite is associated with a volume expansion of the order of 1–3%. At the surface, this initially results in a compressive stress, as in the case of copper, but when the underlying layer finally cools and expands, the surface is put into tension. These tensile stresses can be reduced by using a steel of lower hardenability so that the interior does not got through a martensitic transformation, but instead transforms to bainite at a higher temperature.

B. Quench Cracking

In a steel, there is always the possibility of immediate quench cracking, due to the level of tensile residual stresses developed when untempered martensite is formed. However, cold cracking (or delayed cracking) is the more probable event, and for this reason, alloy steel parts that have been quenched are quickly transferred to tempering furnaces or salt baths in order to mnimize the time available for cold cracking to develop. If cracks develop on quenching into water and the part is then transferred to a salt bath for tempering, an explosive reaction can occur as the water trapped in the cracks transforms into steam. Therfore, personnel carrying out this operation need to wear protective face masks and protective clothing.

It is also possible to develop quench cracks above the quench temperature if, during the quench of a material with low ductility, the tensile surface stresses that develop due to the temperture gradient in the material exceed the resistance to cracking of the material, as indicated in Fig. 7-8.

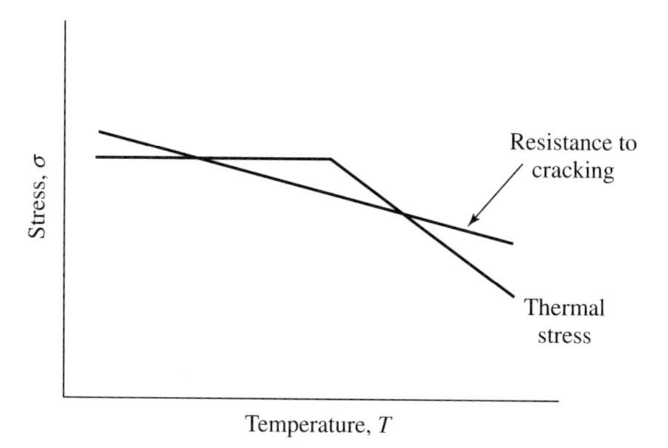

Fig. 7-8. A schematic of thermal stress versus resistance to cracking as a function of temperature.

To avoid the development of residual stresses, heat-treating procedures are used that minimize the temperature gradients responsible for the residual stresses. Such methods are:

(a) *Martempering:* The steel is rapidly cooled to a temperature above the M_s and held to allow a uniform temperature to develop before further cooling to martensite.

(b) *Austempering:* The steel is rapidly cooled to a temperature above the M_s and held there until the transformation to bainite is complete before further slow cooling to room temperature.

V. RESIDUAL STRESS TOUGHENING

If the outer surface of the bowl discussed above in the section on thermal shock had been treated to have a residual compressive stress in the surface region, the resistance to thermal shock would have increased. Glass can be toughened by diffusing large atoms into the surface to develop residual compressive stresses on cooling as, for example, in Corningware. Tempered glass is created by cooling rapidly from an elevated temperature to develop compressive surface residual stresses, much as in the case of a block of copper. This type of glass is used in the side windows of automobiles. When such glass breaks, it fractures into many small fragments. However, the glass used in windshields is not tempered. A windshield consists of a sandwich of two sheets of glass between which is bonded a sheet of plastic. In a crash, the windshield is intended to be flexible enough to reduce head injuries, but be resistant enough to overall fracture to prevent front seat occupants from being ejected through the windshield. In this case the glass generally breaks into large shards that remain attached to the plastic membrane.

VI. RESIDUAL STRESSES RESULTING FROM CARBURIZING, NITRIDING, AND INDUCTION HARDENING

A. Carburizing

Carburizing is carried out in the austenitic range and can lead to the development of a compressive residual stress at the surface in a low-alloy steel during quenching for the following reason (3, 4). Carbon is one of the elements that depresses the M_s temperature of a steel. The M_s temperature of the carburized surface layer, therefore, can be much lower than that of the interior because of the differential in carbon contents. On quenching, the interior, even though at a higher temperature than the surface, is the first to transform due to its higher M_s temperature. Later on, the surface transforms and tries to expand but is now restrained by the already transformed interior. As a result, the surface is left in a state of residual compression, as indicated in Fig. 7-9.

B. Nitriding

Nitriding is carried out at an elevated temperature below the eutectoid temperature for time periods of the order of 9–24 hours, and during the nitriding process any prior residual stresses are relaxed. The purpose of nitriding is to improve the surface wear and fatigue properties. Since the temperatures are lower than in carburizing and no phase transformation is involved, problems with distortion are mini-

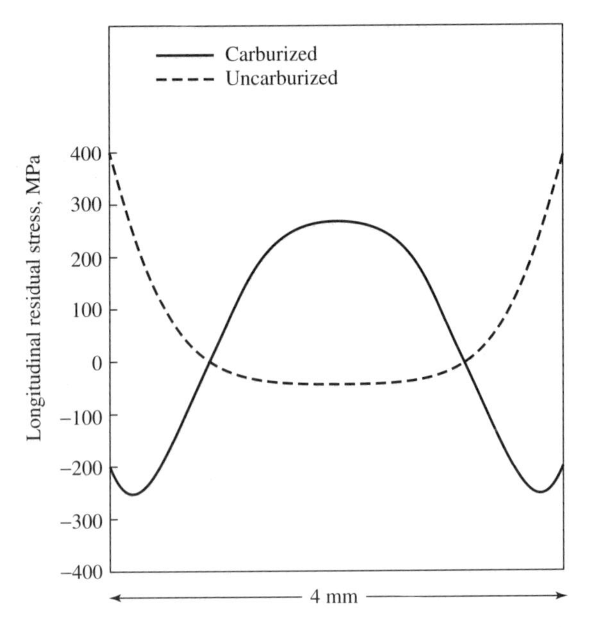

Fig. 7-9. Effect of carburization on the residual stresses in a quenched and tempered (1 hr at 180–200°C) 0.93 Cr-0.26 C steel (After Ebert, 3, and Krause, 4).

mized, an important consideration when heat treating carefully machined parts such as crankshafts. The formation of nitrides leads to a beneficial compressive residual stress in the surface even after the usual slow cooling because of a lower coefficient of expansion of the nitrides.

C. Induction Hardening

Induction hardening is a surface-hardening process in which only the surface layer of a suitable ferrous workpiece is heated by electrical induction to above the transformation temperature and immediately quenched. Compressive residual stresses develop as the surface layers transform from austenite on quenching.

VII. RESIDUAL STRESSES DEVELOPED IN WELDING

In welding operations, the parts being joined often provide a large heat sink, and therefore cooling rates are rapid. As a result, untempered martensite may form in a residual tensile stress field and lead to weld cracking, either through cold cracking or quench cracking. The susceptibility to weld cracking increases as the number of unfavorable welding conditions increases. For example, a medium-carbon steel welded with an electrode not of the low-hydrogen type, and without preheat or postheat, may perform satisfactorily even though there may be a HAZ containing a region of martensite about $\frac{1}{16}$ inch thick if joint restraint is low and cyclic loading in service is limited. However, if the section thickness is doubled, the level of residual stress rises due to greater constraint, and the thickness of the martensitic zone is increased due to a higher cooling rate, and cracking may develop. Even the use of low-hydrogen electrodes may not prevent cracking of the thicker section, but the use of a 315°C (600°F) preheat will prevent cracking by retarding the rate of cooling, and thereby reducing the level of the residual stresses and the amount of martensite formed.

On cooling below the M_s temperature, two forms of martensite can form, depending upon carbon content. In low-carbon steels, the martensite is made up laths that contain a high density of dislocations. In higher carbon steel, the martensite is made up of plates that also contain a high dislocation content, but in addition may be twinned. Preheating is used to reduce the cooling rate in order to minimize the likelihood of the formation of brittle martensite, particularly the twinned martensite. The preheat temperature increases with carbon content and with the thickness of the plates being welded, with recommended preheat temperatures ranging from 100°F for a 0.2 wt % C steel up to 600°F for a 0.6 wt % C steel.

Postweld heating of a weldment serves two purposes. One is to relieve the residual stresses that may have been developed during the welding process. The second is to temper both the weld deposit and HAZ to improve their fracture toughness. It is now mandatory that weldments in all nuclear components and in most pressure vessels be postweld heat treated (PWHT). These postweld heat treatments are carried out at a temperature of the order of 650°C (1200°F) for one hour.

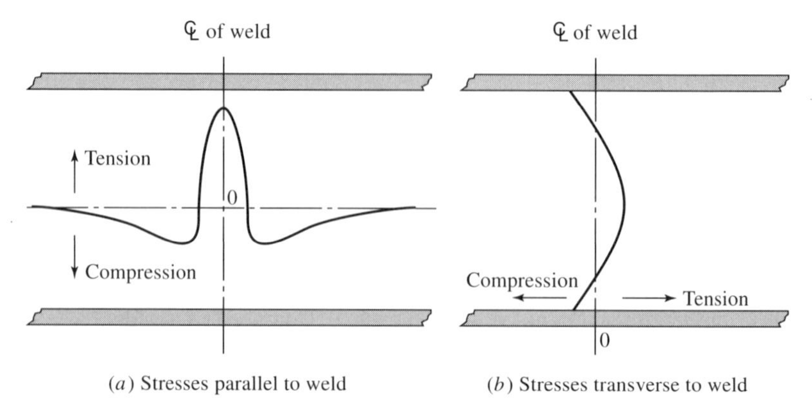

(*a*) Stresses parallel to weld (*b*) Stresses transverse to weld

Fig. 7-10. Transverse and longitudinal residual stresses developed at a butt weld. (From Gurney, 5. With permission of the Cambridge University Press.)

The residual stresses that are developed both parallel to and transverse to a butt weld due to shrinkage of the weld metal on cooling are shown in Fig. 7-10 (5). To minimize distortion, weld passes were made from each edge of the plate toward the center, which accounts for the transverse residual stress distribution, since the last weld metal to solidify was in the central region of the butt weld. The residual stress developed during welding can sometimes be beneficial. For example, the fatigue strength of steel lap welded joints for automotive use was improved by using low transformation temperature (M_s 200°C, M_f 20°C) welding wire (10Cr-10Ni), which induced compressive residual stresses in the surface layers. The fatigue strength at 10^8 cycles was increased from 300 MPa to 450 MPa by this procedure (6).

VIII. MEASUREMENT OF RESIDUAL STRESSES

The two main methods for the determination of the magnitude of residual stress, the x-ray method and the metal removal method, were described in Chapter 5. Another example of the metal removal method is known as the boring-out method. This method can be used to determine residual stress levels in cannon or other cylindrical bodies. In the case of a cannon barrel, as a last step in the manufacturing process, a pressure is developed within the barrel of sufficient magnitude to expand the interior of the barrel into the plastic range. When the pressure is removed, a beneficial state of residual compressive stress is developed on the inner surface, a process known as autofrettage. The determination of the magnitude of the residual stresses by the boring-out method involves the machining away of successive layers from the interior of the barrel. As each layer is removed, some of the residual stress is relieved. As a result, on the outer surface of the barrel, longitudinal and circumferential strains develop which are measured by strain gauges. From these measurements the magnitude of the initial residual stress state can be deduced.

IX. SUMMARY

THe thermal stresses that arise in engineered items, such as the turbine disks of jet aircraft engines, are no longer a major problem because of improvements in materials as well as better design procedures. Nevertheless, their presence must be taken into account. Problems with thermal cracking during heat treatment continue. Residual stresses are still a problem because their presence may not be recognized until a cracking problem associated with them develops.

REFERENCES

(1) L. Remy, in Low Cycle Fatigue and Elasto-Plastic Behavior of Materials, ed. by K.-T. Rie and D. P. Portella, Elsevier, Oxford, UK, 1998, pp. 119–130.

(2) W. F. Hosford and R. M. Caddell, Metal Forming, 2nd ed., Prentice-Hall, Engelwood Cliffs, NJ, 1993.

(3) L. J. Ebert, The Role of Residual Stresses in the Mechanical Performance of Case Carburized Steel, Met Trans. A, vol. 9A, 1978, pp. 1537–1551.

(4) G. Krauss, Principles of Heat Treatment of Steel, ASM, Materials Park, OH, 1980.

(5) T. R. Gurney, Fatigue of Welded Structures, Cambridge University Press, Cambridge, UK, 1968, p. 58.

(6) A. Ohta, Y. Maeda, and N. Suzuki, in Proc. 25th Symp. on Fatigue, Japan Soc. Mats. Sci., 2000, pp. 284–287.

8

Statistical Distributions

I. INTRODUCTION

This chapter deals briefly with two types of statistical distributions, the normal, or Gaussian, distribution and the extreme value distribution. Normal distributions are useful in dealing, for example, with the scatter encountered in fatigue test results. Extreme value distributions are useful in dealing with early failures in a population and in analyzing the influence of inclusions on fatigue strength.

II. DISTRIBUTION FUNCTIONS

There are two types of distribution functions. One is the relative distribution function, which gives the number of observations that fall between two limits, for example, in determination of the yield strengths of 449 tensile bars, the percentage of the results that fall between 828 and 841 MPa. The other distribution function is the cumulative distribution function, which gives, for example, the total number of specimens of the 449 tested with yield strengths less than 841 MPa.

The central tendency of a relative distribution is often of interest, and it can be determined by one of three measures. These are:

1. *Arithmetic Mean (Average),* μ*:* The sum of the values in the distribution divided by their number
2. *Median:* The value in an ordered set of values below which and above which there are an equal number of values
3. *Mode:* The value that occurs with the highest frequency

In a normal distribution, these three measures of central tendency are equal, whereas in an extreme value distribution they generally differ. The standard deviation is a measure of the spread in a population of n items, and is expressed as

$$\sigma = \left[\frac{\sum_{i=1}^{i=n} (X_i - \mu)^2}{n - 1} \right]^{1/2}, \tag{8-1}$$

where X_i is the ith value in an ordered series. The variance is the square of σ, and is also used as a measure of scatter. The coefficient of variation v is expressed as

$$\mathrm{v} = \frac{\sigma}{\mu}. \tag{8-2}$$

The normal distribution focuses on the central characteristics of a distribution, and is characterized by a symmetric bell-shaped curve and two parameters, the mean of the population μ, and the standard deviation σ. The Weibull extreme value distribution (1) is used in the assessment of the probability of early failure in a distribution, the left-hand side of a distribution, and is widely used in the bearing industry. The Gumbel extreme value distribution (2) is used to predict items on the right-hand side of a distribution such as the size of the largest inclusion to be found within a given volume.

III. THE NORMAL DISTRIBUTION

The equation derived by Gauss for the bell-shaped, normal distribution is

$$f(x) = \frac{1}{\sigma\sqrt{2\pi}} e^{-(x-\mu)^2/2\sigma^2}, \tag{8-3}$$

where $f(x)$ is the height of the frequency distribution curve at a particular value of x. An example of this distribution, the standardized normal frequency distribution, is given in Fig. 8-1a. The distribution is symmetrical about $z = 0$, and extends from $-\infty$ to $+\infty$. (The German ten-mark banknote issued in October 1993 features a portrait of Gauss, the above equation, and a bell-shaped distribution curve.) The relative frequency of a value of $(x - \mu)/\sigma$ falling between $-\infty$ and a specified value is given by the area under the curve between these limits, with the area under the entire curve being unity. The relative frequency is also known as the cumulative distribution function, and is obtained by numerically integrating Eq. 8-3 between the limits of $-\infty$ and $(x - \mu)/\sigma$. The following Table 8-1 lists the areas under the curve for several ranges of σ. The area between $\pm\sigma$ of the mean is equal to 0.68, that is 68% of the values fall between these limits in the case of the normal distribution.

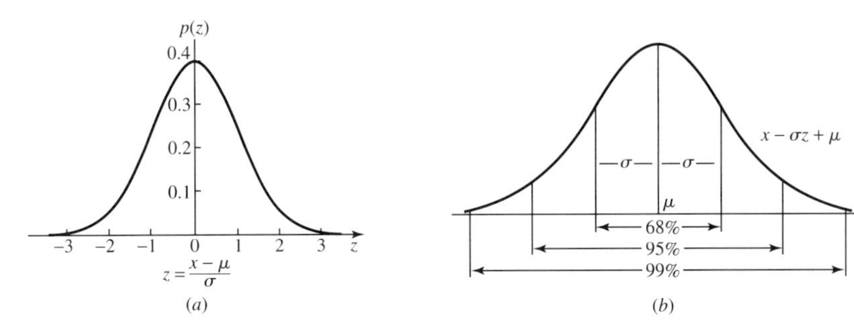

Fig. 8-1. (*a*) The bell-shaped curve associated with the standardized normal frequency distribution. (*b*) One, two and three σ ranges about the mean of a normal distribution.

Similarly, 95% of the values are between $\pm 2\sigma$ of the mean, and 99.7% of the values are between $\pm 3\sigma$ of the mean as indicated in Fig. 8-1*b*. One commonly stated production goal in industry is the 6σ goal, meaning that of 99.7% of the products should have properties that fall within defined limits.

To determine if a distribution can be treated as a normal distribution, the cumulative frequency distribution is plotted on normal probability paper. In this type of plot, the cumulative frequency (in percent) is plotted against the parameter of interest, for example, the yield strength. If the distribution is indeed normal, it will plot as a straight line. The frequency distribution of the yield strengths of a steel is shown in Table 8-2, and is used as an illustration.

Figure 8-2 (4) shows the above data plotted on normal probability data, and it is seen that the distribution can be fitted by a straight line, indicating a normal distribution. The mean value and the standard deviation can also be determined from Fig. 8-2. The mean is the value of the abscissa corresponding to a cumulative frequency of 50%. The standard deviation is equal to one-half of the difference between the abscissa value at the 84% cumulative frequency level and the abscissa value at the 16% cumulative frequency level.

TABLE 8-1. Areas Under the Normal Frequency Curve for Given Ranges of σ

Range	Area under Curve
$\pm 0.5\sigma$	0.3830
$\pm 1.0\sigma$	0.6826
$\pm 2.0\sigma$	0.9744
± 3.0	0.9974

TABLE 8–2. Frequency Tabulation of Yield Strengths of a Steel (3)

Yield Strength, ksi	Frequency	Cumulative Frequency	Cumulative Frequency, %
114–115.9	4	4	0.9
116–117.9	6	10	2.2
118–119.9	8	18	3.8
120–121.9	26	44	9.6
122–123.9	29	73	16.1
124–125.9	44	117	25.9
126–127.9	47	164	36.4
128–129.9	59	223	49.5
130–131.9	67	290	64.5
132–133.9	45	335	74.5
134–135.9	49	384	85.4
136–137.9	29	413	91.9
138–139.9	17	430	95.7
140–142.9	9	439	97.7
142–143.9	6	445	99.0
144–145.9	4	449	99.9

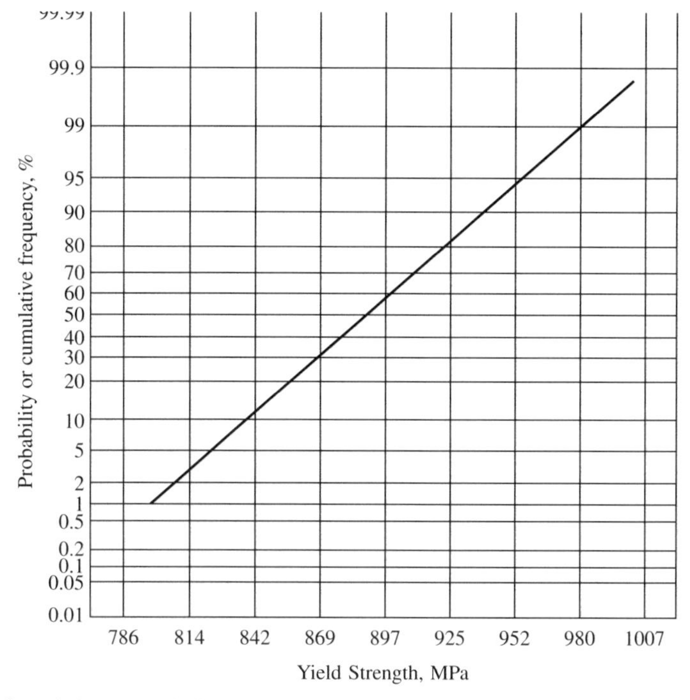

Fig. 8-2. Cumulative normal distribution plot of yield strengths plotted on normal probability paper. (From Dieter, 4.)

IV. THE WEIBULL DISTRIBUTION (1)

In a Weibull distribution of yield strengths, for example, the probability that the yield strength σ_Y will fall in the range from 0 to σ_Y (i.e., the cumulative probability distribution function) is denoted by $G(\sigma_Y)$, and is given as

$$G(\sigma_Y) = 1 - e^{-[(\sigma - \sigma_{Y_0})/\sigma_0]^m}, \tag{8-4}$$

where σ_0 is called the scale parameter, m is called the shape parameter, a measure of the degree of scatter, and σ_{Y_0}, the lowest yield strength in the population, is called the location parameter. The quantity $G(\sigma_Y)$ is estimated from a group of N samples by noting the number of samples n that have a yield strength of σ or less; thus

$$G(\sigma_Y) = \frac{n}{N}. \tag{8-5}$$

However, because N is generally small, it is preferable to define $G(\sigma_Y)$ as

$$G(\sigma_Y) = \frac{n}{N + 1}. \tag{8-6}$$

Equation 8-4 can also be written in a more useful form as

$$\ln \ln \left[\frac{1}{1 - G(\sigma_Y)} \right] = -m \ln \sigma_0 + m \ln(\sigma_Y - \sigma_{Y_0}). \tag{8-7}$$

If the distribution is of the Weibull type, a plot of ln ln $\{1/[1 - G(\sigma_Y)]\}$ versus ln $(\sigma_Y - \sigma_{Y_0})$ will result in a straight line of slope m. σ_0 is called a scale parameter because a change in its value will result in a corresponding shift in the position of the line on the plot. For $G(\sigma_Y)$ equal to 0.632, ln ln $\{1/[1 - G(\sigma_Y)]\}$ is zero, and ln σ_0 is equal to the corresponding value of the abscissa. The slope m provides an indication of the skewness of the frequency distribution. As shown in Fig. 8-3 (5), an m value of 3.5 corresponds to a normal, bell-shaped distribution. As m decreases below 3.5, the spread of the data increases and there is a large percentage of "infant mortalities." For m values greater than 3.5, there is a benefit in that the dispersion of the data is low and the reliability of prediction is increased. However, the value of σ_0 is another consideration, for this value fixes the position of the Weibull distribution with respect to the horizontal axis.

In some cases, it may be possible to fit a straight line only through lower portion of the cumulative distribution curve, yet this may be sufficient for the particular application.

Figure 8-4 shows the yield strength data of Table 8-1 plotted in accord with Eq. 8-5 for various values of σ_{Y_0}. It is seen that the best straight line fit to the data at low values of G, the usual region of interest, is obtained for a value of σ_{Y_0} equal to 105 ksi.

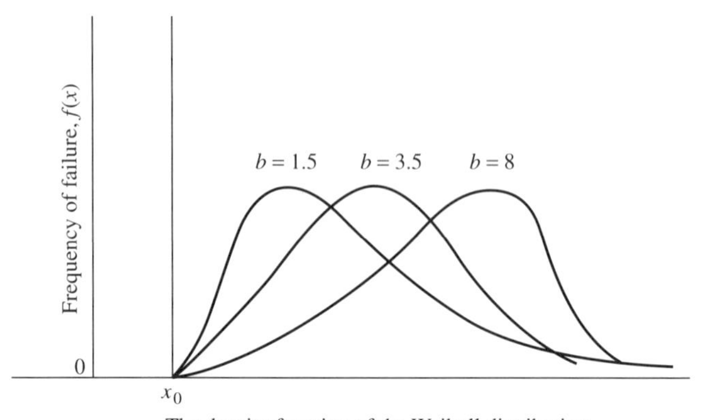

Fig. 8-3. Weibull relative frequency distribution as influenced by slope parameter b. (After Lipson and Sheth, 5.)

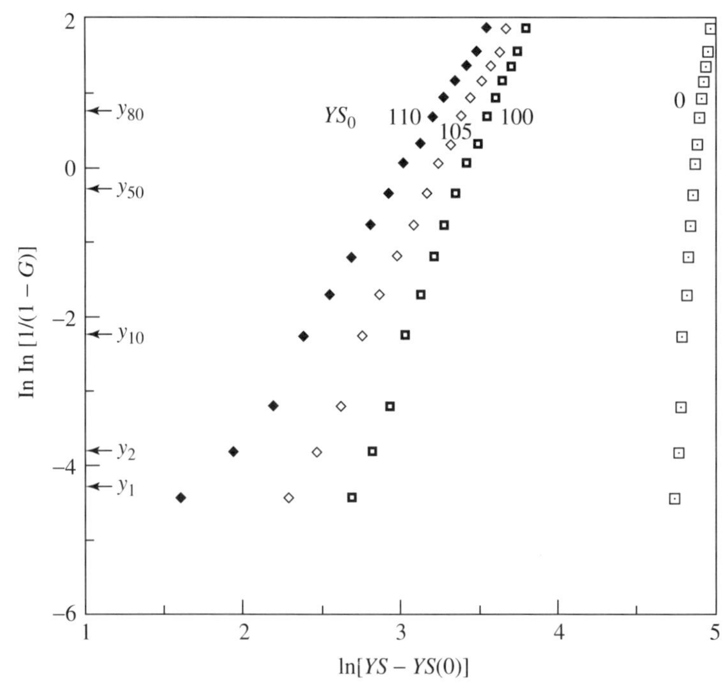

Fig. 8-4. Weibull plot of yield strengths. YS_0 is the expected minimum value of the yield strength.

The slope of this line is 4.6, and σ_0 is equal to 27.9 ksi. Special coordinate paper, known as Weibull probability paper, is commercially available to expedite the plotting of data.

V. THE GUMBEL DISTRIBUTION (2)

Another type of extreme value distribution is due to Gumbel. A plot of the Gumbel distribution differs from that of Weibull in that the ordinate scale is equal to $-[\ln(-\ln(G)]$, and the abscissa scale is linear rather than logarithmic. Whereas the Weibull distribution is used in estimating the probability of early failure in a population, the Gumbel extreme value distribution is useful in predicting the likelihood of an event such as a hurricane or flood of a given magnitude.

Murakami and coworkers (6) have applied Gumbel's extreme value distribution method to predict the maximum size of inclusion present within a given volume of metal, an important consideration with respect to quality control and the fatigue strength. In the procedure developed by Gumbel, the cumulative distribution function F_j is defined as

$$F_j = \frac{j}{n+1}, \tag{8-8}$$

where n is the total number of observations, and j is the jth observation when the n observation are arranged in order from the smallest to the largest.

A reduced variate y_j is then related to Eq. 8-8 by

$$y_j = -\ln\left[-\ln\left(\frac{j}{n+1}\right)\right] = -\ln[-\ln(F_j)]. \tag{8-9}$$

This implies that

$$e^{-y_j} = -\ln\left(\frac{j}{n+1}\right), \tag{8-10}$$

and that

$$e^{-e^{-y}} = \frac{j}{n+1} = F_j, \tag{8-11}$$

that is, the cumulative distribution function F_j is a double exponential function.

The concept of the *return period* is introduced. If an event has a probability p, on average $1/p$ trials are needed for the event to happen once. The return period T_j is defined as

$$T_j = \frac{1}{p} = \frac{1}{1-F_j}. \tag{8-12}$$

The return period for the median of a distribution is 2, for the upper quartile it is 4, and so on. Scales other than time may be used, for example, a length scale is used in considering the cumulative probability that an inclusion of a given size will be found in examining n numbers of metallographic sections.

The return period T_j can be expressed as

$$T_j = \frac{1}{1 - [j/(n+1)]} = \frac{n+1}{n+1-j}. \tag{8-13}$$

Therefore

$$j = \frac{(n+1)(T_j - 1)}{T_j}. \tag{8-14}$$

Upon substituting Eq. 8-14 into Eq. 8-9, we obtain

$$y_j = -\ln\left[-\ln\frac{(n+1)(T_j-1)}{(n+1)T_j}\right] = -\ln\left[-\ln\frac{T_j-1}{T_j}\right]. \tag{8-15}$$

We will be dealing with inclusion distributions, and in particular with the largest sized inclusion found in an area S_0. Equation 8-13 is therefore modified to the following form:

$$T_j = \frac{n+1}{n+1-j}\,\frac{S_0}{S_0} = \frac{S+S_0}{S+S_0-jS_0}. \tag{8-16}$$

If $j = n$, then the return period for an inclusion of the largest size observed is

$$T_n = \frac{S+S_0}{S_0}, \qquad \text{or } (T_n = n+1), \tag{8-17}$$

and if $S >> S_0$, then $T_n = S/S_0$. (8-18)

A. Example

An example of the use of the extreme value distribution to predict the size of the largest inclusion to be found in a given volume of metal is now given. The first step is to prepare a polished section of the metal and to determine the square root of the area of the largest inclusion found within an area S_0, typical values for S_0 being 0.075–0.482 mm^2. This operation is repeated on n different sections, where n is often set at 40. The values of $\sqrt{\text{area}_{max,j}}$ are classified, starting with the smallest, and indexed with $j = 1 \ldots n$. The cumulative distribution function F_j and the reduced variate y_j are then calculated from Eqs. 8-8 and 8-9 above. Next the data are plot-

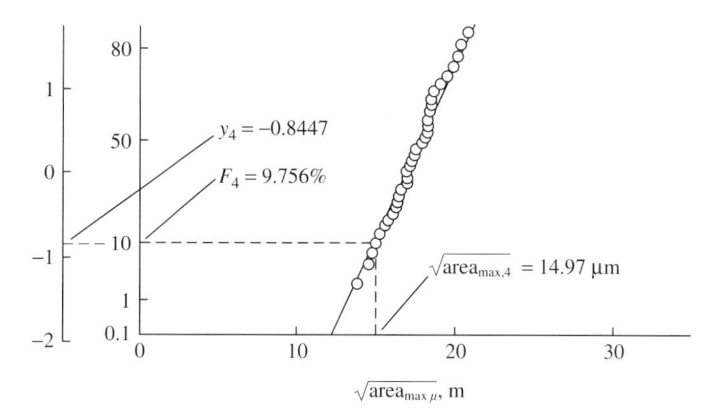

Fig. 8-5. Cumulative distribution of laregest inclusion sizes (n = 40) plotted according to Gumbel's statistics of the extreme. (From Murakami et al., 6.)

ted on extreme value distribution probability paper with the reduced variate as the abscissa and the $\sqrt{\text{area}_{\text{max}}}$ as the ordinate, as in Fig. 8-5. (The value of F_j, expressed in percent, may also be used as the ordinate, since F_j and y_j are related through Eq. 8-9. The return period T_j can also be used as the ordinate, since T_j is a function of F_j, that is, $F_j = (T_j - 1)/T_j$ from Eqs. 8-8 and 8-12.)

If a straight line can be drawn thorough the data, as in Fig. 8-5, the distribution can be considered to be doubly exponential in accord with Gumbel's statistical method of the extremes. In this example, Fig. 8-5 indicates that the cumulative probability that the largest inclusion found will be of size 14.97 μm or less is 9.756% (corresponding to a j value of 4, $F_j = 4/40 + 1 = .09756$).

If a straight line is found as in Fig. 8-5, it can be extended beyond the range of the data to make predictions concerning the probability of finding an inclusion larger than those already found within a given volume of metal. For example, consider the distribution in size of graphite nodules in ductile iron. Assume that S_0, the inspected area in an examined section, is 1 mm^2, and that 50 sections are examined to determine the size of the largest graphite nodule in each section and its associated $\sqrt{\text{area}_{\text{max}}}$. With this information a plot of the cumulative probability distribution as a function of the square root of the $\sqrt{\text{area}_{\text{max}}}$ similar to the plot in Fig. 8-5 can be established. In this case F_{50} is equal to 0.98 and the return period, T_{50}, is 51. Therefore in the examination of 50 sections there is a 98% probability that a graphite nodule of the maximum size from the plot corresponding to a cumulative frequency of 0.98 will be found. Now suppose 5 fatigue samples are to be tested in bending. Each specimen has a diameter of 10 mm and a test length of 18 mm, and only graphite nodules at or just below the surface are of concern as nucleation sites for fatigue cracks. In this case S would be 5 × 10 mm × π × 18 mm or about 2830 mm^2, corresponding to a return period, T, of 2830/1 or 2830. If the cumulative frequency plots as a straight line in the Gumbel representation it can be extrapolated to this T value. The largest sized graphite nodule in the five specimens to be expected can then be determined from the graph corresponding to this T value. Such information is useful in assessing the effect of the size of graphite nodules on fatigue strength as well as for quality control purposes.

VI. SUMMARY

The statistical treatment of data is useful in quality control and in establishing lower limits on properties such as the yield strength and fatigue strength. An awareness of the scatter in such properties is important in design as well as in failure analysis.

REFERENCES

(1) W. Weibull, J. Appl. Mech., vol 18, 1951, pp. 293–297; vol. 19, 1952, pp. 109–113.

(2) E. J. Gumbel, Statistics of Extremes, Columbia University Press, New York, 1958.

(3) F. B. Stulen, W. C. Schulte, and H. N. Cummings, in Statistical Methods in Materials Research, ed. by D. E. Hardenbergh, Pennsylvania State University, University Park, 1956.

(4) G. E. Dieter, Jr., Mecchanical Metallurgy, 1st ed., McGraw-Hill, New York, 1961.

(5) C. Lipson and N. J. Sheth, Statistical Design and Analysis of Engineering Experiments, McGraw-Hill, New York, 1973.

(6) Y. Murakami, T. Toriyama, and E. M. Coudert, Instructions for a New Method of Inclusion Rating and Correlations with the Fatigue Limit, J. of Testing and Evaluation, JTEVA, vol. 22, no. 4, July 1994, pp. 318–326.

9

Creep Failure

I. INTRODUCTION

The time-dependent deformation of materials under stress is known as creep. In metals, it is primarily an elevated-temperature diffusion-controlled process that is terminated by rupture. This chapter reviews the creep mechanisms and creep lifetime prediction procedures. The several creep failure mechanisms are also discussed.

II. BACKGROUND

At temperatures above $0.4T_M$, where T_M is the melting point in K, time-dependent deformation, that is, creep and rupture processes of materials are primary design considerations. In many applications, for example, the aircraft gas turbine engine, the operating temperature, and hence the efficiency, are limited by the creep characteristics of materials, so that there is therefore an economic incentive to develop alloys of superior creep resistance. Designs are usually made on the basis of a maximum permissible amount of creep, such as 0.1 or 1% during the expected lifetime of a particular component. These service lives can vary from minutes for a rocket engine component, to tens of thousands of hours for jet engine components, to hundreds of thousands of hours for high-pressure steam lines. At elevated temperatures, creep deformation may not be the only mechanism governing service lifetime, for alloys have been found to fracture after very limited extension (1) or after unexpectedly short service lifetimes. However, today creep rupture is not a common occurrence. This has come about because of advances in alloy development, particularly in the aircraft field, inspection procedures, and the development of codes such as the American Society of Mechanical Engineers (ASME) Boiler and Pressure Vessel Code and the recommended rules of the American Petroleum Institute (API). Creep problems with carbon steels can be avoided by limiting their use of to temperatures below $0.4T_m$, that is, below the creep range. Therefore, in boiling water and pressurized nuclear reactors, which operate at 288°C (550°F), creep is not a problem. However, some pressurized boiler components, such as superheater headers, superheater tubes, and steam pipes, do operate in the creep range using creep-resistant alloys such as $1\frac{1}{4}$ Cr $\frac{1}{2}$ Mo and $2\frac{1}{4}$ Cr -1 Mo steels. Gas turbine engines also operate above $0.4T_m$, and therefore creep is of concern. When creep failures occur, it is often because a power plant or engine has been operated beyond its safe lifetime, or because the safe operating conditions have exceeded prescribed limits. In addition, unanticipated creep failures have occurred during normal service at welds and other locations in high-pressure steam lines, as will be discussed subsequently.

In a review of failures in fossil-fired steam power plants, it was found that 81% of the failures were mechanical in nature, and that the remainder were due to corrosion. Of the mechanical failures, 65% were classified as short-time, elevated-temperature failures. Only 9% were due to creep, with the rest being due to such causes as fatigue, weld failures, erosion, and so on (2).

III. CHARACTERISTICS OF CREEP

Figure 9-1 shows typical creep extension plots for constant stress conditions. The characteristic regions of the load extension curve are:

(a) The primary creep region in which the creep strain rate decreases with time.

(b) The secondary creep region in which, under constant true stress, the creep strain rate has a minimum and steady-state value $\dot{\varepsilon}_{ss}$. However, under constant load conditions, no steady-state value may be evident.

(c) The tertiary creep region in which, even under constant true stress conditions, the creep strain rate accelerates with time.

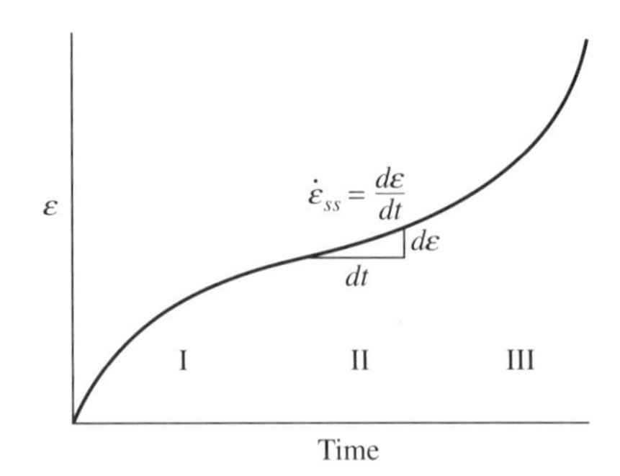

Fig. 9-1. Creep extension as a function of time for constant stress loading conditions.

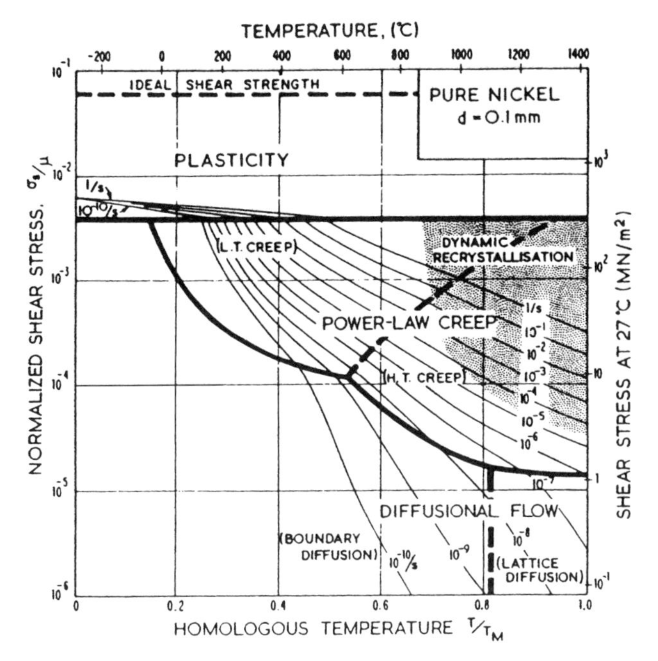

Fig. 9-2. An example of a creep deformation map for pure nickel of grain size 1 mm, work hardened. (LT = low temperature, HT = high temperature.) (From Frost and Ashby, 3. Reprinted by permission.)

A major development in recent years has been the introduction of creep deformation maps (3). These maps indicate, for a given alloy, the creep deformation and rupture mechanisms as a function of temperature and stress level, and permit creep analyses to be carried out on a more rational basis. Figure 9-2 is an example of a creep deformation map. The ordinate is σ/E, and the abscissa is T/T_m. Such maps indicate the mechanism of deformation corresponding to a given combination of stress level and temperature. At temperatures below $0.4T_M$, the deformation is either elastic or plastic, whereas above this temperature under constant loading conditions, creep deformation occurs by either one of two mechanisms, power law creep or viscous creep. Both of these mechanisms are diffusion controlled. At low temperatures in the creep range, creep occurs by diffusion along short circuit paths, that is, the grain boundaries or the cores of the dislocations. At high temperatures, the diffusion path is through the bulk of the material.

In power law creep, the rate-limiting step is known as climb. In this process, blocked glide dislocations move perpendicular to their glide planes rather than along them as atoms diffuse away from or to the core lines of the dislocations. By climbing, the dislocations are able to circumvent obstacles such as precipitates in their glide paths. Power law creep occurs in components such as turbine blades and high-pressure steam pipelines. In the latter case, stresses may be quite low, that is, 45.5 MPa (6.6 ksi). Viscous creep occurs at still lower stress levels. The viscous creep mechanism does not involve the climb of dislocations, but only the diffusion of atoms in such a way as to lead to elongation accompanied by a decrease in cross-sectional area under tensile stress. The equations governing these two types of creep are as given below.

In power law creep, the steady-state creep rate is proportional to the stress raised to a power n, where n takes on values that are commonly in the range 3–6. The creep strain rate is related to the diffusion-controlled climb of dislocations, and is expressed as

$$\dot{\varepsilon}_{ss} = G\sigma^n e^{-Q/RT}, \tag{9-1}$$

where $\dot{\varepsilon}_{ss}$ is the steady-state strain rate, G is a material constant, Q is the stress-modified activation energy, R is the gas constant equal to 8.314 J mol^{-1} K^{-1}, and T is the absolute temperature.

The value of the exponent n has a strong influence on the shape of a bar during creep. Let l denote the length of the bar and A the cross-sectional area. As in plastic deformation, volume is conserved during creep; then:

$$Al = \text{constant}, \tag{9-2}$$

$$A\,dl + l\,dA = 0, \tag{9-3}$$

$$\frac{dA}{A} = -\frac{dl}{l} = -d\varepsilon, \tag{9-4}$$

$$\frac{dA}{A\,dt} = -\frac{d\varepsilon}{dt} = -H\sigma^n = -H\left(\frac{F}{A}\right)^n, \tag{9-5}$$

$$A^{n-1}\,dA = -HF^n\,dt, \tag{9-6}$$

which after integration leads to

$$A_t = (A_0^n - nHF^n\,\Delta t)^{1/n}, \tag{9-7}$$

where A_0 is the initial cross-sectional area, and A_t is the cross-sectional area after a time, t. If $n = 1$ as in diffusion creep or linear viscous flow, then a reduced cross section draws down only slightly faster than the rest of the bar, and this is why molten glass can be drawn down into a thin fiber. For example, consider the creep deformation of a bar with an initial cross section A_{01}, which initially contains a reduced section A_{02} equal to $0.9A_{01}$. When, under viscous creep deformation, the reduced section reaches $0.5A_{01}$, the remainder of the bar would have an area of $0.6A_{01}$. That is, the ratio of the two areas at that time would be only 0.833 as compared to the initial ratio of 0.9. However, if n were equal to 3–5 as in power law creep, then as can be seen from Eq. 9-5, the reduced cross section would draw down much faster than the rest of the bar, with the result that a sharp neck would be developed.

IV. CREEP PARAMETERS

A number of parametric relationships have been developed for design purposes in the attempt to relate stress, creep rate, time to rupture, and temperature in order to extrapolate available data to longer rupture times at lower stresses, under the assumption that the mechanism of creep remains the same in the range of interest. One of the more widely used of these parameters is the Larson-Miller parameter. In the Larson-Miller approach, data on the steady-state creep rate or the creep rupture life at a given stress level at one temperature are used to estimate the creep rates and lifetimes at the same stress level at other temperatures. The starting point for the development of this parameter is the Arrhenius type of equation:

$$\dot{\varepsilon}_{ss} = Ae^{-Q(\sigma)/RT}, \tag{9-8}$$

or

$$\mathrm{T}(\log \dot{\varepsilon}_{ss} - C) = -2.3\frac{Q}{R} = -f(\sigma), \tag{9-9a}$$

where $C = \log A$, or

$$\log \dot{\varepsilon}_{ss} = C - f(\sigma)\frac{1}{T}\cdot \tag{9-9b}$$

If the log of the steady-state creep rate data for a series of stress levels is plotted versus $1/T$, such a plot will produce a series of straight lines with a common intercept C, as in Fig. 9-3a, provided that the data conform to Eq. 9-8. Once the value of C is determined, a master plot of $\log \sigma$ versus $T(\log \dot{\varepsilon}_{ss} - C)$, the Larson-Miller parameter, can be constructed, as shown in Fig. 9-4.

When the extension in tertiary creep is small compared to the extension in steady-state creep, the empirical Monkman-Grant law,

$$\dot{\varepsilon}_{ss} t_r = \text{constant} = B = \varepsilon_r, \tag{9-10}$$

where t_r is the time to rupture and ε_r is the strain to rupture, can be used to develop an alternate form of the Larson-Miller parameter,

$$T\left(\log \frac{B}{t_r} - C\right) = -f(\sigma), \tag{9-11}$$

or

$$T(\log B - \log t_r - C) = -f(\sigma),$$

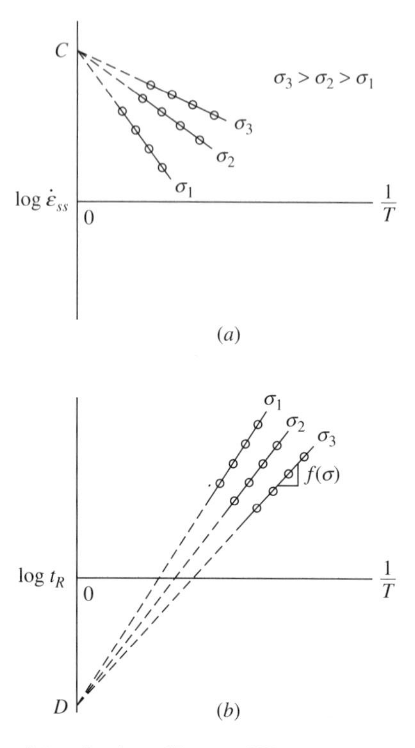

Fig. 9-3. Plots used in the determination of Larson-Miller constants. (*a*) log $\dot{\varepsilon}$ vs. $1/T$ (*b*) log t_r vs. $1/T$.

or

$$T(D + \log t_r) = f(\sigma), \tag{9-12a}$$

or

$$\log t_r = -D + f(\sigma)\frac{1}{T}\cdot \tag{9-12b}$$

The constant D is determined by plotting experimental data for a series of stress levels, as shown in Fig. 9-3*b*. Once D is determined, a master plot of log σ versus $T(D + \log t_r)$ can be constructed, as shown in Fig. 9-4.

To estimate the time to failure under variable loading and temperature conditions during the creep lifetime, a linear damage rule, known as Robinson's life fraction rule, is used. To carry out the calculations, data are needed on the time to rupture under constant loading conditions at the temperatures of interest. This rule is

$$\sum_{i=1}^{i=i} \frac{t_i}{t_{fi}} = 1, \tag{9-13}$$

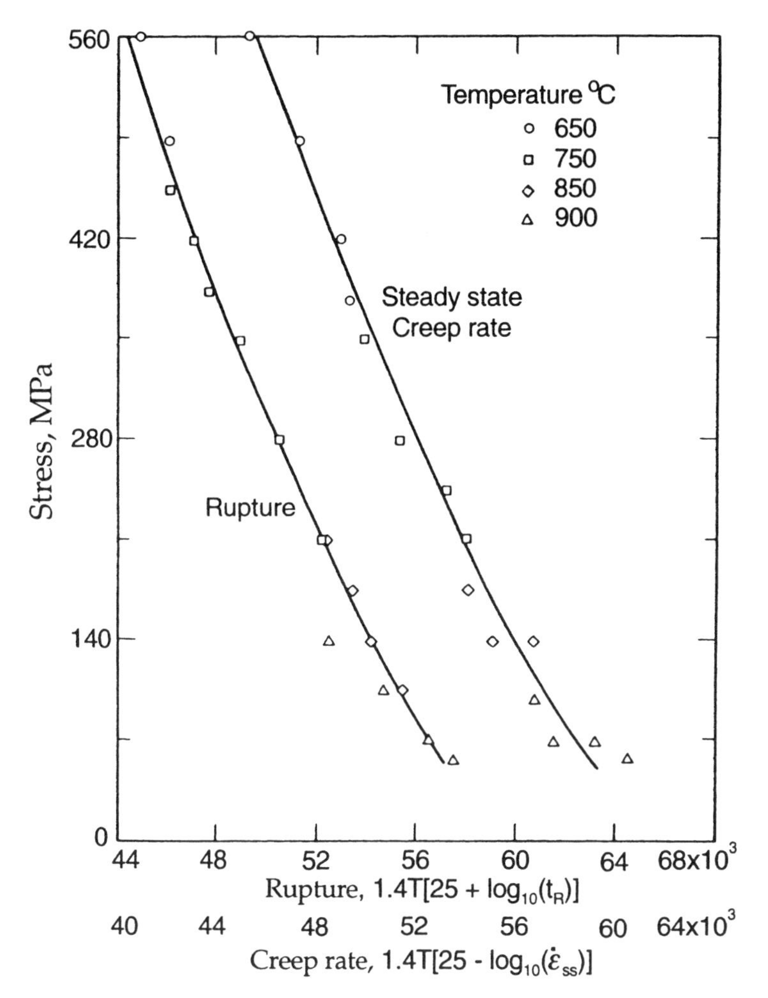

Fig. 9-4. Larson-Miller plot for Inconel X. (After Tetelman and McEvily, Fracture of Structural Materials, 1967, John Wiley & Sons.)

where t_i is the time spent at the ith stress level at a given temperature, and t_{fi} is the time to failure under the same loading conditions.

In using any extrapolation approach to predict long-time behavior, it is assumed that the microstucture remains essentially unaltered. However, this may not be the case, and the predictions may be nonconservative. For example, at temperatures above 425°C (800°F), carbon steels in long-time service are subject to graphitization, a process that leads to loss of strength and overall ductility. In this process, Fe_3C is replaced by iron and carbon. Above 470°C (875°F), carbon-molybdenum steels are similarly affected. In welded structures, this transformation from carbide to graphite takes place preferentially along heat-affected zones adjacent to the welds and can lead to creep failures requiring repair or replacement (4). For service above 425°C (800°F), the use of carbon-molybdenum steels containing $\frac{1}{2}$% or more chromium eliminates the danger of graphitization. It is also reported that carbon-molybdenum steels deoxidized with silicon are not prone to graphitization after prolonged exposure to elevated temperatures (5).

V. CREEP FRACTURE MECHANISMS

The modes of creep failure are compared to the modes of low-temperature failure in Fig. 9-5 (6). The creep failure modes are:

(a) *Intergranular Creep Failure:* This is the most likely mode of creep fracture in long-time exposure to temperature and stress. In this failure process, voids nu-

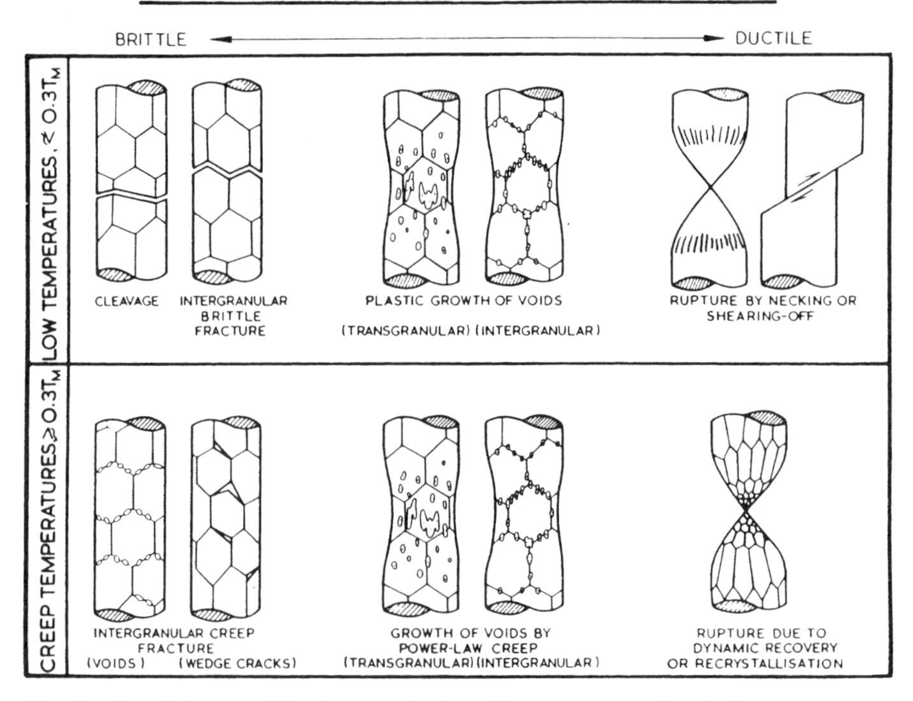

Fig. 9-5. Broad classes of fracture mechanisms. The upper row refers to low temperatures ($<0.3\ T_M$) where plastic flow does not depend strongly on temperature or time; the lower row refer to the temperature range ($>0.3\ T_M$) in which materials creep. (After Ashby et al., 6. Reprinted by permission).

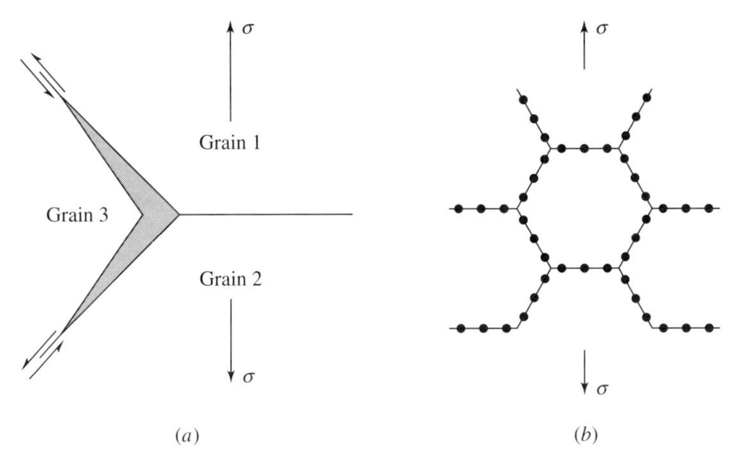

Fig. 9-6. (*a*) Example of a wedge crack. (After Chang and Grant, 11. Reprinted with permission.) (*b*) Example of grain boundary voids (After Kennedy, 12. Reprinted with permission).

cleate along grain boundaries and wedge cracks appear at triple points (1). Figure 9-6 shows examples of a wedge crack and grain boundary voids. In time, these defects grow and link up, and as a result, there is little reduction in cross-sectional area and a "flat" or thick-walled fracture results. This mode of failure is associated with components that have been in service for many years. Because grain boundaries are weak areas in creep, due to the grain boundary sliding that contributes to creep elongation, and also because they are sites for void nucleation, means have been developed to minimize the effects of grain boundaries. Directional solidification is used to grow columnar grains that are parallel to, rather than at an angle to, the principal creep stress. Single crystals are used in order to completely eliminate the negative contributions of grain boundaries to creep. The use of directionally solidified and single-crystal components has permitted gas turbine engine operating temperatures to be raised significantly. Single crystals were first used in the hot stage of aircraft gas turbine engines. Advances in casting technology now permit the use of single crystals in industrial gas turbines where the blades are much larger than in aircraft engines.

(b) *Transgranular Creep Fracture:* This mode of failure can occur in short-time creep failures, and the failure process is similar to tensile rupture below the creep range. The ductility and reduction in area are usually large and much greater than at room temperature, with the result that a heat exchanger tube will locally balloon and lead to a thin-walled fracture.

(c) *Point Rupture Fracture:* At sufficiently high temperatures and low stresses, dynamic recrystallization, that is, recrystallization during creep, can remove microstructural creep damage. As a result, voids do not nucleate, and necking down to a point can occur.

VI. FRACTURE MECHANISM MAPS (6)

These maps are analogous to creep deformation maps, and indicate the mode of failure as a function of stress and temperature. An examination of a creep fracture surface together with the information contained in these maps can give an indication of the operating conditions at the time of failure. Figure 9-7 gives an example of a fracture mechanism map.

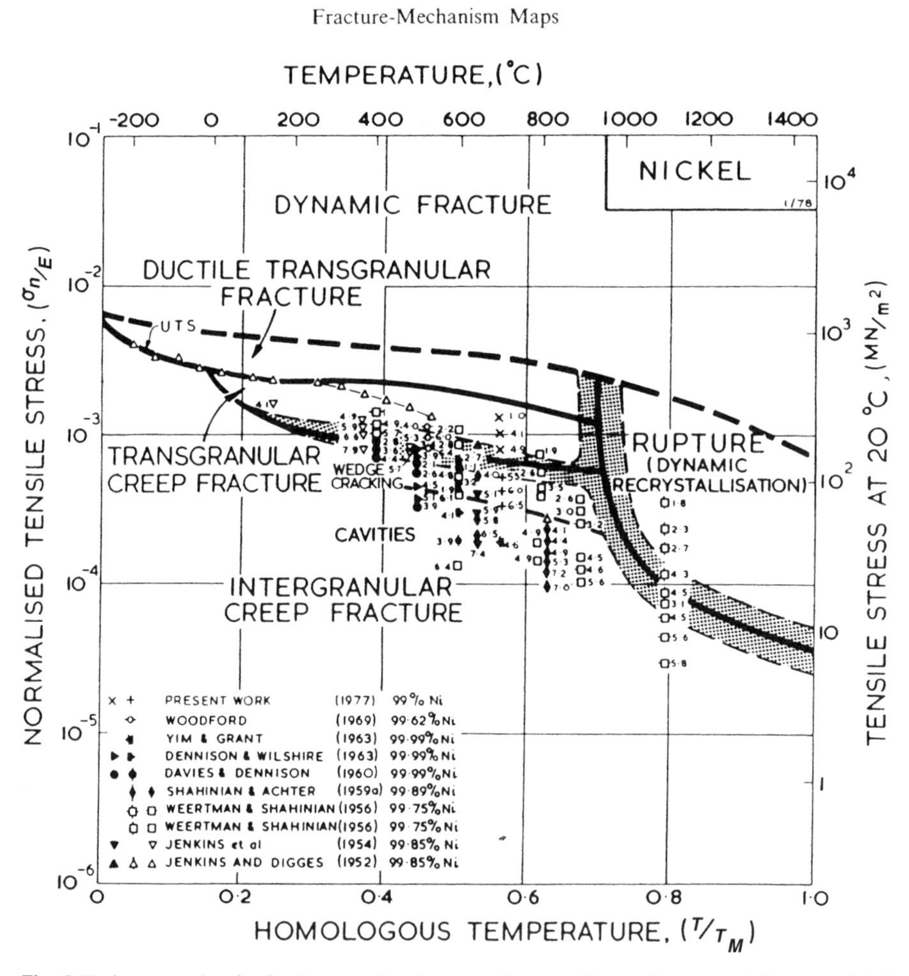

Fig. 9-7. An example of a fracture mechanism map for round bars of commercially pure nickel tested in tension. The map shows four fields corresponding to four modes of failure. Individual tests are marked by symbols and (for creep tests) labeled with the logarithm, to base 10, of the time to fracture in seconds ($\log_{10} t_f$.) Solid symbols mean that the fracture was identified as intergranular. Shading indicates a mixed mode of fracture. (After Ashby et al., 6. Reprinted with permission).

VII. CASE STUDY: FAILURE AT LONGITUDINALLY WELDED PIPE (7)

A thick-walled failure at a longitudinal submerged arc weld in 25.4-mm thick power-plant low-chromium alloy steel pipelines operating at 538°C occurred after 12 years of service. The pipeline should have lasted over 30 years based upon creep rupture analyses, which assumed that the weld and base metals had the same creep properties and were stressed at 45.5 MPa; in this case, however, the weld metal was less creep resistant than the base metal. It is noted that pipelines are nominally in a state of plane strain. The nominal stresses in a pipeline are the hoop stress $\sigma_1 = \sigma$ (hoop), $\sigma_2 = \sigma/2$ (longitudinal), and $\sigma_3 \approx 0$ (through-thickness). The strain in the longitudinal direction is given as

$$\varepsilon_2 \propto [\sigma_2 - \tfrac{1}{2}(\sigma_1)] = 0. \tag{9-14}$$

For constancy of volume, $\dot{\varepsilon}_1 = -\dot{\varepsilon}_3$. Therefore, during creep a pipeline expands in the circumferential direction and grows thinner in the radial direction, and there is no change of length in the longitudinal direction.

The weld was symmetric double-V weld, Fig. 9-8, and because of its lower creep resistance compared to the base metal, the creep extension in the narrow portion of the weld lagged the creep extension in the wider parts of the weld. This differential led to the development of an increased tensile stress in the narrow portion of the weld, which helped to explain the premature failures. In the low-chromium alloy steel, a high density of spherical inclusions associated with the submerged arc welding process was present along the fusion line. A sulfur coating on the inclusions prevented good bonding with the matrix, and the high tensile stress at the root of the weld promoted the early formation and growth of cavities at the inclusions. As these cavities grew, they linked up to form a crack along the fusion line within 1–2 years of service. The crack then extended along the fusion line toward the inner and outer surfaces of the pipe as additional cavities formed at grain boundary inclusions in the high strain field ahead of the crack. These cavities then grew and coalesced with the crack tip. A detailed analysis involving power law creep behavior, $\dot{\varepsilon} = B\sigma^n$, and elastic-plastic creep crack growth relationships led to a predicted lifetime of ten years, a value close to the time involved in the actual failure.

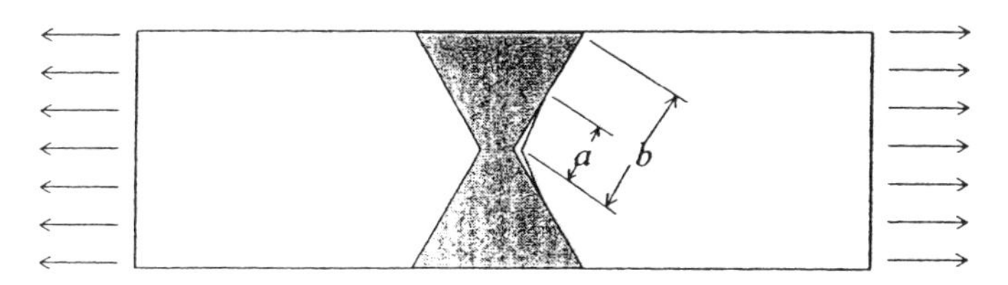

Fig. 9-8. An idealized model of a longitudinal pipe weld. The curvature of the pipe is neglected and plane strain conditions are assumed. The arrows represent the uniform hoop stress in the pipe away from the weld (After Stevick and Finnie, 7. Reprinted with permission.)

VIII. CASE STUDY: FAILURE OF A HEAT EXCHANGER TUBE

An oil-fired power plant had been in service for a number of years when, during the 1970s, an oil shortage developed. Since the shortage appeared to be of long duration, management decided to convert the plant to a coal-fired operation. The plant was therefore modified to accommodate the use of coal. During the modification, a vertical bank of new heat exchanger tubes was installed. In operation, the tubes are heated externally by hot gases from the furnace, and there is a flow of steam within the tubes. The tubes were made of 1018 steel, and were 50 mm in diameter with a wall thickness of 5 mm. The steam pressure was 10 MPa, resulting in a hoop stress of 40 MPa. The plant had been in operation only a day or so following the modification when rupture in a transgranular, thin-wall mode of one of the tubes occurred, see Fig. 9-9, necessitating a costly shutdown of the plant. The fact that the failure was of a thin-walled type was a direct indication of a short-time fracture due to overheating. The reduction in wall thickness is much less pronounced in long-time creep failures, since they occur due to the linking-up of grain boundary voids. In addition, in long-time failures, there is usually considerable intergranular cracking in regions adjacent to the thick-walled fracture, which of course did not occur in the case under consideration.

The tube was repaired by welding in a replacement section, but again, after the plant had been in operation only a short while, the replacement section failed, also in a thin-walled manner. During the second repair operation, one of the repair crew happened to look upward into the failed tube and noticed that an obstruction was present. The obstruction turned out to be a wrench that had been accidentally dropped into the tube during the plant conversion and had wedged itself midway down the tube. The wrench had disrupted the flow of steam in the tube, causing a hot spot to develop. As a result, the tube temperature had risen to a point where the flow strength of the metal was exceeded, and rupture occurred in the observed short-time, thin-walled fashion.

The microstructure in the vicinity of the break can give an indication of the maximum temperature reached as well as of the subsequent cooling rate. For example, in one instance when an attempt was made to remove the upper end of an oil drum made of a low-carbon steel by flame cutting with an oxyacetylene torch, an explosion occurred. The procedure involved in flame cutting is to preheat an area to a cherry red color with the oxygen supply to the cutting torch set at a low level, and then to increase the oxygen supply for the cutting operation itself. In this case, as soon as the cutting flame penetrated the metal, the fumes in the drum ignited with explosive force. Subsequent metallographic examination of the preheated area revealed an acicular microstructure, rather than a ferritic-pearlitic microstructure, thereby indicating that the area had been heated into the austenitic range during preheating and then cooled rapidly after the explosion. In the case of the failed heat exchanger tube, the microstructure at the break was ferritic-pearlitic, an indication that it had been heated to a temperature below the transformation temperature but high enough to allow softening and rupture to occur.

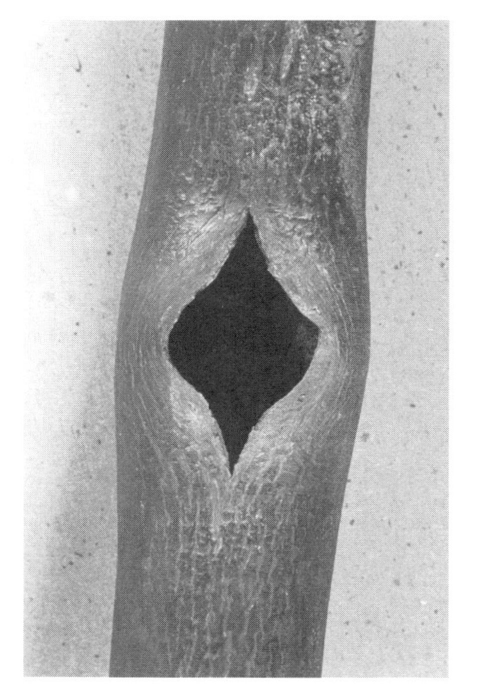

(*a*)

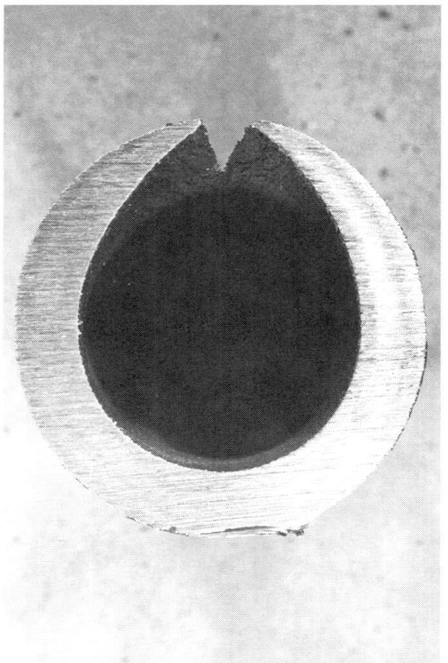

(*b*)

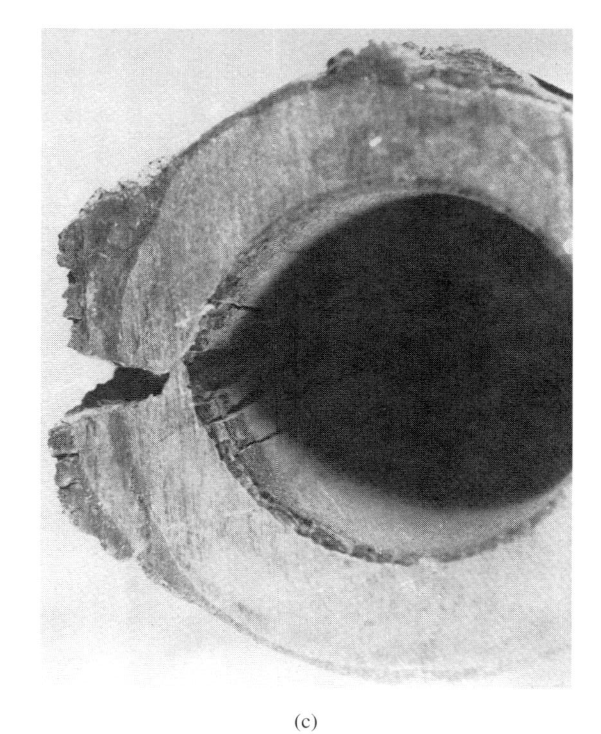

(*c*)

Fig. 9-9. (*a*). Longitudinal view of a thin-wall creep fracture, (*b*). Cross-sectional view of a thin-wall creep fracture. (*c*) Cross-sectional view of a thick-wall creep fracture (After French, 2. Reprinted with permission.)

When excessive amounts of creep are found in jet aircraft engines, the microstructure can be examined to determine if high thermal excursions due to improper engine operation had occurred. Similarly, the examination of the microstructure in any creep failure can lead to an indication of the thermal history of the component.

IX. CASE STUDY: AN OVALIZED TUBE (8)

When a tube of oval cross-sectional shape is subjected to internal pressure, in addition to the hoop stress, a bending stress will also be introduced. Under pressure, the tube cross section tries to become more circular, thereby introducing the bending stresses, which are a maximum at the end of the major axis of the ellipse. At this location, the bending stress adds to the hoop stress on the inside surface and subtracts from the hoop stress on the outside surface. Ovalization occurs at tube bends with attendant pipe wall thickening at the inner radius of the bend and thinning at the outer radius of the bend. To minimize ovalization, it is standard practice to hot form the bend after first packing the pipe with sand.

In one instance, a Cr-Mo tube with an outer diameter of 44.5 mm and a nominal wall thickness of 7.5 mm failed by cracking longitudinally at a bend after operating for only 3500 hours at 538°C at a pressure of 13.7 MPa. The nominal hoop pressure was calculated to be 33.8 MPa. Measurement of the tube wall thickness revealed that the wall thickness was 8.5 mm at the inner radius of the bend, and 6.1 mm at the outer radius of the bend. There was also significant ovalization, with the maximum tube diameter exceeding the minimum by 4 mm. It was confirmed that the tube bending had been carried out without packing with sand, and without the use of proper bending equipment.

The allowable ASME code stress for the material at the operating temperature is 53.8 MPa. When the draft methodology of API (American Petroleum Institute) 579 Recommended Practice was used to calculate the combined bending and hoop stress, the resultant value for the observed degree of ovality was found to be well in excess of the allowable limit, and the reason for the premature failure is readily seen.

X. CASE STUDIES: FAILURES IN FOSSIL-FUEL FIRED BOILERS (2)

Oil-Fired Steam Boiler Tube

An oil-fired steam boiler had been in service for two years when a furnace sidewall tube ruptured. The steam temperature was 540°C (1005°F), and the steam pressure was 14.7 MPa (2150 psig). The fracture occurred at the downstream edge of a circumferential weld, and was of the short-time, knife-edge type. The microstructure at the failure location was in a spheroidized condition, whereas, further away from the failure, it contained pearlite. It was deduced that the weld upset the smooth steam flow and formed an area of turbulence, leading to the formation of an insulating steam blanket. As a result, the metal temperature rose to 650–700°C, at which point the metal was not able to withstand the internal pressure, and the rupture occurred.

In other failures, insulating scales have been found to form on the interior of tubes, due to the reaction between steam and iron to form magnetite. These scales have led to catastrophic increases in metal temperature. The reduction of tube wall thickness due to corrosion or erosion can also be a contributing factor.

B. Waterwall Tube

A waterwall tube failed in a short-time, knife-edge manner. The nominal temperature was 540°C and the steam pressure was 15.4 MPa. The microstructure at the point of failure consisted of ferrite and bainite, whereas away from the failure the microstructure consisted of ferrite and pearlite. The presence of bainite indicated that the tube had been heated above the A_1 temperature into the two-phase $\alpha - \gamma$ region and rapidly cooled. From the relative amounts of these two phases, an estimate of the maximum temperature reached can be made thorough the use of the lever rule. The lever rule states that the % α-phase present at a given temperature is given by

$$\% \, \alpha = \frac{(\% \text{ carbon in } \gamma - \% \text{ carbon in alloy})}{(\% \text{ carbon in } \gamma - \% \text{ carbon in } \alpha)} \times 100. \tag{9-15}$$

In this case, the lever rule indicated that the highest temperature had been in the range 790-820°C. The cause of the failure was attributed to hot spots, which developed as the result of poor circulation of the furnace gases.

In another failure, the microstructure was entirely bainitic, indicating that a temperature in excess of 870°C had been reached.

C. Superheater Tube

A 2.25Cr-1Mo steel superheater tube experienced a nominal temperature of 540°C at a pressure of 13.5 MPa. The tube outside diameter (OD) was 44 mm and the wall thickness was 6.6 mm. After some time in service, the tube failed in a classic creep type of fracture, as shown in Fig. 9-9c, with a 38-mm long longitudinal split. The wall thickness at the fracture showed virtually no reduction. The wall thickness 180° from the fracture was 7.6 mm, which included an outer scale of 1.3 mm and an inner scale of 0.8 mm. It was estimated that, with the insulating scales present, the outer skin temperature had risen to about 650°C, whereas the safe upper temperature limit for the avoidance of severe oxidation was 605°C. The cause of the creep failure was due to prolonged exposure to the 650°C temperature.

Overheating can also result in the formation of methane (CH_4) gas as hydrogen from the flue gas atmosphere combines with carbon in the steel and decarburizes it. If the methane forms at subsurface sites, the gas can create an internal pressure leading to blister formation on the OD.

XI. RESIDUAL LIFE ASSESSMENT

Design codes have traditionally been about design only, and have not been intended to cover the assessment of operating plant that has been in service for some time. In the case of boilers, pressure vessels, and tanks, periodic inspections are carried out, but they do not provide a complete engineering evaluation and have not been governed by a regulatory code in terms of engineering assessment of damage and the prediction of remaining safe life (9). An immediate reaction upon finding defects such as cracks or pores during an evaluation is to condemn the equipment as not meeting code requirements and indicating that repair is necessary. Recently, however, "Fitness for Service" codes have been and are being developed by the ASME and the API, codes that provide rules and guidelines for repairs and alterations based upon risk-based inspection procedures. Three levels of analysis may be involved: qualitative, semiquantitative, and fully quantitative. As an example, cracks in a tee connection weld that operated at 585°C at 4.5 MPa were discovered during an unplanned shutdown. After a qualitative analysis of the risks involved, and as a temporary expedient, weld repairs were made, even though grinding prior to welding did not completely eliminate the cracks. Later, however, it was found to be desirable to extend the period of operation before a major repair was undertaken, and a more detailed analysis, including the use of computer codes for creep crack growth, was therefore undertaken to evaluate the safety of the system. Based upon the size of the crack, an evaluation of the loading during operation and shutdown, and the time required to grow the crack by creep to a critical size, it was concluded that the tees needed to be replaced much sooner than desired by plant management.

This example relates to the important question concerning a plant that has been in service for some time: "How much longer can the plant operate before major problems start to develop?" Answers to this question are needed in order to plan for major overhauls and perhaps replacement of the plant.

It turns out that, for components that operate in the creep range, this is not an easy question to answer. However, the following procedure has been developed to provide a partial answer. It relies upon the observation of grain boundary voids. The procedure involves the cleaning, polishing, and etching of suspect areas in situ, and then making replicas for observation under the microscope. If voids are found, and depending on their stage of development into cracks, an estimate of residual life of the component can be made that is usually of the order of several years or less. Care in polishing the surface is needed, for it has been shown (10) that grain boundary cavities in Cr-Mo low-alloy steels are enlarged by the polish-etch procedures used, and may not be apparent upon more careful polishing. In fact, the apparent voids may have been locations at which grain boundary carbides were removed during polishing. However, if voids are not found, a definitive life assessment cannot be made, except for the fact that the component will last at least as long as the time required for observable voids to develop, grow and link up, and cause failure. This may require at least five years, so that at least this amount of time is available to plan for repairs and replacement.

An alternative procedure is to examine the microstucture for the degree of spheroidization that may have occurred in service as an indication of residual life (8). A

Cr-Mo superheater stub failed after 175,000 hours at a temperature of 515°C and extensive spheroidization had taken place with little evidence of the original pearlite being evident. However, when a Cr-Mo steam tube failed after 70,000 hours at 490°C, extensive void formation and cracking on grain boundaries could be observed, but with no detectable change in the original ferrite-pearlite microstructure. Longer times and higher temperatures may be needed before spheroidizaton is apparent.

XII. STRESS RELAXATION

The substitution of elastic strain by plastic strain (creep) at elevated temperatures leads to a loss of tension in bolted connections and is therefore is a problem. This process occurs at constant displacement and is known as stress relaxation. The equations governing this process are as follows:

$$\varepsilon_t = \varepsilon_{el} + \varepsilon_{pl} \tag{9-16}$$

where ε_t is the total strain, ε_{el} is the elastic strain, and ε_{pl} is the inelastic strain. Since the total strain is fixed in time,

$$\Delta\dot{\varepsilon} = 0 = \Delta\dot{\varepsilon}_{el} + \Delta\dot{\varepsilon}_{pl}, \tag{9-17}$$

and

$$\dot{\varepsilon}_{el} = -\dot{\varepsilon}_{pl}, \tag{9-18}$$

which leads to

$$\frac{1}{E}\frac{d\sigma}{dt} = -A\sigma^n. \tag{9-19}$$

The variables are then separated to yield

$$\sigma^{-n}\,d\sigma = -AE\,dt, \tag{9-20}$$

and upon integrating

$$\left.\frac{1}{-n+1}\sigma^{-n+1}\right]_0^{\sigma(t)} = -AE\,\Delta t. \tag{9-21}$$

The stress varies in time from its initial value σ_0 to a lower value $\sigma(t)$, which leads to

$$\frac{1}{\sigma(t)^{n-1}} - \frac{1}{\sigma_0^{n-1}} = (n-1)AE\,\Delta t. \tag{9-22}$$

The time $\Delta t_{1/2}$ required for the stress to relax to one-half of initial value is obtained by setting $\sigma(t) = \frac{1}{2}\sigma_0$, so that

$$\Delta t_{1/2} = \frac{(2^{n-1} - 1)}{(n - 1)AE\sigma_0^{n-1}}. \tag{9-23}$$

XIII. SUMMARY

This chapter has dealt with the nature of the creep process and creep failure modes. Design procedures were reviewed, and the topics of residual life determination and stress relaxation were dicussed. Several case studies have served to illustrate the variety of problems that can arise because of creep.

REFERENCES

(1) H. E. Evans, Mechanisms of Creep Fracture, Elsevier Applied Science, Oxford, UK, 1984.

(2) D. N. French, Metallurgical Failures in Fossil Fired Boilers, Wiley, New York, 1983.

(3) H. J. Frost and M. F. Ashby, Deformation-Mechanism Maps, Pergamon Press, Oxford, 1982.

(4) Fossil Power Systems, ed. by J. G. Singer, Combustion Engineering, Windsor, CT, 1981, p. 10–22.

(5) H. Thielsch, Defects & Failures in Pressure Vessels & Piping, Reinhold Publishing Co., New York, 1965.

(6) M. F. Ashby, C. Ghandi, and D. M. R. Taplin, Fracture Mechanism Maps and Their Construction for F.C.C. Metals and Alloys, Acta Met., vol. 27, 1979, pp. 699–729.

(7) G. R. Stevick and I. Finnie, Crack Initiation and Growth in Weldments at Elevated Temperature, in Localized Damage IV, ed. by H. Nisitani, M. H. Aliabadi, S.-I. Nishida, and D. J. Cartwright, Computational Mechanics Publishers, Southhampton, 1996, pp. 473–484.

(8) H. C. Furtado, J. A. Collins, and I. Le May, in Risk, Economy and Safety, Failure Minimization and Analysis, 1998 edition, ed. by R. K. Penny, Balkema, Rotterdam, 1998, pp. 75–80.

(9) H. C. Furtado and I. Le May, in Risk, Economy and Safety, Failure Minimization and Analysis, 1996 edition, ed. by R. K. Penny, Balkema, Rotterdam, 1996, pp. 151–157.

(10) T. L. da Silveira and I. Le May, Mater. Char., vol. 28, 1992, pp. 75–85.

10

Fatigue

I. INTRODUCTION

Fatigue is a leading cause of failure in mechanical components and structures that are subjected to repeated loads. This chapter presents a general background concerning the initiation and propagation of fatigue cracks, together with consideration of design procedures and case studies.

II. FATIGUE: BACKGROUND

A fatigue failure is the result of the repeated application of stress below the tensile strength of the material. The failure process consists of the initiation of one or more cracks, the propagation of a dominant crack, and final separation. Fatigue cracks usually initiate at the surface, and for this reason, the condition of the surface and the environment are important factors influencing fatigue behavior. Processes such as shot peening or carburizing can improve high cycle fatigue performance by developing surface compressive residual stresses that make the initiation of surface cracks more difficult. However, when the surface is made highly resistant to the initiation of fatigue cracks, subsurface crack initiation can occur.

In the first half of the nineteenth century, a number of train wrecks occurred as the result of the fatigue failure of railroad car axles. These accidents brought attention to this new failure mode, and as a result, a number of investigations concerning the nature of fatigue were initiated. In 1839, Poncelet spoke of the failure of metals under the repeated action of tension and compression, and used the term "fatigue." Rankine (1843) observed that the growth of fatigue cracks in axles was an important aspect of the fatigue process. McConnell in England (1849) claimed, probably because of the brittle appearance of fatigue fracture surfaces he observed, that "a change from fibrous to crystalline character" occurred during fatigue, an unfounded idea that has long appealed to shop mechanics. Fairbairn (1864) carried out fatigue tests of components and was able to propose allowable stresses for bridge structures subject to fatigue loading. Some of the most important work aimed at characterizing fatigue as a material property was done by Wöhler in Germany during the period from 1858 to 1870. He was the first to carry out extensive fatigue test programs on test specimens rather than actual components, and to use both rotating bending and axial loading test methods. Because of his strong contributions to the field, plots of the stress amplitude versus the log of the fatigue life are commonly referred to in Europe as "Wöhler curves." In the United States, such curves are known as *S/N* curves.

The shape of these *S/N* curves is material-dependant, particularly in the high cycle regime. For low-carbon steels, the *S/N* curves exhibit a well-defined "knee" at about 10^6 cycles, as shown in Fig. 10-1. Beyond the knee, the curves are horizontal, and we can speak of a fatigue limit. As indicated in this figure, a useful relationship for steels is that the fatigue limit is about one-half of the tensile strength. For most other alloys, however, as indicated in Fig. 10-2, the knee is absent and the *S/N* curve at high numbers of cycles continues to decline slowly with increase in the number of cycles to failure. For such materials, it is customary to define the fatigue strength or endurance limit as the stress amplitude corresponding to a specified number of cycles, often 10^7. Further, there is generally no correlation with other properties such as the tensile strength.

Some components will experience huge numbers of fatigue cycles in their lifetimes, and for reliable operation, will be designed such that the stress amplitudes experienced are below the fatigue curve by a margin of safety. Such components are designed for infinite life. Other components, such as aircraft structures, will experience stresses above the fatigue strength, and are designed for finite life. How-

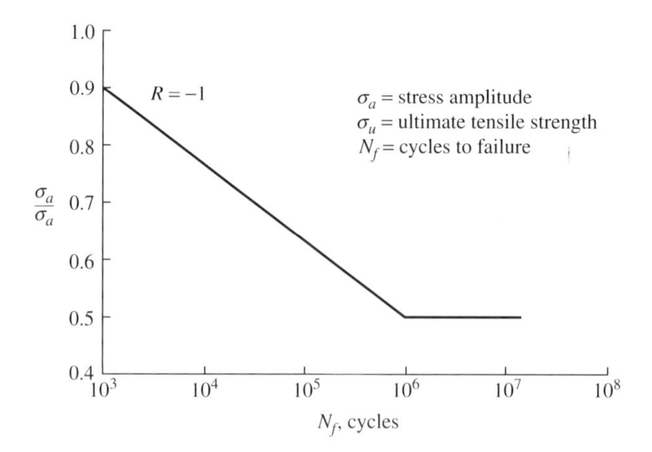

Fig. 10-1. Typical S/N behavior for steels.

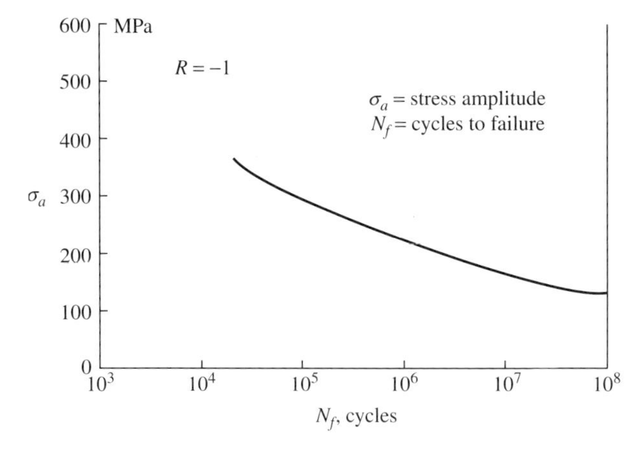

Fig. 10-2. S/N curve for an aluminum alloy.

ever, it is not unusual for the actual life to exceed the original design life. In such cases, careful inspection and the replacement of fatigue-cracked components is necessary to insure continued reliable performance. When repeated stresses are applied such that some of the stresses are above the endurance limit and some below, it is not safe to assume that the stresses applied below the endurance limit do no damage. This is because the stresses that are above the endurance limit may initiate cracks, which are then able to propagate at stress amplitudes that lie below the original endurance limit.

III. STATISTICS OF FATIGUE

A. Statististical Distributions

An important consideration in dealing with the reliability of a structure is the scatter in in-service loads encountered in practice compared to the scatter in a material's ability to resist the imposed loading conditions, Fig. 1-1. Figure 10-3 is a schematic diagram that indicates the distribution both of fatigue lives and of fatigue strengths. An example of the degree of scatter encountered in testing laboratory samples is given in Fig. 10-4 (1). This figure is a logarithmic-normal probability plot showing individual fatigue lifetimes obtained at different stresses for the aluminum alloy 7075-T6 under zero-tension loading. An even larger scatter may be found in service due to differences in nominal loading and environmental conditions. A straight line on this plot corresponds to a log-normal distribution of data points, and the steeper the slope, the smaller the standard deviation in lifetimes. Note that the scatter increases with decrease in $\sigma_{\max}$. The line corresponding to the highest stress amplitude is steep, and the scatter is relatively small. However, at the lowest $\sigma_{\max}$, 207 MPa (30,000 psi), the line is much less steep and there is more than an order of magnitude difference in lifetimes.

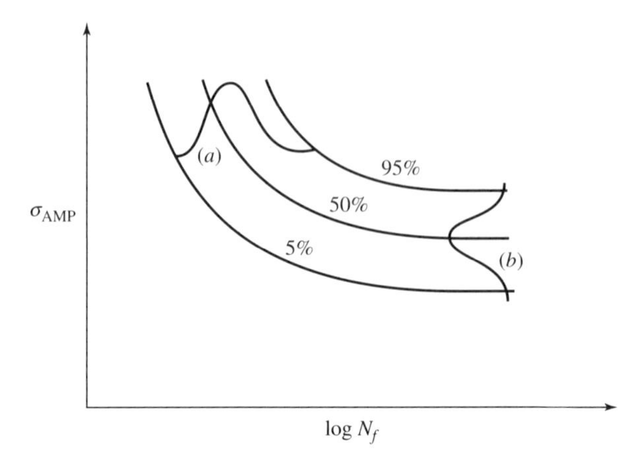

Fig. 10-3. (*a*) The scatter in fatigue lives and (*b*) the scatter in fatigue strengths. The probability of failure is indicated.

The statistical distribution of the fatigue strengths indicated in Fig. 10-3 is usually analyzed either as a normal distribution rather than a log-normal distribution or as an extreme value distribution.

B. An Application of Statistics to Fatigue

Extreme value distributions are important considerations in fatigue. Murakami et al. (2) have made use of the statistics of the extreme to analyze the distribution of inclusion sizes in an SAE 9254 spring steel, and Fig. 10-5 shows the distribution of inclusion sizes plotted in terms of the statistics of the extreme for this steel. The value of S_0, the standard inspection area, in this case was 0.0309 mm^2, and n, the number of sections examined, was 40. Murakami and coworkers have shown that the lower bound on fatigue strength, σ_{wl}, is related to the size of the largest inclusion expected at or near the surface in a rotating bending fatigue test. As the number of specimens increases, the probability of encountering larger defects also increases, with the result that σ_{wl} is a function of the number of rotating beam specimens tested, N, expressed as

$$\sigma_{wl} = \frac{1.41(H_v + 120)}{(\sqrt{\text{area}_{\max(N)}})^{1/6}}, \tag{10-1}$$

where σ_{wl} is in MPa, H_v is the Vickers hardness number expressed in units of kgf/mm^2, and $\sqrt{\text{area}_{\max(N)}}$ is in microns. For N values of 10 and 100, the expressions for the fatigue strength are

$$\sigma_{wl(10)} = \frac{1.41(H_v + 120)}{(29.2)^{1/6}} = 0.80(H_v + 120), \text{ MPa}, \tag{10-2}$$

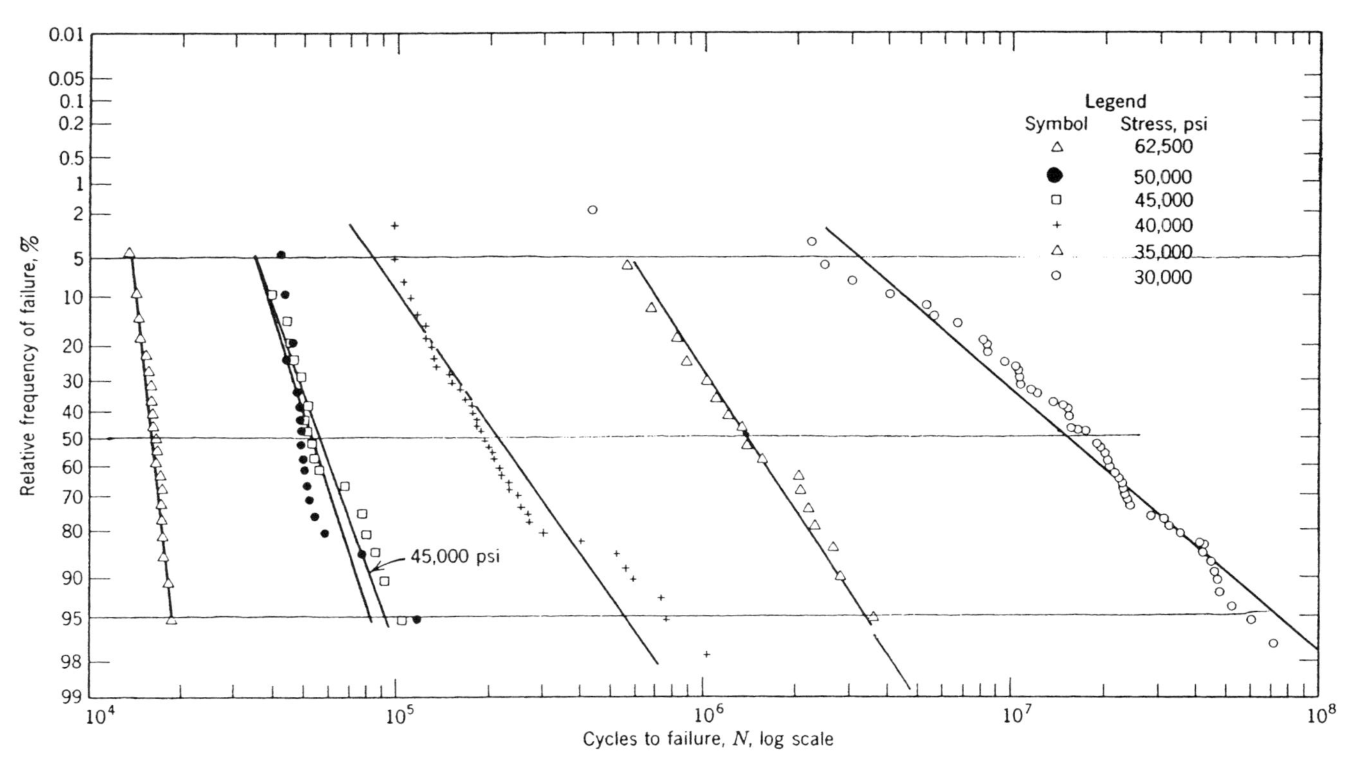

Fig. 10-4. Logarithmic normal probability diagram showing individual fatigue lifetimes obtianed at different stress ranges at $R \approx 0$. (From Sinclair and Dolan, 1. Reprinted with permission of the ASME.)

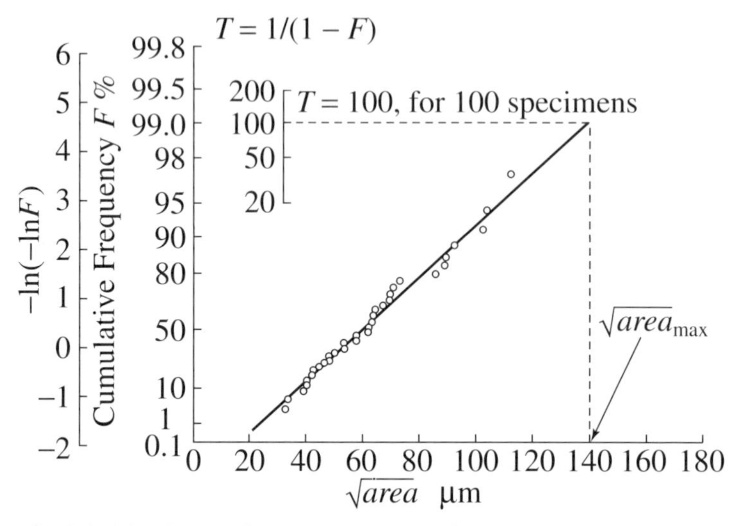

Fig. 10-5. Statistical distribution of the extreme values of the maximum size of inclusion at the center of the fatique fracture origin for SKH51 tool steel. (From Murakami, 2, NIST.)

and

$$\sigma_{wl(100)} = \frac{1.41(H_v + 120)}{(32.9)^{1/6}} = 0.79(H_v + 120), \text{ MPa.} \tag{10-3}$$

These equations are plotted in Fig. 10-6 and compared with experimental results. It is noted that the type of inclusion, whether it be an oxide, sulfide, or silicide, for example, did not affect the endurance limit, only the $\sqrt{\text{area}_{max}}$ parameter did. However, at stress amplitudes in the finite fatigue life range, the nature of the inclusion becomes a more important consideration.

IV. DESIGN CONSIDERATIONS

The fatigue lifetime is strongly affected by a number of factors such as the type of loading, for example, bending, axial, multiaxial, or torsion. The presence of a notch or hole will reduce the fatigue resistance, but not simply as a direct function of the notch stress concentration factor, for the size of the notch or hole is also a consideration. For example, a plate of steel containing a hole of 1 mm diameter will have a higher fatigue strength than a plate containing a hole of 1 cm diameter, despite the fact that in both cases the stress concentration factor is 3.0. Materials in engineering usage are often cyclically loaded in corrosive environments that degrade the fatigue properties. For example, ship propellers and drilling platforms operate in sea

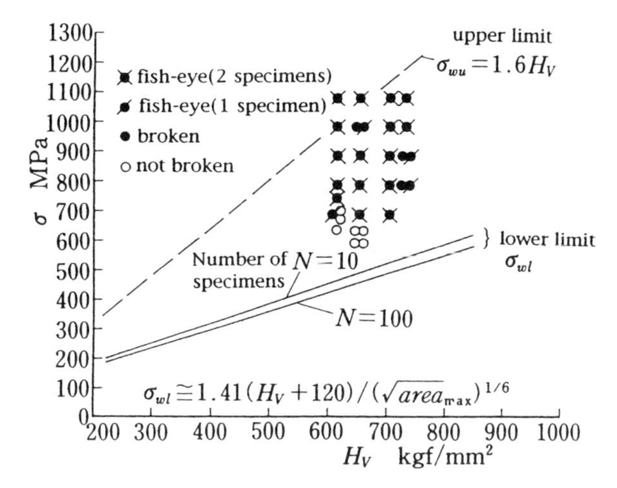

Fig. 10-6. A comparison of the predicted and experimental lower bounds of the fatigue strength of SKH51 tool steel as function of hardness level, H_V. (From Murakami, 2, NIST.)

water, which accelerates both the crack initiation process and the crack growth process. The term "corrosion fatigue" is used to describe the combined action of cyclic loading and corrosion. A related form of fatigue is known as fretting corrosion fatigue, with fretting being due to the relative motion of surfaces under pressure, as for example between a rivet head and the clamped sheet metal in an aircraft-structure.

Aircraft turbine blades experience huge numbers of vibratory cycles during their lifetimes and for this reason there is interest in fatigue at ultra-high cyclic lifetimes, such as 10^{10} cycles. In some of these ultra-high fatigue lifetime studies, it has been observed that for certain steels at high numbers of cycles a drop in fatigue resistance can occur, due to the growth of subsurface fatigue cracks. In assessing the fatigue performance of turbine blades, the possibility of foreign object damage (FOD) due to the ingestion of hard particles into an aircraft turbine engine must be taken into account. The effect of bird strikes on turbine blade performance is another consideration, one which, until recently, precluded the use of composite materials as first stage turbine blades because of their low resistance to bird-impact damage.

Much of the early testing in fatigue was carried out under fully reversed loading, that is the mean stress was zero as in the case of a railroad car axle. However, components are often cycled about a mean stress other than zero, and the designations R and A are used to describe such loading conditions. These designations are defined as

$$R = \frac{\sigma_{min}}{\sigma_{max}}, \tag{10-4}$$

and

$$A = \frac{\sigma_a}{\sigma_m}, \tag{10-5}$$

where σ_{min} and σ_{max} are the minimum and maximum stresses in a loading cycle, respectively, σ_a is the stress amplitude, and σ_m is the mean (average) stress. The stress range $\Delta\sigma$ is equal to $\sigma_{\max} - \sigma_{\min}$. The stress amplitude is equal to $\Delta\sigma/2$.

For certain components, the value of R may change significantly during use due to changes in the mean stress. For example, consider a point on the lower side of an aircraft wing. While the plane is taxiing on the ground, the mean stress is in compression and R is negative, but in flight the same point is in tension and R is positive.

Gerber (1873) and Goodman (1899) proposed relationships to deal with the dependence of allowable stress amplitudes on positive mean stresses, and their relationships are depicted in Fig. 10-7. The Gerber relation is parabolic and is given by

$$\sigma_a = \sigma_{a,R=-1}\left[1 - \left(\frac{\sigma_m}{\sigma_{UTS}}\right)^2\right], \tag{10-6}$$

where σ_a is the stress amplitude corresponding to a given lifetime, $\sigma_{a,R=-1}$ is the stress amplitude for a given fatigue lifetime under fully reversed loading, which is known from experiments, σ_m is the mean stress, and σ_{UTS} is the ultimate tensile strength. The Goodman relation is linear, and is given by

$$\sigma_a = \sigma_{a,R=-1}\left(1 - \frac{\sigma_m}{\sigma_{UTS}}\right). \tag{10-7}$$

Alloys differ in their conformity to either of these empirical relationships. Today, such relationships are still important in design, although fracture mechanics considerations are also becoming more widely used as well, as will be discussed.

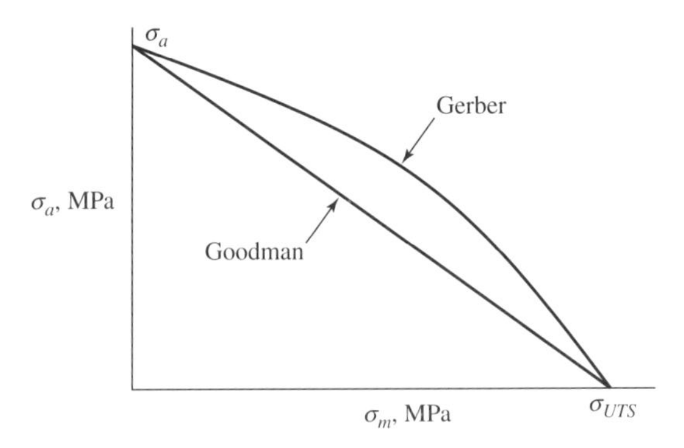

Fig. 10-7. The Goodman and the Gerber diagrams for the prediction of the effect of mean stress on the allowable stress amplitude for a given fatigue life.

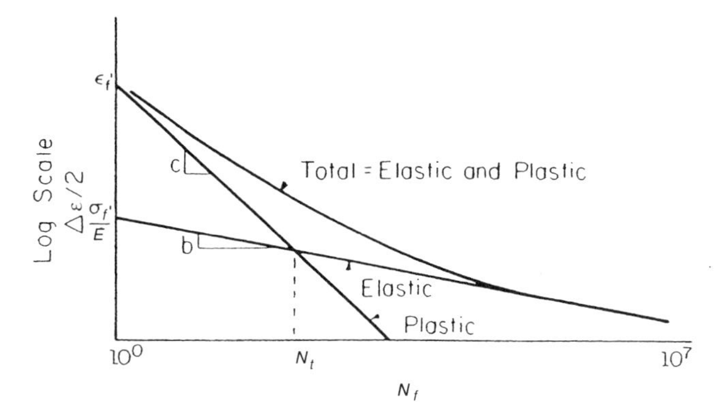

Fig. 10-8. The log of the number of cycles to failure as a function of logs of the elastic, the plastic and the total strain amplitudes, where b and c are respectively the slopes of the elastic and the plastic components of the total strain versus the number of cycles to failure.

Two analytical relationships are commonly used in fatigue analyses. One is due to Basquin, who in 1910 proposed that a linear relationship exists between the log of the stress amplitude and the log of the number of cycles to failure, or

$$N_f = C\sigma_a^{-g}, \tag{10-8}$$

where C and g are constants, and σ_a is the stress amplitude. The other is due to Coffin and Manson, who in the 1950s independently proposed that the log of the plastic strain range, $\Delta\varepsilon_p$, is linearly related to the log of the number of cycles to failure, or,

$$N_f = D(\Delta\varepsilon_p)^{-d}, \tag{10-9}$$

where D and d are constants. This is an important relationship for cases in which the plastic strain amplitude is large with respect to the elastic strain amplitude, that is, the low cycle fatigue regime.

A strain-life curve for fully reversed strain is shown in Fig. 10-8. The Coffin-Manson and Basquin relationships plot as straight lines as a function of either the elastic or plastic component of strain in Fig. 10-8. That is

$$\frac{\Delta\varepsilon_p}{2} = \varepsilon_f'(N_f)^c, \tag{10-10}$$

and

$$\frac{\Delta\varepsilon_e}{2} = \frac{\sigma_f'}{E}(N_f)^b, \tag{10-11}$$

where the empirical constants are: σ_f', the fatigue strength coefficient; ε_f', the fatigue ductility coefficient; b, the fatigue strength exponent; and c, the fatigue ductility exponent.

The fatigue life where the two straight lines cross is known as the transition life. At this life, the plastic strain amplitude is equal to the elastic strain amplitude, and it can be considered that the fatigue lives to the left of the intersection represent the low cycle fatigue range, and that the fatigue lives to the right of the intersection represent the high cycle fatigue range. The transition life can be obtained by setting the elastic and plastic strains in the above equations equal to each other and solving for N_f to obtain

$$N_t = \left(\frac{\varepsilon_f' E}{\sigma_f'} \right)^{1/(b-c)}. \tag{10-12}$$

The principal variables are ε_f' and σ_f'. As ε_f' increases, σ_f' usually decreases, with the result that the value of N_t is much higher for low-strength alloys (perhaps 10^5 cycles) than for high-strength alloys (perhaps ten cycles).

In using either the Basquin relation or the Coffin-Manson relation for calculating the fatigue life, the average fatigue life is considered. In designing to allow for scatter in fatigue strength, for corrosion, and for in-service surface damage, the design allowable stress for a critical part, such as an aircraft propeller, may be a small fraction of the smooth bar average fatigue strength. The determination of such low design allowables has often stemmed more from experience than from analysis.

Components in service are generally subjected to variable amplitude loading rather than the constant amplitude loading used to generate *S/N* curves, and the Palmgren-Miner law is used to make an approximate prediction of the fatigue lifetime under variable amplitude loading conditions. The law is referred to as a linear damage summation rule and is expressed as

$$\sum_{i=1}^{m} \frac{n_i}{N_{fi}} = 1, \tag{10-13}$$

where n_i is the number of cycles applied at the ith loading level, N_{fi} is the number of cycles to failure at that level, and m is the number of load levels. This law is used in estimating the fatigue life where not only the stress amplitude varies but the mean stress does as well. An example would be the case of aircraft structures, which experience repeated ground-air-ground (GAG) cycles associated with takeoff, cruise, and landing. In a such a GAG cycle, the mean stresses vary significantly, as do the loads resulting from maneuvers, gusts, and accelerations.

Jet aircraft engine components, such as the disks into which the turbine blades are inserted, are subjected to a type of cyclic loading known as thermal-mechanical fatigue (TMF). During engine startup, the temperature at the surface of a disk increases more rapidly than does that of the interior of the disk. As a result, a compressive stress is developed in the surface, and compressive plastic strains can result. On shutdown, the reverse situation can develop, with tensile stresses being developed in the surface of the disk. It is possible for these cyclic stresses and strains to result in low-cycle fatigue crack initiation in the disk, with possible disastrous consequences, unless the disks are carefully inspected at appropriate intervals.

In dealing with notches and their effect on fatigue, a relation due to Neuber (3) is often used. The relation is

$$K_T^2 = K_\sigma K_\varepsilon. \tag{10-14}$$

This relation states that the square of the stress concentration factor, K_T, is equal to the product of the stress concentration factor, K_σ, defined as $\sigma_{loc}/\sigma_{app}$, times the strain concentration factor, K_ε, defined as $\varepsilon_{loc}/\varepsilon_{app}$, where the subscripts refer to the local conditions at the notch and to the applied loading conditions, respectively. This relation was derived for Mode III loading in the elastic-plastic range, but has been used for other modes of loading as well. This relationship is obviously true in the elastic range, but appears to hold reasonably well for cases where the material at the tip of a notch is in the plastic range as well, for when plastic deformation occurs at a notch under a continuously rising load, the stress concentration factor will decrease and the strain concentration factor will increase. For applied stress within the elastic range, the Neuber relation can be written as

$$K_T^2 = \frac{\sigma_{loc}}{\sigma_{app}} \frac{\varepsilon_{loc}}{\varepsilon_{app}} = \frac{\sigma_{loc}\varepsilon_{loc}}{\sigma_{app}^2/E}, \qquad \text{or } K_T\sigma_{app} = \sqrt{E\sigma_{loc}\varepsilon_{loc}}. \tag{10-15}$$

The last of these equations is used as follows. Unnotched specimens are cycled to failure at different strain amplitudes under fully reversed strain conditions, the cyclic amplitude stress associated with each amplitude is noted, and the square root quantity on the right-hand side of the equation is calculated. Then it is assumed that when the magnitude on the left-hand side equals the magnitude on the right-hand side, both sides of the equation represent equal fatigue lifetimes. On this basis, fatigue data obtained with unnotched specimens can be used to estimate the fatigue life for a notched specimen of any K_T value. However, as mentioned above, it is observed in fatigue testing that K_f, the stress concentration factor determined under cyclic loading conditions, is less than K_T, the theoretical stress concentration factor. This effect has been attributed to the crack length emanating from the stress raiser and to crack closure (4). To account for it, empirical expressions such as the following, which is also based upon Neuber's (5) work, have been used:

$$K_f = 1 + \frac{K_T - 1}{1 + \sqrt{\rho'/\rho}}, \tag{10-16}$$

where ρ' is considered to be a material constant, which in the case of steel decreases in magnitude with increase in tensile strength.

A design procedure that has been developed by the Japanese Society of Mechanical Engineers (6, 7) takes into account the effect of part geometry and tensile strength in determining the fatigue strength, σ_w, of a component. The basic equation is

$$\sigma_w = \frac{\sigma_{w0}\zeta_s}{\beta}, \tag{10-17}$$

where σ_{w0} is the fatigue strength of a polished specimen, ζ_s is a an empirically determined fatigue strength reduction factor for surface finish and material, and β is determined with the aid of a design chart. For example, consider the bending of a shaft with a shoulder. To use the design chart for this case, Fig. 10-9, information is needed on: the tensile strength of the material, σ_B; the diameter of the shoulder, D; the diameter of the shaft, d; and the radius of the transition from shaft to shoulder, ρ. Let us assume a tensile strength of 50 kg/mm^2 (490 MPa), D = 50 mm, d = 25 mm, and ρ = 2.5 mm. Therefore d/ρ = 10, and $1 - d/D$ = 0.5. From the chart $\xi_1 = 1.45$, $\xi_2 = 0.92$, $\xi_3 = 0.48$, and $\xi_4 = 0.93$. The notch factor β is defined as

$$\beta = 1 + \xi_1\xi_2\xi_3\xi_4 = 1 + 1.45 \times 0.92 \times 0.48 \times 0.93 = 1.60. \tag{10-18}$$

Therefore, the fatigue strength of the shaft is estimated to be $0.625\sigma_{w0}\zeta_s \cdot \sigma_{w0}$ can be estimated to be one-half of the tensile strength, so that σ_w = 15.6 kg/mm^2 (153

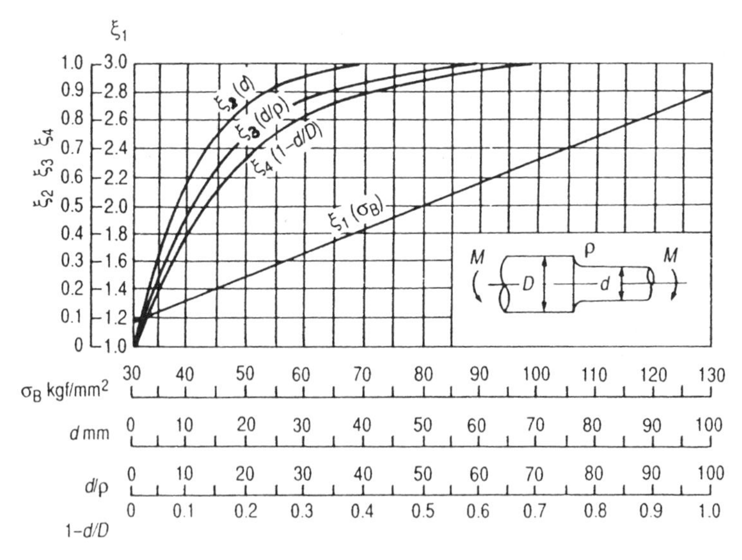

Fig. 10-9. The notch factor for a steel shaft with a shoulder in bending as a function of the tensile strength, σ_B, the major and minor diameters, and the transition radius, ρ. (From Nishida 6, based upon 7. Reprinted with permission of the Japan Society of Mechanical Engineers.)

MPa) $\times \zeta_s$. Additional design charts are available to cover other geometries and loading conditions.

Although design analysis precedes the manufacture of a fatigue critical component, fatigue testing is an important part of the certification process, as in the case of aircraft. For the Boeing 717, parts were tested under simulated in-service loading (spectrum loading) to five times the planned number of operational cycles (8). The fatigue testing of components is an essential part of vehicle development in the ground transportation industry as well. In the case of a part that has failed in service, it may be possible to carry out an analysis to determine why the part did not have the requisite strength to sustain the applied loads.

It is of interest to see how knowledge of fatigue is used in the development of design rules. Albrecht and Wright (9) have reviewed the procedures used in the design of welded steel bridges against fatigue. For each of eight different types of weld detail, a basic *S/N* plot is constructed that consists of two straight lines on a log-log plot. The finite-life line is drawn through the data points at a slope of 1/3. This line intersects a horizontal line representing the fatigue limit at what is referred to as the transition life. The design curve in the finite life range is set at two standard deviations to the left of the mean *S/N* line, and the design fatigue limit is set at one-half of the constant amplitude fatigue limit to account for variable amplitude loading, since if cracks are nucleated at stresses above the fatigue limit, then stresses applied below the fatigue limit are significant. The standard deviation ranges from 0.221 for plain rolled I beams to 0.101 for cover plates welded to the flange of an I beam, and the transition lives range from 14×10^6 to 180×10^6 cycles, depending upon the detail.

V. MECHANISM OF FATIGUE

Plastic deformation plays an essential role in the initiation and propagation of fatigue cracks. An important step in understanding the mechanism of fatigue was taken by Ewing and Humphrey (1903), who were among the first to use the metallurgical microscope in making their observations. In fatigue tests of polished specimens of Swedish iron, they noted that slip bands that developed under cyclic loading differed in appearance from those that formed under tensile loading. Under cyclic loading, the bands formed in uniformly spaced groups within a grain at low stress amplitudes, whereas under tensile loading, the slip bands were more uniformly distributed across the grains. Most importantly, Ewing and Humphrey noted that fatigue cracks originated within the fatigue slip bands. Later work by Gough in the 1920s further demonstrated that plastic deformation was essential to fatigue crack initiation. Fatigue slip bands are favored sites for continuing plastic deformation. If a surface containing these bands is lightly polished and the specimen then recycled, fresh slip bands will appear at the same site as fatigue slip bands were previously observed. Hence the name persistent slip bands (PSBs). In the 1950s, with the advent of the transmission electron microscope (TEM), the fatigue process within fatigue slip bands was observed in greater detail by Forsyth, who found that thin ribbons of material were sometimes extruded from these slip bands prior to the development of cracks. Today, the scanning electron microscope (SEM) is a main tool for such observations.

Until the 1960s, the only information available to the design engineer was often an *S/N* curve. However, the *S/N* curve does not distinguish between the crack initiation and crack growth portions of the fatigue process. Interest in the fatigue crack growth increased in the 1950s with the recognition that macroscopic cracks could form and grow in aircraft structures due to cyclic loading. The Comet aircraft disasters in particular emphasized this point. Two other developments also promoted interest in this topic. One was the observation by Forsyth and Ryder (10) that fatigue cracks advanced an increment in each cycle. The other was the emergence of fracture mechanics as a method for dealing with fatigue cracks in an analytical manner and led to the formulation of Paris's law for the rate of fatigue crack growth:

$$\frac{da}{dN} = C(\Delta K)^m \tag{10-19}$$

where a is the crack length, ΔK is the range of the stress intensity factor for tension-tension loading, and m and C are constants. Figure 10-10 is a plot in log-log coordinates of the rate of fatigue crack growth as a function of ΔK. The overall curve is sigmoidal in shape, bounded at the lower end by the threshold for fatigue crack growth, and at the upper end by the fracture toughness of the material. Region I is known as the near-threshold range, and region II is the region in which Paris's law, Eq. 10-19, is applicable. The fractographic markings that are associated with fatigue crack growth are known as striations and they are found in the upper portion of this region. Region III involves both tensile and fatigue modes of fracture.

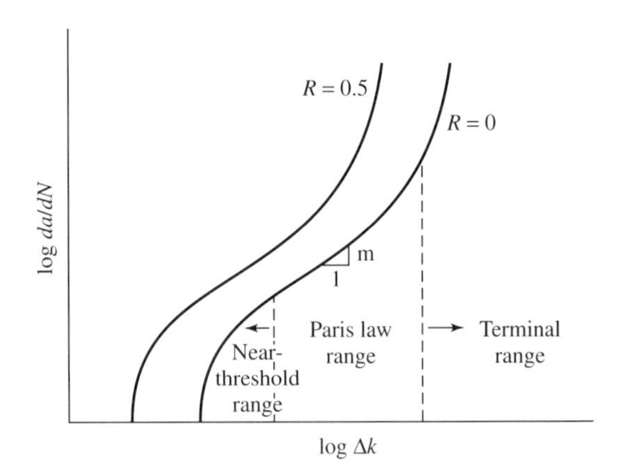

Fig. 10-10. The rate of fatigue crack growth as a function of ΔK.

In 1970, Elber (11) introduced the concept of crack closure, which has since figured importantly in considerations of fatigue crack growth. When a fatigue crack is loaded in tension under plane stress conditions, there will be a contraction of material at the crack tip in the through-thickness direction due to the Poisson's effect. When the specimen is unloaded, this sucked-in material will cause contact of the crack surfaces at the tip to be made before complete unloading, and a compressive stress is developed in the wake of the crack tip over the region of premature contact. In the next loading cycle, these compressive stresses must be overcome before the crack tip can again open to allow the crack to propagate. The effective range of the tensile cycle is thus that range between the opening load and the maximum load of the cycle, that is,

$$\Delta K_{eff} = K_{max} - K_{op}. \tag{10-20}$$

Similar considerations apply to plane strain loading, where the mismatch of rough fracture surfaces leads to premature contact and the crack closure phenomenon. The following relationship based upon crack closure concepts has been found to describe the fatigue crack growth behavior of a wide variety of alloys:

$$\frac{da}{dN} = A(\Delta K_{eff} - \Delta K_{effth})^2, \tag{10-21}$$

where A is a material constant whose magnitude is influenced by the environment, and ΔK_{effth} is the range of the stress intensity factor at the threshold level. In dealing with very short cracks, that is, of the order of 10 μm, the endurance limit of the material exerts more control on fatigue crack growth behavior than does the threshold level. Because of the large ratio of plastic zone size to crack length, the effective crack length is taken to be the actual crack length plus one-half of the Dugdale plastic zone size. In addition, the crack closure level for a newly initiated fatigue crack is zero since it does not have a crack wake, but the crack closure level increases up to the level of a macroscopic crack, K_{opmax}, as the crack length increases. To take such factors into account, Eq. 10-21 is rewritten as follows:

$$\frac{da}{dN} = A\left\{\left[\sqrt{\pi r_e\left(\sec\frac{\pi}{2}\frac{\sigma_{max}}{\sigma_Y} + 1\right)} + Y\sqrt{\frac{\pi}{2}a\left(\sec\frac{\pi}{2}\frac{\sigma_{max}}{\sigma_Y} + 1\right)}\right]\Delta\sigma - (1 - e^{-ka})(K_{opmax} - K_{min}) - \Delta K_{effth}\right\}^2, \tag{10-22}$$

where r_e is a material constant, of the order of 1 μm, that provides a bridge between the endurance limit and the effective threshold level. The constant is obtained by setting a in the second term equal to r_e, $\Delta\sigma$ equal to the endurance limit, and then equating the first two terms to ΔK_{effth}. The empirical material constant k governs the rate of crack closure development. For a medium strength steel, it has a value of 6000/m.

VI. FACTORS AFFECTING CRACK INITIATION

As mentioned above, plastic deformation is essential to the nucleation and growth of fatigue cracks. Anything that promotes plastic deformation will decrease the fatigue resistance of a material. The surface is the region at which fatigue cracks usually initiate, although when the surface is harder than the interior, subsurface fatigue cracks may initiate. Stress concentrators such as notches and corrosion pits will degrade fatigue resistance. The handbook values for fatigue strength are generally given for highly polished specimens tested under fully reversed loading conditions ($R = -1$). Care should therefore be used in applying these values to actual components. The degree of surface roughness, which is a function of the machining or processing method, will be a factor, since roughness acts as a stress concentrator. Figure 10-11 indicates how varying the surface finish, as a function of the mode of processing, affects the fatigue strength (12). The decarburization that occurs during the hot-forging of a steel will lead to a soft surface layer in which resistance to plastic deformation is reduced with a corresponding decrease in fatigue strength. An increase in strength properties may increase fatigue strength, at least in steel, where it is observed that the endurance limit is one-half of the tensile strength. However, in high-strength aluminum alloys, this is not the case, that is, the fatigue strength does not increase with tensile strength but can be rather insensitive to the strength level. The behavior of aluminum alloys indicates that, despite a high strength level, local soft spots that exist in the microstructure permit plastic deformation at relatively low stress levels. Compressive residual surface stresses are useful in preventing the development of fatigue cracks. On the other hand, tensile residual surface stresses are to be avoided.

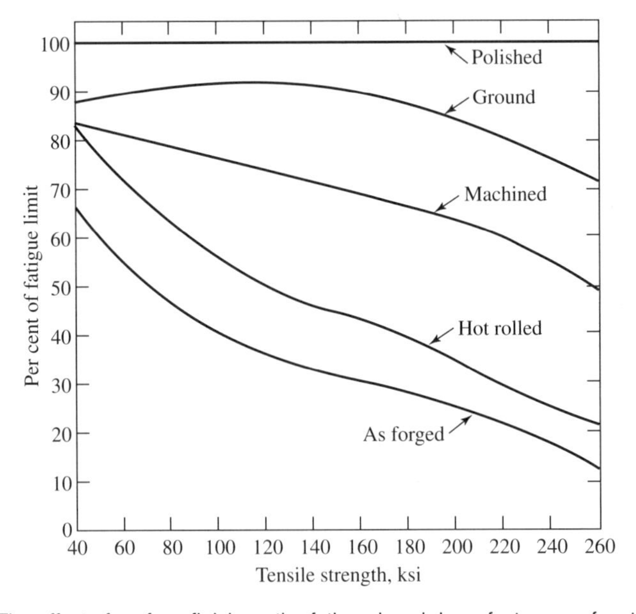

Fig. 10-11. The effect of surface finish on the fatigue knockdown factor as a function of tensile strength, 36. (Reprinted with permission of John Wiley.)

VII. FRETTING FATIGUE (13)

Fretting is a wear process that occurs when two components are held together and undergo a small localized relative displacement, typically of the order of only 10 nm. In contrast to normal wear processes, not only are the displacements much smaller, but also the fretting debris remains trapped between the two surfaces in contact. Under certain cyclic loading conditions, the resultant surface wear can promote the initiation of fatigue cracks, which can then propagate and perhaps lead to failure. In aircraft structures, the contact area between a rivet head and the aluminum skin is a typical location at which fretting fatigue cracks can develop. The fir tree or dovetail slots in a turbine disk into which the turbine blade roots fit is another location of concern, particularly in the low-pressure stages of the compressor, where the temperatures are below the creep range. Fretting fatigue problems can arise whenever two components are tightly clamped together and undergo rotation, as in the case of train wheels press-fitted on axles, or vibration, as in the case of tubing passing through a support plate in a steam generator. In one such case, there was initially a clearance between tubing and plate, but over time, corrosion occurred, the tubing became in effect clamped in place, and fretting fatigue cracks developed.

In the attempt to avoid fretting fatigue failures, limits are set on the contact stress and the peak stresses, corners are rounded, and stress relief grooves may be machined to relieve points of high contact stress. In addition, shot peening (see below) and graphite antifriction lubricants applied as an aerosol are used to improve resistance to fretting fatigue. Rolls-Royce is applying a new copper-nickel-indium antifriction coating, which is flame sprayed onto Trent 800 fan blade roots to supplement a graphite based antifriction lubricant, to eliminate a cracking problem in 92,000-lb. maximum thrust engines, which resulted in the loss of fan blade on a Boeing 777 during a take off run (34). As a result of this failure, the fan blades of Trent 800 engines, which do not have the new coating, are required to be ultrasonically inspected once the engine reaches 600 taekoff and landing cycles, and then every 100 cycles thereafter. The new coating should allow the inspection interval to be increased to 1,200 cycles.

Periodic inspections of turbine blade roots can also be carried out, but on occasion, the large number of fretting cycles rapidly accumulated due to high-frequency turbine blade oscillations have led to fatigue crack initiation in such a short time, that periodic inspections were not effective.

VIII. SHOT PEENING

Shot peening is used as a means to remove the brittle oxide scales that form on steel during heat treatment, but Almen and Black at General Motors (14) noted that shot peening also improved fatigue resistance by imparting a residual compressive stress to the peened surface. The shot peening process involves the use of a stream of metal or glass shot of a selected size, moving at a chosen velocity and impinging upon the surface being treated. The peening conditions are standardized by means of test strips (Almen strips). Shot peening a thin strip leaves the shot-peened surface in compression, the mid-portion of the strip in tension, and, so that there be no net moment, the surface opposite to the peened surface in compression. This state of stress leads to a curvature of the strip as in the bending of a beam such that, if the peening is from above, the strip adopts a convex upward shape, that is, it rises in the Almen test. During a check calibration, the Almen strips are peened and the height of the strip as a function of time is noted. After some time, the height will not change signficantly. If the height differs from that specified, the peening conditions are altered accordingly.

IX. FACTORS INVOLVED IN FATIGUE CRACK PROPAGATION

Figure 10-12 schematically indicates the formation of a fatigue crack within a slip band at the surface and its subsequent propagation under cyclic tensile loading. In clean material, the initial crack propagates along a slip plane within the slip band and is referred to as a Stage I crack. At some point, usually within the first grain or two, the crack will change orientation and begin Stage II (Mode I) propagation in a direction perpendicular to the principal tensile stress. In material containing inclusions, a crack may initiate at an inclusion, and the slip band process of initiation may be bypassed. In calculations of the fatigue life, the cycles spent in crack initiation are sometimes neglected, and the lifetime is computed on the basis of the growth of micron-sized cracks to a critical size for failure in Mode I.

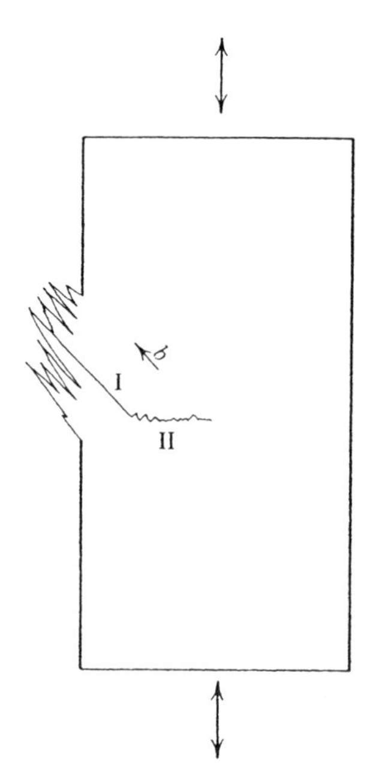

Fig. 10-12. A schematic diagram of the Stage I and Stage II fatigue crack growth. (b is the Burgers vector.)

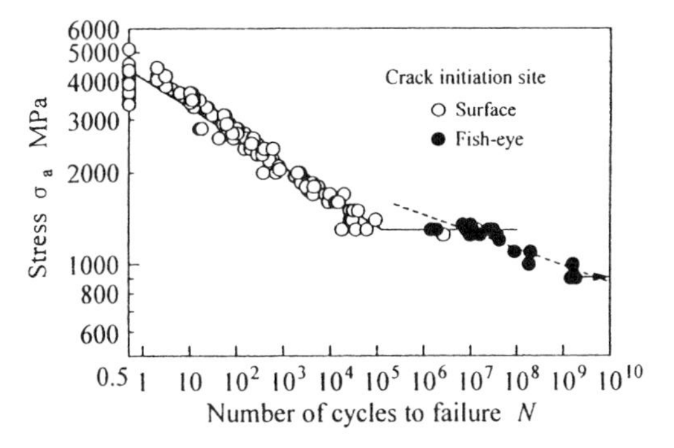

Fig. 10-13. S/N diagram for the bearing steel SUJ2. (From Sakai et al., 15. Reprinted with permission from the Society of Materials Science of Japan.)

In the ultra-long-life fatigue regime of steels, that is, 10^6–10^9 cycles to failure, fatigue cracks can originate at subsurface inclusions rather than at the surface. Of particular concern is that these ultra-high cycle fatigue failures can occur at stresses below the nominal fatigue limit, as indicated in Fig. 10-13 (15). In some cases, subsurface cracking occurs because the surface had been strengthened by residual compressive stresses due to shot peening, heat treatment, machining, a protective oxide, and so on. In other cases, the stress raising effect of the largest subsurface inclusion within the most highly stressed volume may lead to subsurface crack nucleation, particularly in axial loading. Small amounts of bainite within a martensitic matrix have also led to subsurface fatigue crack formation.

In the case of an inclusion-nucleated subsurface fatigue crack, optical examination of the region immediately adjacent to the inclusion has revealed the presence of a faceted, optically dark area (ODA) (16). The ODA can vary in radial extent from a few microns to as large as 50 μm, increasing in size the longer the fatigue lifetime. Secondary ion mass spectroscopy (SIMS) has shown that hydrogen is present in the ODA, and the initiation of the fatigue crack fatigue is thought to be promoted by hydrogen that had segregated to the inclusion. The fact that the ODA increases in size with increase in fatigue lifetime indicates that the hydrogen was able to diffuse along with the crack and promoted crack growth at the lowest crack growth rates. As the fatigue crack propagates beyond the ODA while still subsurface, it propagates in vacuum without the aid of hydrogen, and the optical image becomes relatively bright. However, once the crack penetrates through to the surface, it propagates in air, and the optical image darkens. The overall effect that is created is known as a fish-eye, Fig. 10-14*a* and *b*. In the case of Fig. 10-14*a*, a subsurface crack grew to the surface before fracture occurred. In the case of Fig. 10-14*b*, the fatigue crack was still a subsurface crack when fracture occurred. The reason for the change from a dark (ODA) to light and then to a dark optical appearance is simply due to changes in fracture surface roughness. The fracture surface is rough in the ODA, but much smoother in the surrounding optically bright area. It then becomes rough again after the crack has penetrated to the surface. The fatigue crack

(*a*)

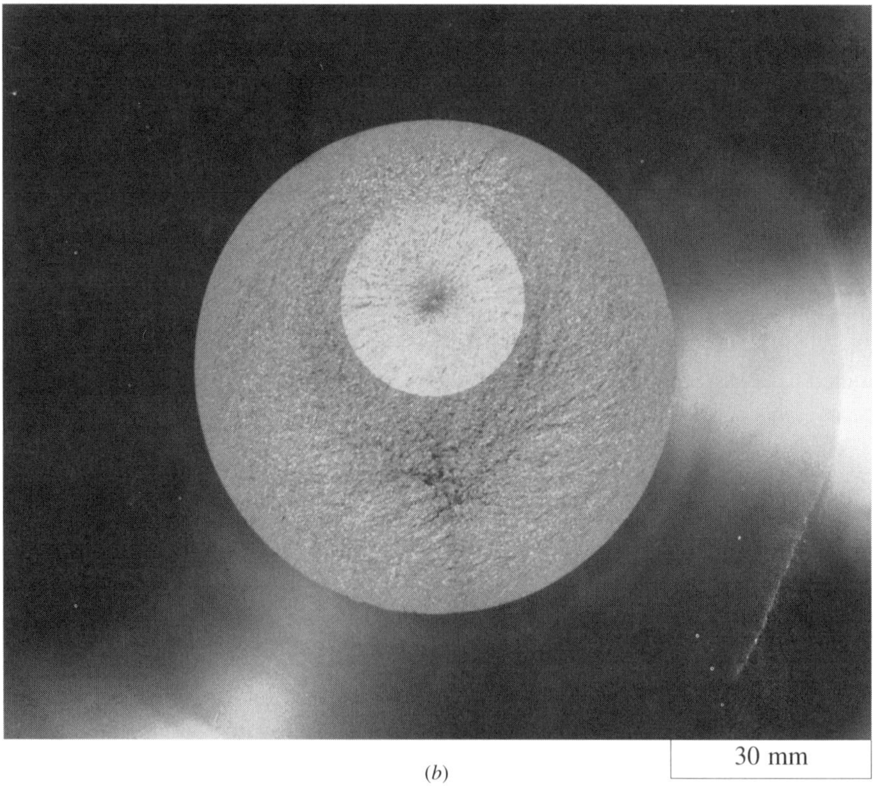

30 mm

(*b*)

Fig. 10-14. (*a*) An example of a fish-eye fracture in SAE 9254 steel. (From Murakami et al., 2). (*b*) A completely subsurface fish-eye fracture (courtesy of Y. Murakami.)

propagation process involves the opening and closing of the crack tip in each loading cycle. The rough appearance created in air is due to the fracturing of the oxide at the crack tip at different levels around the periphery of the crack front during the loading portion of a fatigue cycle. In vacuum, the crack advance along the crack front is much more co-planar, and an optically bright region results. It is also noted that similar differences in roughness occur, for example in elevated temperature fatigue tests conducted in either air or vacuum.

Figure 10-15 provides examples of fatigue striations that are created during fatigue crack propagation. These markings are most easily observed in ductile metals and may be difficult to observe in high-strength alloys of limited ductility. They are created on a cycle-by-cycle basis and are usually observed at crack growth rates ranging from of 0.1 μm per cycle up to more than 10 μm per cycle. The striations often have a staircase profile, which is created in the manner shown in Fig. 10-16. Their presence or absence may be critical in deciding the nature of a fracture. When striations are present on a fracture surface, they can at times be used to trace the loading history of the failed part. For example, in the case of the fatigue failure of a support beam for a wing landing flap, the fatigue crack initiated at a bolt hole, and striations were observed to be in pairs. The flap beam saw two dominant loads in each flight, a high load during landing (flap out) and a lower load during takeoff (flap partially out), and the width of the striation spacing created on landing was

(*a*)

Fig. 10-15. Examples of fatigue striations observed in aluminum alloys.

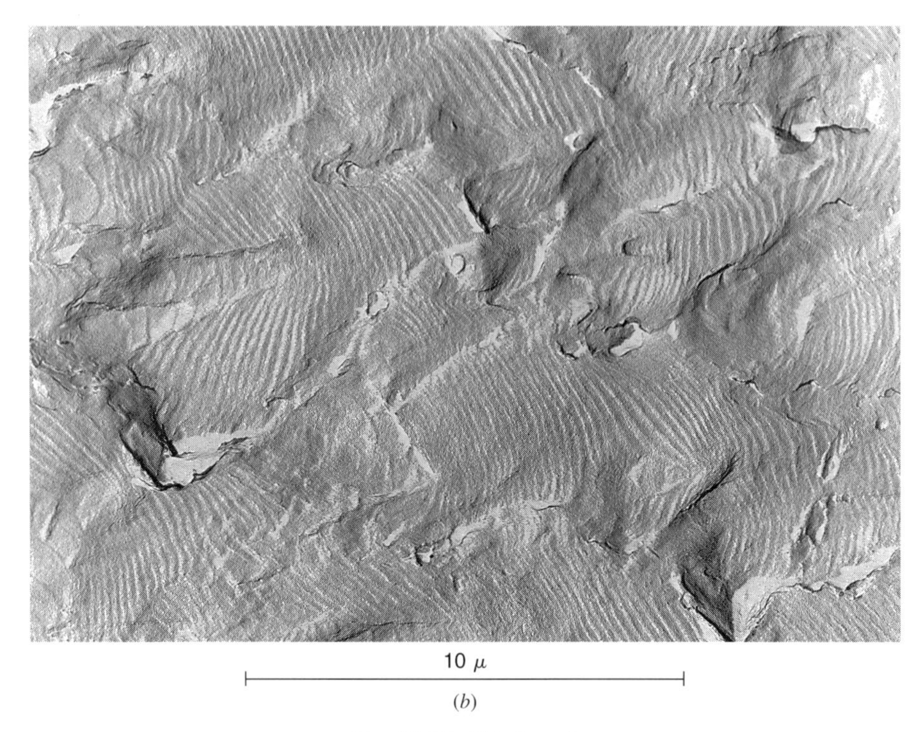

Fig. 10-15. *(Continued)*

twice that created on takeoff. A count of these striations indicated that the crack growth life was about 5000 flights, long enough for the crack to have been detected by periodic inspection (17). In the case of an aluminum helicopter blade containing a fatigue crack, the main cyclic loading occurs as the blade is brought from rest up to speed and then back to rest again. A count of the striations created by this loading can also indicate the number of flights spent in crack propagation. This is true of other rotating parts such as aircraft engine crankshafts.

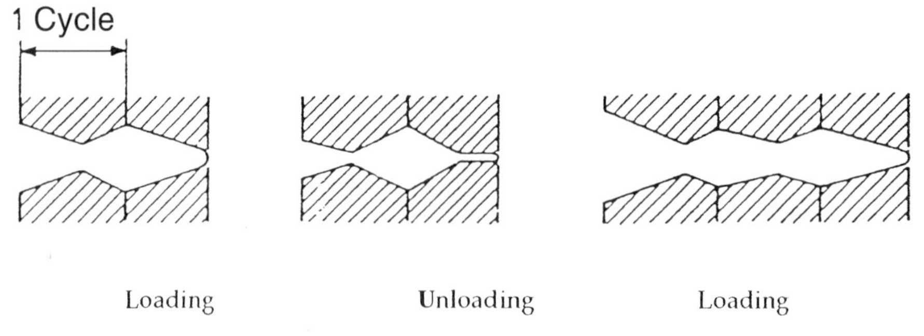

Fig. 10-16. A model for the formation of staircase-profile type fatigue striations. (From Schivje, 32. Copyright ASTM. Reprinted with permission.)

X. FATIGUE FAILURE ANALYSIS

A. Macroscopic

When a component fails in fatigue, an examination of the fracture surfaces may provide substantiating evidence such as well-defined "beach mark" patterns. These patterns can be created by environmental interactions with the crack front, and the term "thumbnail" fracture is often used to describe their shape around the origin of fatigue cracking. In steel parts that are subjected to reversed loading, the fatigue fracture surfaces can rub together as in $R = -1$ loading and develop a highly polished appearance, particularly near the origin of fatigue cracking. In such cases, no fine details will be observed, but the macroscopic beach markings may be apparent, as in Fig. 10-17. Such markings can develop when the part experiences periods of cycling followed by periods of rest. During the rest period, corrosion occurs at the crack tip to outline the crack front, thereby creating the beach mark. Variable amplitude loading may also create beach marks, particularly if there are many cycles of small amplitude mixed with a few high amplitude cycles. In this case the texture of the fracture may vary from smooth for the low amplitude cycles to rough for the high amplitude cycles, and it is the roughness that leads to the creation of the beach marks. A well-defined transition between the fatigue region and the region of final overload failure can also establish that the fracture was due to fatigue.

Fig. 10-17. Fatigue beach markings on a steel component. The markings are visible to the eye, and the fatigue crack initiated at a weld defect at the top of the figure.

B. Microscopic

An examination of a fatigue fracture surface using light microscopy or scanning electron microcopy (SEM) will reveal whether the fracture is intergranular or transgranular. If the fracture is transgranular, which is the usual mode, the fracture may be faceted, indicating a crystallographic path for crack growth. This mode is observed in aluminum alloys at low growth rates and in single crystals of nickel-base superalloys, for example. Transgranular fractures are often noncrystallographic in that faceted growth is not observed. It is on such fracture surfaces that fatigue striations are observed.

At higher ranges of the stress intensity factor corresponding to growth rates greater than 0.1 μm per cycle, fatigue striations may be visible, and their presence can be related to the magnitude of the stress amplitudes that had been applied. They are observable only over a limited range of growth rates because at too low a growth rate, that is, usually below 0.1 μm per cycle, they are not resolvable in an SEM, and at high growth rates, that is, above 10^{-3} mm/cycle, the fracture surface contains increasing amounts of what are called static fracture patches, which are the precursors to final separation. Because of these limitations, striations may provide only limited information about the past in-service history. However, even this information can be critical in trying to resolve whether a particular fracture occurred by fatigue or not. The spacing of striations is da/dN and if a plot of da/dN versus ΔK is available for the appropriate R condition, the applied ΔK values can be determined. However, the presence of high residual stresses, such as may be present in a welded joint, can make this type of analysis more difficult. Striations can also provide information as to the number of cycles or flights that a part experienced during the crack propagation phase of a fatigue failure. In some cases only the major loadings may be reflected, as in the case of an aluminum helicopter blade where a relatively few start-up cycles of high stress amplitude create a different fracture surface texture than do the cycles experienced during flight. In such a case it may not be possible to resolve individual minor striations because of rubbing of the fracture surfaces. Similarly, a failed crankshaft of a propeller-driven aircraft may exhibit two different but repetitive textures, one associated with takeoff and the other with cruise conditions. In such a case, it may be possible to determine the number of flights since the last inspection, for example.

Since fatigue striations are not usually observed below a crack growth rate of 0.1 μm per cycle, it is not possible to reconstruct the loading history of a failed part on the basis of spacing of striations in this range of crack growth rates. However, an alternative procedure has been developed in which the fracture surface roughness is measured with a profilometer either transverse to (18) or parallel to (19) the direction of fatigue crack propagation. As shown in Fig. 10-18, the fracture surface roughness of a 12CrMo steel increases with increase in crack growth rate, and from this information the corresponding ΔK values and loading history can be estimated. The method has potential applicability in the range of fatigue crack growth rates of from 10^{-9} to 10^{-7} m/cycle, but a reference fracture surface of known loading history is needed for comparison with the fatigue fracture surface of a failed component. As

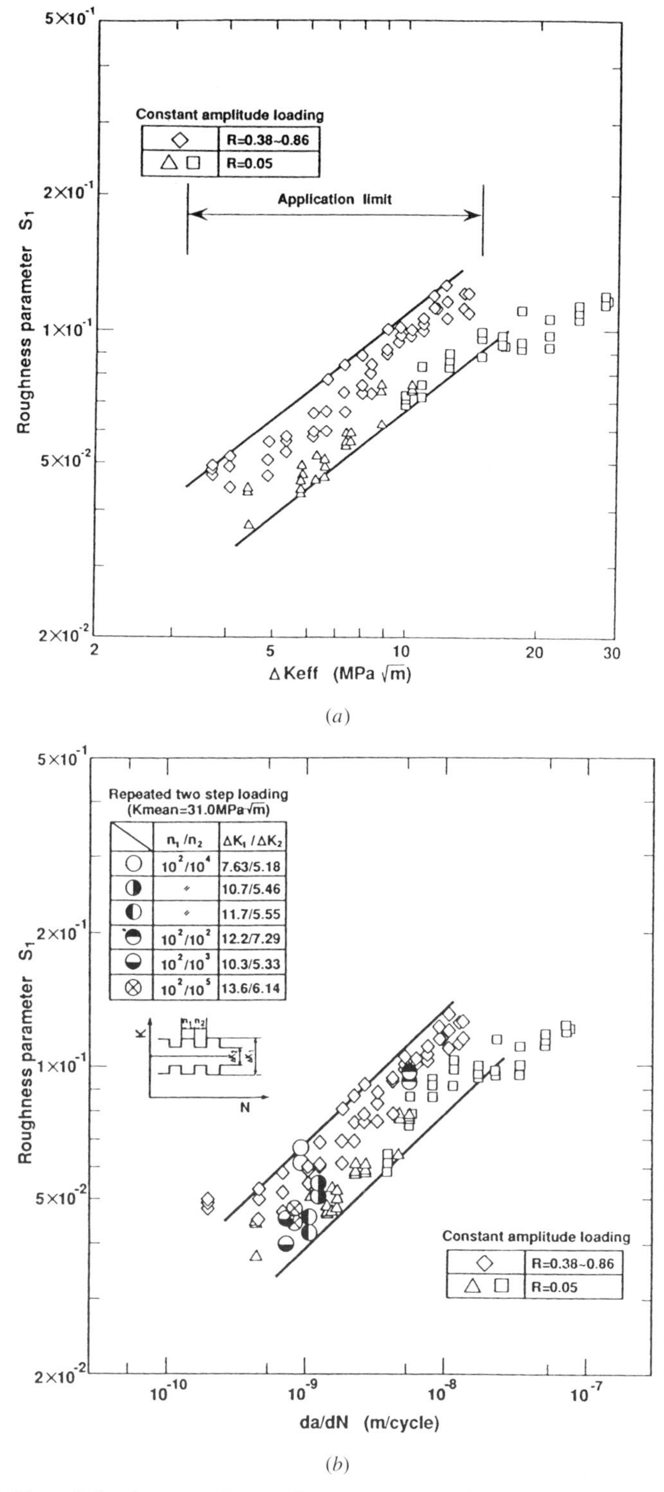

Fig. 10-18. The relation between the roughness parameter S_1 and da/dN for a 12CrMo steel. (From Fujihara et al., 18. Reprinted with permission of the Society of Materials Science, Japan.)

in the case of striations, corrosion of the fracture surfaces will increase the difficulty of establishing a good correlation between the roughness parameter and the rate of fatigue crack growth.

For materials in which fatigue striations are difficult to observe, it may be worthwhile to conduct laboratory fatigue and fracture tests to establish standards that can be used to determine the characteristics of the fracture surfaces associated with each type of fracture. Knowledge of the loading history of a failed part is helpful in the examination of a fracture surface, for it is important to establish that a set of "striationlike" markings under investigation are the result of cyclic loading, since markings that are similar in appearance to fatigue striations have been observed in cases of creep failure, stress corrosion cracking, and overload failures. Such markings may correspond to an intermittent mode of crack growth or to a hesitation in growth rate of the crack front. In brittle materials such as glass, "Wallner lines," which resemble striations, may be present on the fracture surface. These lines are the result of interaction between the expanding crack front and reflected elastic waves in the body.

In some cases, it may not be feasible to examine the actual fractured component in the laboratory. In such instances, replicas of the fracture surface can be made in the field and examined later in the laboratory, as described in Chapter 5.

XI. ENVIRONMENTAL EFFECTS

The environment can have a strong effect on fatigue behavior, as shown in Fig. 10-19. Corrosion has two effects: (*a*) it can lead to the formation of corrosion pits,

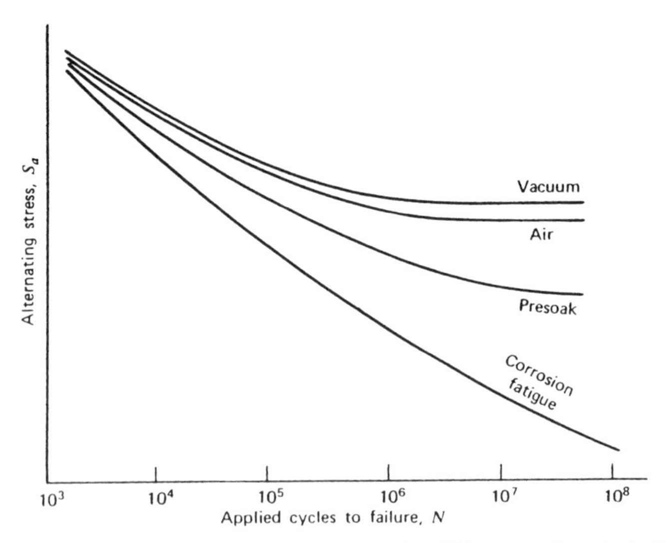

Fig. 10-19. Effect of various environments on the S/N curve of a steel. (From Fuchs and Stephens, 33.)

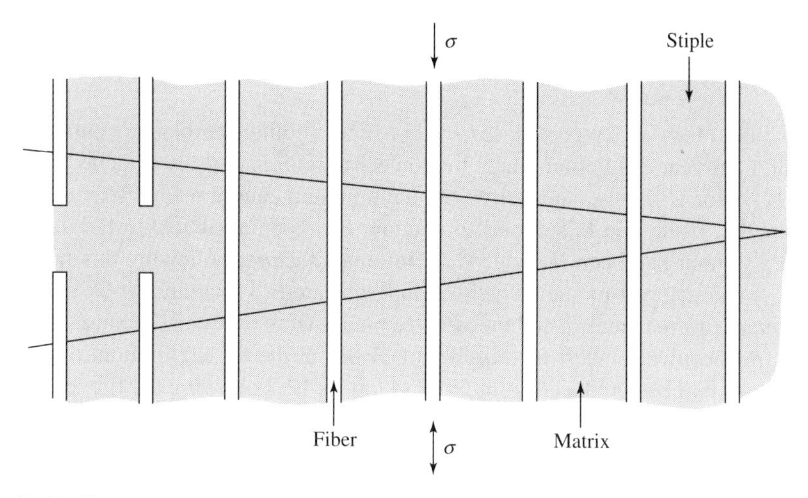

Fig. 10-20. The bridging of a crack in a composite material by unbroken fibers in the wake of the crack tip.

which are stress raisers that degrade the fatigue resistance of a material, and (*b*) it accelerates the crack propagation process. These corrosion processes are of particular concern with respect to aging aircraft. A review of the current understanding of pitting corrosion and corrosion fatigue in aluminum airframe alloys has been presented by Wei and Harlow (20).

XII. COMPOSITE MATERIALS

Properly made composites can be quite resistant to fatigue; for example, the fatigue strength at 10^7 cycles can be as high as 90% of the tensile strength of the composite. However, environmental degradation of polymer matrix composites as well as mechanical damage to fibers can occur to degrade the fatigue resistance. The fatigue crack propagation process can be reduced by the phenomenon known as bridging, which is illustrated in Fig. 10-20. Unbroken fibers in the wake of the crack tip serve to reduce the level of the crack tip opening displacement (CTOD), and thereby reduce the rate of crack propagation.

Composites can be reinforced also by particles and by short fibers known as whiskers, and, there is current interest in strengthening aluminum with either SiC particles or fibers for automotive engine applications. Another type of composite, known as Glare, is made of alternating layers of aluminum and fiberglass bonded together. Its key structural property is better fatigue resistance, for in fatigue critical areas it can take up to 20–25% higher loads than conventional aluminum. This composite will be used for fuselage components of a 555 passenger plane, the Airbus A380 (35).

XIII. CASE STUDY: HYDROELECTRIC GENERATOR POWER FAILURE (21)

The failure of an epoxy-coated, low-carbon steel cooling fan blade occurred in one unit of a ten-year-old Pelton-wheel hydroelectric station, and the detached blade cut into the stator windings, short circuited the unit, and caused a fire. Six months earlier, another blade had failed, and this failure had been attributed to fatigue initiating at a pit that had been introduced during manufacture. Following this failure, all blades were stripped of their coatings and subjected to examinaton. A number of small cracks were detected, and the affected blades were replaced. Because of a shortage of replacement blades, the number of blades at the top and bottom of the generator units had been reduced from 24 to as low as 12. Following the fire, cracks that had developed in a one-month period were found in two blades of another generating unit. These blades, as well as the failed blade that led to the fire, were adjacent to gaps that had been created where blades had been removed to provide them for another unit. Examination of the fracture surfaces showed that the cracks were high-cycle fatigue cracks, and it was clear that blade vibration was a major problem.

Both FEM analysis and an experimental nodal analysis were made, and both indicated the first natural frequency of the blades in the regions where the cracks initiated was 70 Hz. An estimate of the cyclic stress range was made based upon the observed striation spacings, and it was found to be high enough to indicate that a resonant condition must have been present. However, this condition must have been intermittent for the blades to have lasted even a month at 70 Hz. It was concluded that the cracking problem was caused by air-induced vibration brought about by the absence of some of the blades and associated with periodic vortex shedding. The problem was solved by installing complete new sets of fan blades that were heavier than the original blades, and that therefore had a higher natural frequency. Prior to installation, the blades were carefully inspected to be sure that no surface defects were present that might trigger fatigue cracking.

XIV. CASE STUDY: FATIGUE OF A B747 FUSE PIN

In 1992, a B747-200 freighter was six minutes out of the Amsterdam airport at an altitude of 2000 m (6500 feet) and climbing, when suddenly the right inboard engine and pylon separated from the wing. When this engine separated, it was under full power, and because of a gyroscopic moment, it moved into the path of the right outboard engine and struck it, causing the outboard engine also to separate from the wing. (These two engines fell into a lake and were not recovered for several months.) The aircraft became unstable, crashed into an apartment complex, and was destroyed. The four persons on board the aircraft, as well as approximately fifty people in the apartment complex, lost their lives. The 747 was 22 years old and had accumulated 45,746 flight hours and 10,107 flight cycles. 257 flight cycles had elapsed since the last ultrasonic inspection of critical parts involved in the accident.

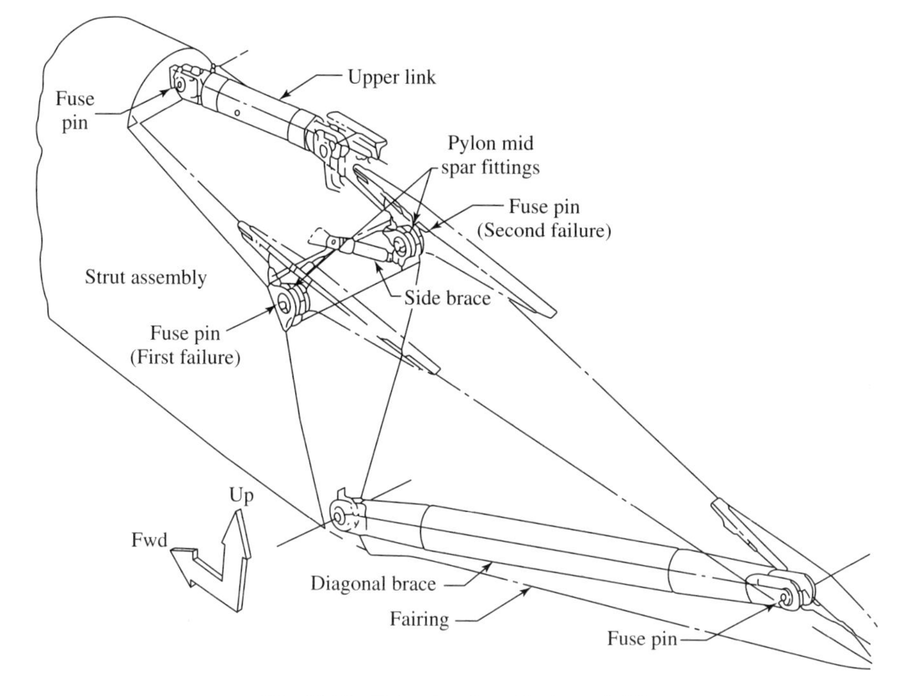

Fig. 10-21. Pylon to wing attachment, 37.

Each engine weighed 8600 lb (38.3 kN) and developed 50,000 lb (222 kN) of thrust at full power, and these forces, as well as another 20,000 lb (89 kN) of force due to the pitching of the engine during takeoff and climb, had to be resisted by the support structure. Figure 10-21 is a sketch of the pylon structure that attached the engine to the wing. The design incorporated 4340 low-alloy steel fuse pins, which were intended to shear off should the engine strike the ground, thereby allowing the engine to separate from the wing, thus minimizing the chance of a fire due to rupture of a fuel tank. A sketch of a fuse pin is shown in Fig. 10-22. Each fuse pin transmitted load from two lugs on the pylon to a clevis that was attached to the wing. The investigation showed the mid-spar fuse pins of the right inboard engine were critical in the sequence of events resulting in the accident. Unfortunately, the inboard mid-spar fuse pin was not recovered. However, because the outer lug on the inboard pylon fitting had broken under combined bending and tensile loading, it was clear that the inboard side of the mid-span fuse pin had failed first due to fatigue, and that, as a result, the load on the outboard mid-spar fuse pin had been greatly increased. The mid-portion of the outboard fuse pin was recovered, and was still within the clevis. On the inboard side, it had failed in shear, but on the outboard side, it had failed by fatigue. The main fatigue crack had initiated at multiple sites on the inner surface of the fuse pin at a reduced section where machining grooves were present, and had grown in a radial direction. Examination of the fatigue frac-

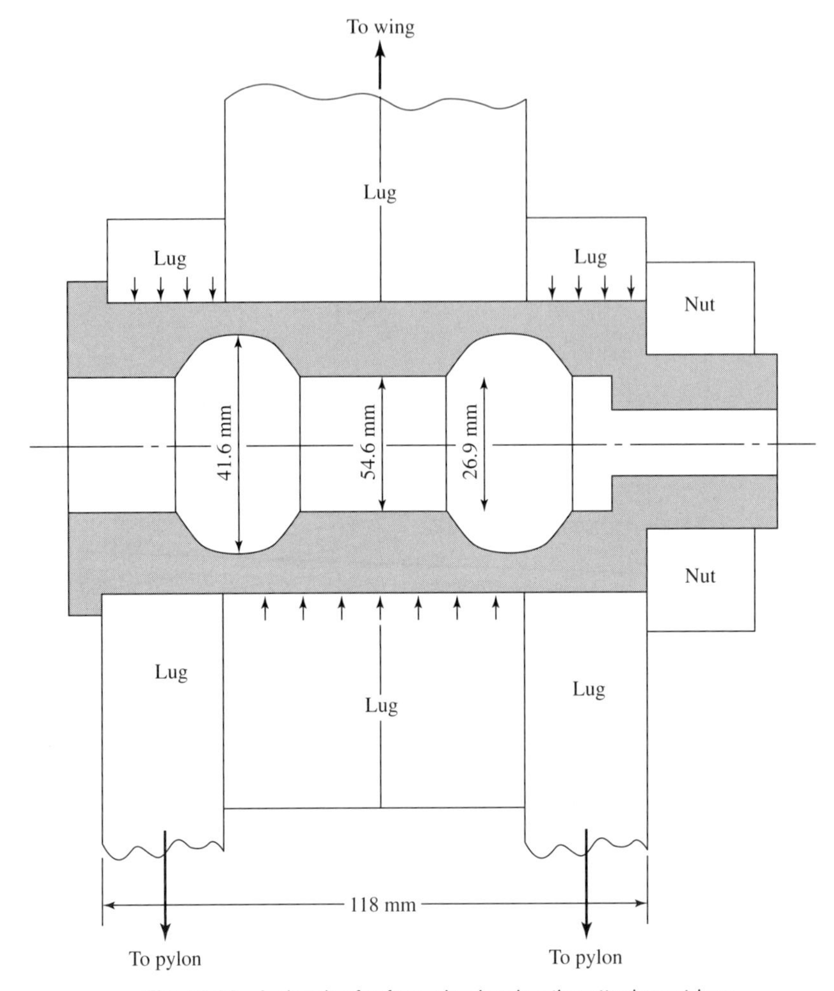

Fig. 10-22. A sketch of a fuse pin showing the attachment lugs.

ture surface revealed the presence of striations spaced at approximately 1–3 μm intervals. An example of these striations is shown in Fig. 10-23. Approximately 1900 major striations were present, and at a depth of 2.2 mm, they were spaced at an interval of 1.0 μm. The striation spacing indicated a fatigue crack growth rate of $1–3 \times 10^{-3}$ mm/cycle, which in 4340 steel corresponds to a ΔK of 50–70 $\text{MPa}\sqrt{\text{m}}$ (22).

A question arose as to whether the major striations represented cycle-by-cycle events or flight-by-flight events. If they were flight-by-flight indications, the fatigue crack should have been detected by the ultrasonic method at the time of the last inspection, 257 flight cycles before the crash. If, on the other hand, they were cycle-by-cycle striations, they must have been created by engine oscillations after failure of the inboard fuse pin and just before the inboard engine separated. Detailed ex-

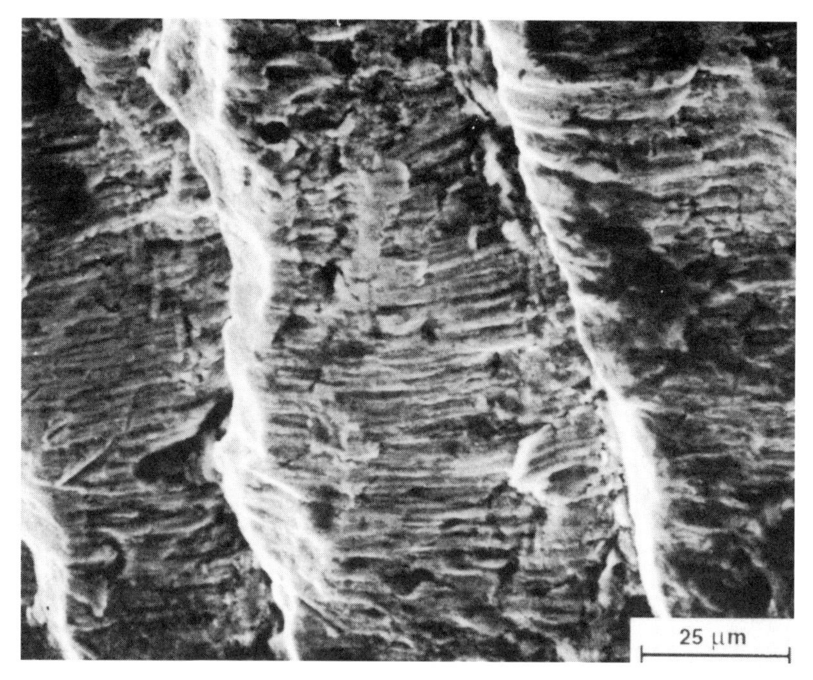

Fig. 10-23. Fatigue striations found on failed outboard fuse pin. (From Oldersma and Wanhill, 22.)

amination of the striations revealed the presence of minor, striationlike markings between striations. In the flight-by-flight interpretation, these striationlike markings were indications of in-flight loading events due to gust and maneuver loads. However, the interpretation of these minor markings was not straightforward, for in auxiliary tests under constant load amplitude conditions, it was observed that, at growth rates of the order of 1–3 μm per cycle, minor markings were created between striations. The Netherlands Aviation Safety Board (23) found that no firm conclusion could be drawn as to whether or not a fatigue crack in the outboard mid-spar fuse pin should have been detectable at the last ultrasonic inspection.

A stress intensity factor can be calculated to obtain an estimate of the range of stress that was actually applied. Since the length-to-depth ratio of the crack was large, a single-edge notched configuration in pure bending is assumed, and at a depth of 2.2 mm and a total thickness of 7 mm, ΔK is equal to 0.03 $\Delta\sigma$ (24). For a nominal load of 60 kips per fuse pin, a maximum stress of 400 MPa has been calculated (25). At a crack depth of 2.2 mm, the calculated value of ΔK is then 12.0 MPa$\sqrt{\text{m}}$. This is too low a value to create an observable striation in steel under the flight-by-flight conditions that existed prior to the failure of the inboard fuse pin. Therefore, the stress intensity must have been higher. It could have been higher due to higher load amplitudes and also if the crack tip deformation had been elastic-plastic rather that linear-elastic. To account for elastic-plastic behavior, Irwin (26) has suggested increasing the actual crack size by one-half of the plastic zone size (PZS). The size of the plane-stress plastic zone, as given by Dugdale (27), is

$$PZS_{pl\sigma} = \left[\sec\left(\frac{\pi}{2}\frac{\sigma}{\sigma_y}\right) - 1\right]a. \tag{10-23}$$

In plane strain, the expression for the plastic zone size using the Irwin plastic constraint factor is

$$PZS_{pl\varepsilon} = \left[\sec\left(\frac{\pi}{2}\frac{\sigma}{\sigma_y}\right) - 1\right]\frac{a}{3}. \tag{10-24}$$

The estimated tensile strength of the 4340 fuse-pin steel is 860 MPa, based upon a Brinell Hardness Number (BHN) of 240, and the yield strength is estimated to be 745 MPa (28). If one-half of the plastic zone size is added to the crack length, the expression for ΔK_{E-P}, the elastic-plastic value of ΔK for plane-strain conditions at a crack depth of 2.2 mm, becomes

$$\Delta K_{E-P(pl\varepsilon)} = 1.11\ \Delta\sigma\sqrt{\frac{\pi}{6}a\left(\sec\frac{\pi}{2}\frac{\sigma}{\sigma_y} + 5\right)}\ \text{MPa}\sqrt{\text{m}}, \tag{10-25}$$

which is equal to 12.2 MPa$\sqrt{\text{m}}$.

The elastic-plastic value results in a small increase in ΔK over the linear-elastic fracture mechanics (LEFM) value, but not enough to create well-defined striations for which a ΔK of at least 30 MPa$\sqrt{\text{m}}$ is needed. Therefore, the observed striations, even on a flight-by-flight basis, could not have been created at the nominal stress amplitude that existed prior to the failure of the inboard fuse pin. For a striation spacing of 1 μm at a depth of 2.2 mm and a corresponding ΔK of 50 MPa$\sqrt{\text{m}}$, the required $\Delta\sigma$, according to Eq. 10-25, for plane strain would be 728 MPa. This value is much higher than the nominal 400 MPa that existed prior to the fatigue failure of the inboard fuse pin. The calculations, therefore, indicate fatigue crack growth in the outboard fuse pin had occurred on a cycle-by-cycle basis under the high load amplitudes that developed after the inboard fuse pin had failed.

As a result of this accident, several changes were made in the pylon-to-wing connections. The original type of fuse pin was replaced by a stainless steel fuse pin that did not have thin-walled regions. In addition, the pylon fittings and bracing were also strengthened.

XV. THERMAL-MECHANICAL FATIGUE

A thermal-mechanical loading cycle was described in Chapter 7. As with other types of loading, the repeated application of such a cycle can result in fatigue. A traditional item of concern has been the thermal fatigue cycle imposed upon the disks in the hot stages of a jet aircraft engine during takeoff and landing. Through a combination of design, testing, and in-service inspection, the modern disks now serve reliably, and for a greater number of flight cycles than in years past, before their fatigue lifetime is expended.

XVI. CAVITATION

Cavitation is a form of liquid erosion that involves the formation and collapse of bubbles within the liquid. Cavitation damage has been observed on ship propellers and hydrofoils, in hydraulic pumps and structures, and infrequently in sliding bearings (29). Examples of cavitation damage in steel pump components are shown in Figs. 10-24*a* and *b*. When the local pressure in a liquid is reduced sufficiently, gas-filled bubble can nucleate and grow. If these bubbles subsequently pass into a region of higher pressure, they can implode in a few milliseconds and form microjets of liquid that attain velocities of from 100 to 500 m/sec. A very small fraction of these jets impinges upon the adjacent surface, and a form of Hertzian contact occurs, leading to the development of subsurface shear stresses sufficiently high to nucleate fatigue cracks. These fatigue cracks then grow to the surface, allowing small particles to break away, and thereby creating observable cavitation damage.

The resistance of a metal to cavitation erosion is influenced by several factors, among which are: hardness, the strain energy to fracture, and the corrosion-fatigue

(*a*)

Fig. 10-24. Examples of cavitation damage in pump components.

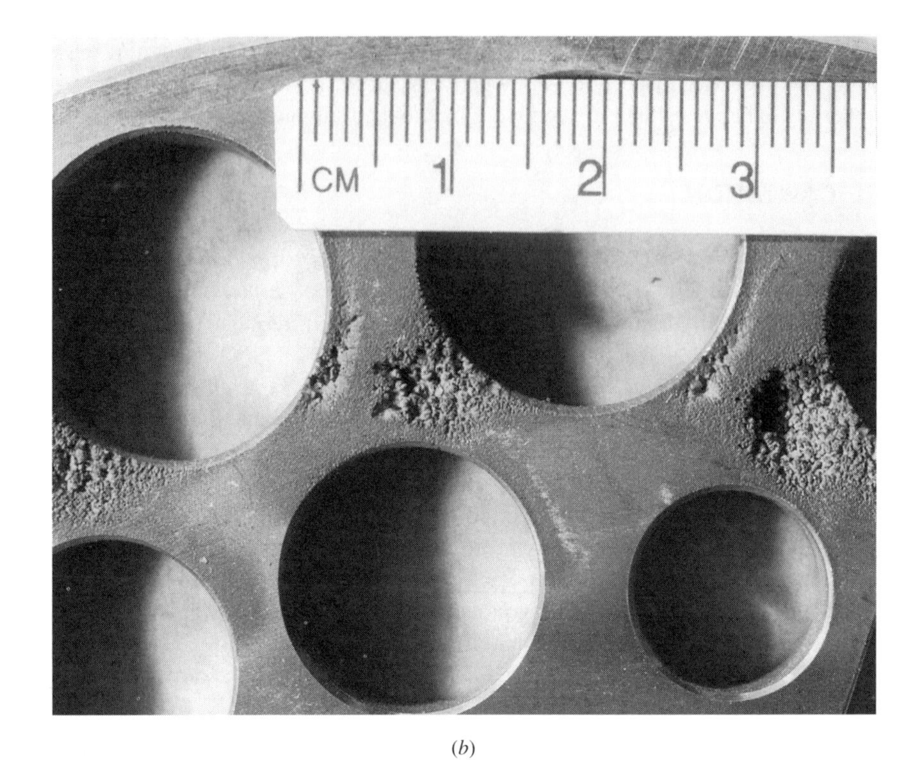

(*b*)

Fig. 10-24. *(Continued)*

strength. Surface treatments such as shot peening are not very effective because they duplicate the processes that occur during the crack nucleation period.

XVII. CASE STUDY: AIRCRAFT GAS TURBINES

Modern gas turbine engines for aircraft feature a significant reduction in the number of engine parts, low operating and maintenance costs, and high reliability, with thrust values per engine varying from 10,000 to over 100,000 lb. For the 100-seat Airbus A318, thrust is in the range of 16,000–24,000 lb, whereas for the long-range B777, thrust will exceed 100,000 lb. The PW6000 engine proposed for the A318 will have an initial on-wing engine time of 15,000 hours, about 6–8 years after entering service, at which time 900 high-temperature turbine blades will be replaced. The life-limiting parts, such as the turbine disks, are designed for 25,000 cycles, meaning that these parts should last until the second overhaul (30).

The PW6000 engine relies upon proven materials, simplified mechanical systems, and advanced aerodynamics. The 56.5-inch-diameter first-stage fan utilizes shroudless (no mid-span supports), solid titanium fan blades, together with a titanium fan case. The inlet vanes in the low-pressure compressor are fabricated from titanium for bird-strike resistance, and the unit's single-piece drum rotor is also forged from titanium. Integrally bladed disks (blisks) are not used in order to facilitate the replacement of damaged blades in the field. Blades in the first two stages of the high-pressure compressor are made of titanium, while the blades in the remaining stages are made from a nickel-based alloy. The unit's disks are also made of a nickel-based alloy. The single-piece turbine exhaust case is made of Howmet C263 alloy. To cope with debris ingestion and erosion, a splitter spacing arrangement is used between fan and low-pressure compressor so that the debris, such as sand, brake dust, and tire remains, can be centrifuged out of the engine's flow path and into the fan duct. This centrifuging also helps the engine shed rain and hail, protecting against flameouts. The addition of a fourth stage to the low-pressure compressor allows for supercharging, which results in a temperature drop of 30°F in the high-pressure turbine. This translates into a significant increase in component life.

Today, aircraft turbines are remarkably durable. For example, fifty years ago an aviation piston engine manufacturer could expect to sell 20–30 times the original cost of the engine in after-market parts. With the advent of the jet engine, this after-market figure dropped to 3–5 times. Despite this marked improvement, fatigue still remains a cause for concern (31). In gas turbine engines, LCF failures are usually caused by large cyclic excursions between zero stress and the operating stress. For example, a rotor disk would experience one LCF cycle as it accelerates from rest to operating speed, then decelerates to shutdown. In the mid-1970s, manufacturers and the U.S. Air Force (USAF) initiated development of fracture-mechanics-based damage-tolerant design criteria, aimed at reducing the risk of unexpected failure from LCF. These criteria were eventually embodied in the Engine Structural Integrity Program. LCF assessments are comprised of an LCF-safe life determination, which is the life to expected crack initiation, and a fracture-mechanics-based assessment to determine the remaining life to failure from an assumed crack, defect, or inspection limit. Both assessments are made and are used to determine an appropriate safety inspection interval, if required, and an eventual retirement life for the component.

On the other hand, high cycle fatigue (HCF) failures generally occur due to numerous, smaller cyclic excursions between two or more stress states. These excursions result from the following drivers:

(a) *Aerodynamic Excitation:* Caused by engine flow path pressure perturbations affecting primarily blades and vanes.
(b) *Mechanical Vibration:* Caused by rotor imbalance that affects external components, plumbing, and static structures; and rub, affecting blade tips and gas path seals.
(c) *Airfoil Flutter:* Caused by aerodynamic instability, affecting blades.
(d) *Acoustic Fatigue:* Affecting mostly sheet metal components in the combustor, nozzle, and augmentor.

HCF has been identified as the cause of 24% of the maintenance problems in aircraft turbine engines, and is the Air Force's number one readiness issue. Nearly half of the affected components are blades and vanes, and since these components can accumulate up to 10^{10} HCF cycles, design methods, criteria, and prediction approaches have been improved to address HCF. However, dramatic increases in engine performance and concurrent reductions in weight have driven up temperatures, stresses, and individual stage loading, and HCF problems have persisted. Although the rate of failure has remained steady, the percentage of HCF problems has actually increased.

Design for HCF tends to be based on experience and safety factors. The Goodman diagram is a basic design tool. Failure is expected to occur for combinations of mean stress and stress amplitude that lie above the line on this diagram corresponding to a given fatigue lifetime. As the number of cycles to failure increases, the allowable stress amplitude correspondingly decreases. The Goodman diagram approach as currently employed is highly empirical and does not adequately address in-service damage or interaction of HCF with other damage modes (fretting, LCF, etc.). Ultimately, HCF problems result from repeated stresses that are higher than the material's ability to resist them, as indicated by the reliability curves shown in Fig. 1-1. The intersection of the tails of these two curves indicates instances where the operating stresses are greater than the material strength—therefore where failures occur. The HCF program is intended to shift these curves away from each other by altering mean values and reducing standard deviations.

The Palmgen-Miner rule is used to estimate the number of cycles to failure under variable amplitude loading conditions under which the components may experience both HCF and LCF loading. Experience has shown that this summation often depends upon load sequence, and the value of the summation can be greater or less than unity. However, the HCF alternating stresses in turbine blades and vanes are often difficult to predict or measure, and as a result, there is significant uncertainty in the actual vibratory stress levels experienced by these components. This is why extensive testing of fatigue prone components is done in the aircraft indus-try to evaluate the influence of edges, holes and attachments, shot peening, and coatings.

Foreign object damage (FOD) is common for fan and compressor blades, with the relatively sharp airfoil leading edges being most susceptible. The size of the damaged areas varies from near microscopic to large tears, dents, or gouges. To assess the effect of FOD, the damage is simulated by V-notches in laboratory test programs, and airfoils are designed assuming reduced HCF capability in FOD-susceptible areas. Blade and disk attachment surfaces are also susceptible to fretting and/or galling surface damage. Both fretting and galling result from relative motion of surfaces in contact with one another; and both are characterized by surface debris, wear, and reduced fatigue capability.

A USAF committee concluded that most titanium failures were attributable to HCF and recommended that the development of a threshold fracture-mechanics approach be undertaken as an alternative to the Goodman diagram approach. An improved HCF design system would consist of:

(a) A crack initiation analysis, improved over the current Goodman approach, to assess basic HCF capabilities. The intent would be to improve prediction approaches, and reduce the level of empiricism. Factors to be considered include: the effects of intrinsic defects, a local stress-strain approach, cyclic hardening and softening, maximum stresses, strain ranges, inelastic stress-strains, tensile and shear failures, HCF/LCF interaction, multiaxial stresses, processing effects, materials anisotropy, and material property variability.

(b) A fracture-mechanics-based system capable of addressing ΔK_{th} behavior, small crack effects, crack closure, and HCF/LCF interaction.

Other goals include (a) the development of nonintrusive methods of accurately measuring the distribution and variation of dynamic stresses and pressures within gas turbine components, and (b) the development of "real-time" measurement capabilities to insure safe operation in the field.

XVIII. SUMMARY

This chapter has provided a broad coverage of the field of fatigue. The main topics were design procedures, including consideration of statistical effects, and factors influencing the initiation and propagation of fatigue cracks. One case study illustrated the use of fracture mechanics in a failure analysis. Another case study dealt with design considerations for gas turbine engines.

XIX. FOR FURTHER READING

S. Suresh, Fatigue of Materials, 2nd ed., Cambridge University Press, Cambridge, UK, 1998.

Fatigue and Fracture, ASM Handbook, vol. 19, ASM, Materials Park, OH, 1996.

J. A. Bannantine, J. C. Comer, and J. L. Hardrock. Fundamentals of Metal Fatigue Analysis, Prentice Hall, Englewood Cliffs, NJ, 1990.

REFERENCES

(1) G. M. Sinclair and T. J. Dolan, Trans. ASME, vol. 75, 1953, p. 867.

(2) Y. Murakami, T. Toriyama, Y. Koyasu, and S. Nishida. Effects of Chemical Composition of Non-Metallic Inclusions on Fatigue Strength of High Strength Steels, J. Iron and Steel Inst. Japan, vol. 79, 1993, pp. 60–66.

(3) H. Neuber, Trans. ASME, J.Appl. Mechs., vol. 28, 1961, p. 544.

(4) A. J. McEvily and K. Minakawa, On Crack Closure and the Notch Size Effect in Fatigue, Eng. Fracture Mechanics, vol. 28, 1988, pp. 519–527.

(5) H. Neuber, Theory of Notch Stresses, Edwards, London, 1946.

(6) Design Handbook for the Fatigue Strength of Metals, Japan Soc. Mechanical Eng. (JSME), vol. 1, 1982, p. 125.

(7) S.-I. Nishida, Failure Analysis in Engineering Applications, Butterworth-Heinemann, Oxford, UK, 1986.

(8) Aviation Week and Space Technology, Sept. 13, 1999.

(9) P. Albrecht and W. Wright, in Fracture Mechanics: Applications and Challenges, ed. by M. Fuentes et al., ESIS Publication 26, Elsevier, Oxford, UK, 2000, pp. 211–234.

(10) P. J. E. Forsyth and D. Ryder, Metallurgia, vol. 63, 1961, pp. 117–124.

(11) W. Elber, Fatigue Crack Closure under Cyclic Tension, Eng. Fracture Mechs., vol. 2, 1970, pp. 37–45.

(12) R. C. Juvinall, Engineering Considerations of Stress, Strain, and Strength, McGraw-Hill, New York, 1967.

(13) C. Ruiz and D. Nowell, in Fracture Mechanics: Applications and Challenges, ed. by M. Fuentes et al., ESIS Publication 26, Elsevier, Oxford, UK, 2000, pp. 73–95.

(14) J. O. Almen and P. H. Black, Residual Stresses and Fatigue in Metals, McGraw-Hill, New York, 1963.

(15) T. Sakai, M. Takeda, N. Tanaka, and N. Oguma, Proc. 25th Symp. on Fatigue, Japan Soc. Material Sci., 2000, Kyoto, pp. 191–194.

(16) Y. Murakami, N. N. Yokohama, and K. Takai, Proc. 25th Symp. on Fatigue, Japan Soc. Material Sci., 2000, Kyoto, pp. 223–226.

(17) J. Schivje, in Fatigue '96, vol. II, ed. by G. Lütjering and H. Nowak, Pergamon, Oxford, UK, 1996, pp. 1149–64.

(18) M. Fujihara, Y. Kondo, and T. Hattori, Japan Soc. Material Sci., vol. 40, no. 453, 1991, pp. 712–717.

(19) K. Furukawa, Proc. 25th Symposium on Fatigue, J. Soc. Material Sci., 2000, Kyoto, pp. 71–73.

(20) R. P. Wei and D. G. Harlow, in Fatigue '99, vol. 4, ed. by X. R. Wu and Z. G. Wang, EMAS, West Midlands, UK, 1999, pp. 2197–2204.

(21) I. Le May and H. C. Furtado, in Technol., Law and Insurance, 1999, vol. 4, pp. 111–119.

(22) A. Oldersma and R. J. H. Wanhill, Netherlands National Aerospace Laboratory, NLR Contract Report 93030 C, Amsterdam, Netherlands, 1993.

(23) Netherlands Aviation Safety Board, 1994, Aircraft Accident Report 92-11, El Al Flight 1862, Boeing 747-258F 4X-AXG, Bijlmermeer, Amsterdam, October 4, 1992, Hoofddorp, Netherlands, 1994.

(24) H. Tada, P. Paris, and G. Irwin, The Stress Analysis of Cracks Handbook, Del Research Corp., Hellertown, PA, 1973.

(25) E. Zahavi, Fatigue Design, CRC Press, New York, 1996.

(26) G. R. Irwin, Fracture, Handbuch der Physik VI, ed. by S. Flugge, Springer, Berlin, Germany, 1958, pp. 551–590.

(27) D. S. Dugdale, J. Mechs. and Physics of Solids, vol. 8, 1960, pp. 557–594.

(28) ASM Metals Handbook, vol. 1, 9th ed., ASM, Materials Park, OH, 1978, p. 426.

(29) F. G. Hammitt and F. J. Heymann, in ASM Metals Handbook, vol. 10, 8th ed., Failure Analysis and Prevention, ASM, Materials Park, OH, 1975, p. 160.

(30) Aviation Week and Space Technology, June 7, 1999, p. 43.

(31) B. A. Cowles, High Cycle Fatigue in Aircraft Gas Turbines—An Industry Perspective, Int. J. Fracture, vol. 80, 1996, pp. 147–163.

(32) J. Schivje, in Fatigue Crack Propagation, ASTM STP 415, Conshohocken, PA, 1967, p. 533.

(33) H. O. Fuchs and R. I. Stephens, Metal Fatigue in Engineering, Wiley, New York, 1980.

(34) Aviation Week and Space Technology, June 4, 2001, p. 52.

(35) M. A. Dornheim, Aviation Week and Space Tech., 2001, vol. 154, No. 25, pp. 126–128.

(36) A. S. Tetelman and A. J. McEvily. Fracture of Structural Materials, Wiley, NY, 1967.

11

Defects

I. INTRODUCTION

Failures of components have often been triggered by defects that were introduced during the manufacturing process. Because of the prevalence of defects, critical parts are inspected in the attempt to prevent defective parts from entering service, sometimes without success. For example, an undetected defect in a Ti-6Al-4V turbine disk led to the crash of a DC-10 aircraft. Since the resolution of any inspection method is limited, it is prudent to consider that defects of a size equal to the limit of resolution are present in assessing the integrity of a structure. Some defects are unavoidable, as in the case of welds, and allowance must be made for such defects to obtain reliable in-service performance. In this chapter, some of the more common types of defects encountered in welds, castings, and rolled or forged products are discussed.

II. WELD DEFECTS

A. General Characteristics

Certain defects, such as the porosity caused by the entrapment of gas evolved during weld metal solidification and an irregular shape, are inherent to the welding process. Figure 11-1 depicts a fillet weld with the toe of the weld being the junction between the face of the weld and the base metal. Irregularities in shape at the toe of a fillet weld act as stress raisers. In addition, irregular surface ripples are formed in the wake of a weld pass, and are concave with respect to the direction of the weld pass, Fig. 11-2. These surface ripples, as well as those formed at the start and stop

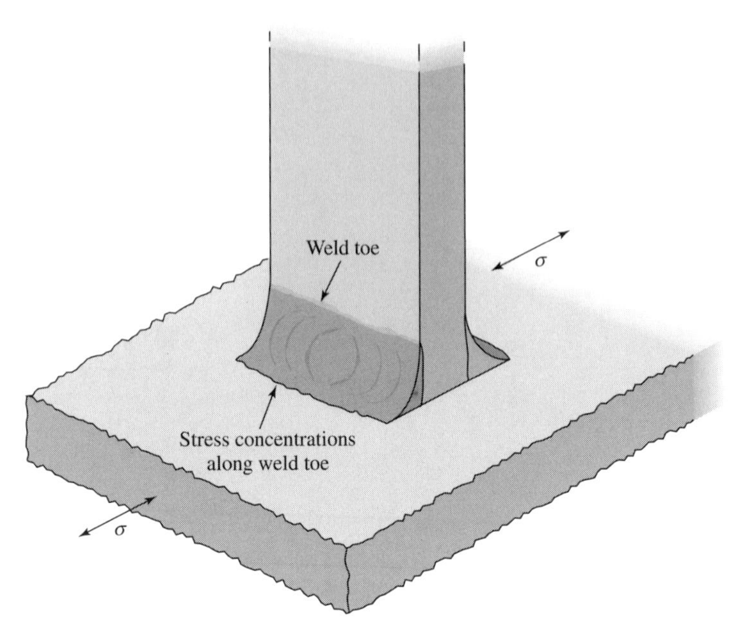

Fig. 11-1. Fillet weld with weld toes indicated. (After the Welding Institute, 6.)

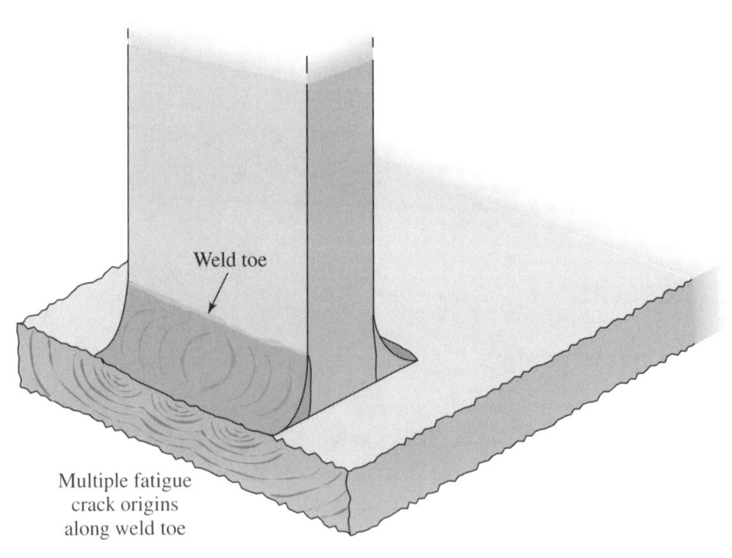

Fig. 11-2. Failure of plate at nonload carrying transverse fillet welds. (After The Welding Institute, 6.)

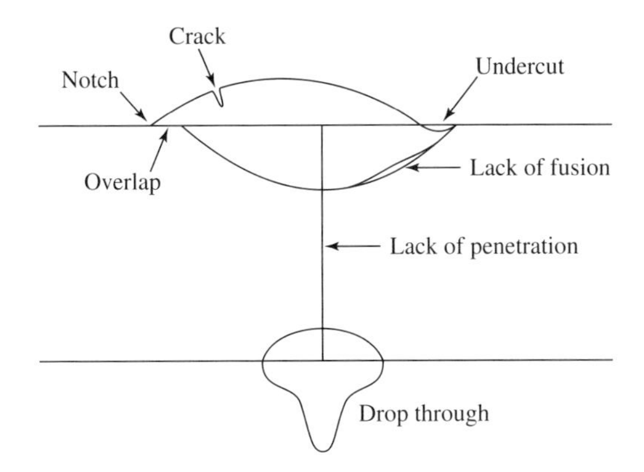

Fig. 11-3. Some typical weld defects.

positions of the welding process, are geometric discontinuities that promote the initiation of fatigue cracks. Although the fatigue properties of many welded structures are strongly influenced by the presence of such defects, under monotonic loading these defects are much less of a problem. No failures due to porosity have been reported in welds where the porosity was within acceptable code requirements (1).

Defects in fillet welds also result from improper welding procedures. A number of common weld defects are shown in Fig. 11-3. These include the undercut, which is a groove melted into the base metal adjacent to the toe of a weld and left unfilled, and the cold lap, which is a protrusion of weld metal beyond the bond at the toe of weld. Lack of penetration indicates that weld metal did not completely fill the gap between the two plates being welded. Lack of fusion can occur if the base metal does not melt, or in a multipass weld, if there is a lack of melting of the previous pass between passes. One type of defect is due to a shoddy work practice known as "slugging a weld." In making a long fillet weld, welders have been known to place a welding rod along the intersection of the two plates being welded and then to cover it over with weld metal, so that to the eye it would appear to be a sound weld. Incredibly, such defects have been found even on booster rockets for space launches.

Weld reinforcement is the excess of weld metal above the plane of the plates being joined by a butt weld. This is actually a misleading term, since the excess creates a stress concentration that lowers the fatigue resistance. Underfill, also referred to as weld cavity or weld crater, denotes the opposite of reinforcement. The fatigue strength of a transverse butt weld is dependent upon the weld profile. The highest fatigue strength of a butt weld is obtained when the weld is flush with the plate surface, that is, when it has zero reinforcement.

Weld porosity in the case of a fillet weld is not a great concern because the geometrical factors discussed above usually exert a stronger influence on fatigue behavior than does the porosity. On the other hand, porosity is of greater concern in butt welds subject to cyclic loading where the degree of porosity influences the fatigue life. To guard against failures due to porosity, radiographic inspection using a radioactive element, such as cobalt 60, together with film is used (see Chapter 13).

Radiographic inspection standards exist to aid in the assessment of the degree of porosity and to determine in a given application whether a weld is acceptable or not.

Some of the other defects associated with welds are:

(a) *Mismatch:* The plates that are being butt welded are offset by transverse or angular misalignment.

(b) *Shrinkage:* Shrinkage of the weld and adjacent area during cooling after welding can result in cracking of the base metal, as well as crater and fusion line cracking in the weld metal.

(c) *Buried Slag Inclusions:* Slag inclusions result from the dissolution of the welding flux in the weld metal. In the case of multipass welding, the slag is normally on the outer portion of a solidified weld and is brushed off. However, if not completely removed, slag inclusions can develop between passes. These defects are generally less harmful than the design details that are associated with fillet and butt welds.

(d) *Weld Spatter:* Weld spatter refers to the metal particles expelled during arc or gas welding. Weld spatter can be particularly dangerous in the case of quenched and tempered steels as it can lead to the formation of untempered martensite in the base metal as well as to surface irregularities, a circumstance that can facilitate the nucleation of fatigue cracks.

In addition, the properties of the heat-affected zone (HAZ), that portion of the base metal that was not melted during the welding operation but whose microstructure and physical properties were altered by exposure to elevated temperature, are also of concern with respect to mechanical behavior.

B. Effect of Cooling Rate

Alloying elements such as carbon, manganese, chromium, and nickel all increase hardenability and decrease the critical cooling rate because they slow down the diffusive decomposition of austenite, but because phases such as hard, untempered martensite may form, these elements also increase the likelihood that weld cracking may occur on cooling. The effect of alloying elements on the tendency for cracks to form on cooling can be assessed from empirical formulas that express the alloy content in terms of a carbon equivalent, CE. An example of such a relationship is

$$\mathrm{CE} = \mathrm{C} + \frac{\mathrm{Mn}}{6} + \frac{\mathrm{Cr} + \mathrm{Mo} + \mathrm{V}}{5} + \frac{\mathrm{Ni} + \mathrm{Cu}}{15}, \tag{11-1}$$

where the concentrations of each element are given in wt %. When the CE exceeds 0.45, the cooling rate must be controlled very carefully to produce sound welds. Since thermal stresses increase both with increase in cooling rate and increase in CE, cracking becomes more likely since the toughness, that is, the resistance to cracking, decreases as the CE increases. To minimize the tendency for cracking, preheating of the weld region just prior to welding is done to reduce the cooling rate after welding, with the specified level of preheat increasing with increase in CE.

C. Laminar Tearing

When a tension-bearing member is fillet welded to a rolled steel plate, a tensile stress is developed normal to the plane of the plate. In this direction the fracture resistance is less than in the in-plane directions because of the anisotropy introduced during rolling. Manganese sulfide particles and other inclusions will be oriented perpendicular to the tensile stress. If this stress is high enough, a type of in-plane fracture, known as laminar tearing, can occur. In critical structures such as submarines, the hull steels are processed in a way so as to reduce this anisotropy in fracture resistance.

III. CASE STUDY: WELDING DEFECT

A. The Alexander Kielland Accident, March 27, 1980 (2)

The Alexander Kielland (AK) was built as a mobile, semisubmersible oil drilling platform of the pentagon type, where five cylindrical columns support the platform; see Fig. 11-4. A major design objective was to minimize platform movement in heavy seas. The AK was built in France, and delivered in 1976, but used as an accommodation platform to house the workers in North Sea offshore-drilling operations rather than as a drilling platform. It originally had a capacity of 80 beds, but over time this was increased to 384.

In the nine months prior to the accident, the AK was anchored close to the drilling platform "Edda 2/7 C." To each of the columns A, B, D, E two anchor wires were attached. Column C was not anchored. Connection between AK and Edda was maintained by a moveable walkway. In bad weather the walkway was lifted on board the AK, and AK was shifted away from Edda. This was done by slackening the anchor wires from the B and D columns and tightening the anchor wires from the A and E columns.

On March 27, 1980, visibility was poor, and because the wind speed was 16–20 meters per second (45 mph), and the wave height was 6–8 meters, the AK was shifted away from Edda by 1750 hours. At about one-half hour later, column brace D-6 failed and quickly other braces failed, allowing column D to break away from the AK. The platform heeled almost at once, and when the angle of heel reached 35° a message "Mayday, Mayday. Kielland is capsizing" was sent out at about 1829.

On board the AK, a strong impact was experienced a few minutes before 1830 hours, followed by some kind of trembling. This seemed to be a wave impact, typical of bad weather. Immediately after the first impact, however, there was another, again accompanied by shaking of the platform, accompanied by sounds of metal tearing. The platform started to heel further and the situation became critical. The angle of heel almost stabilized at 30–35°, but then continued slowly to increase past

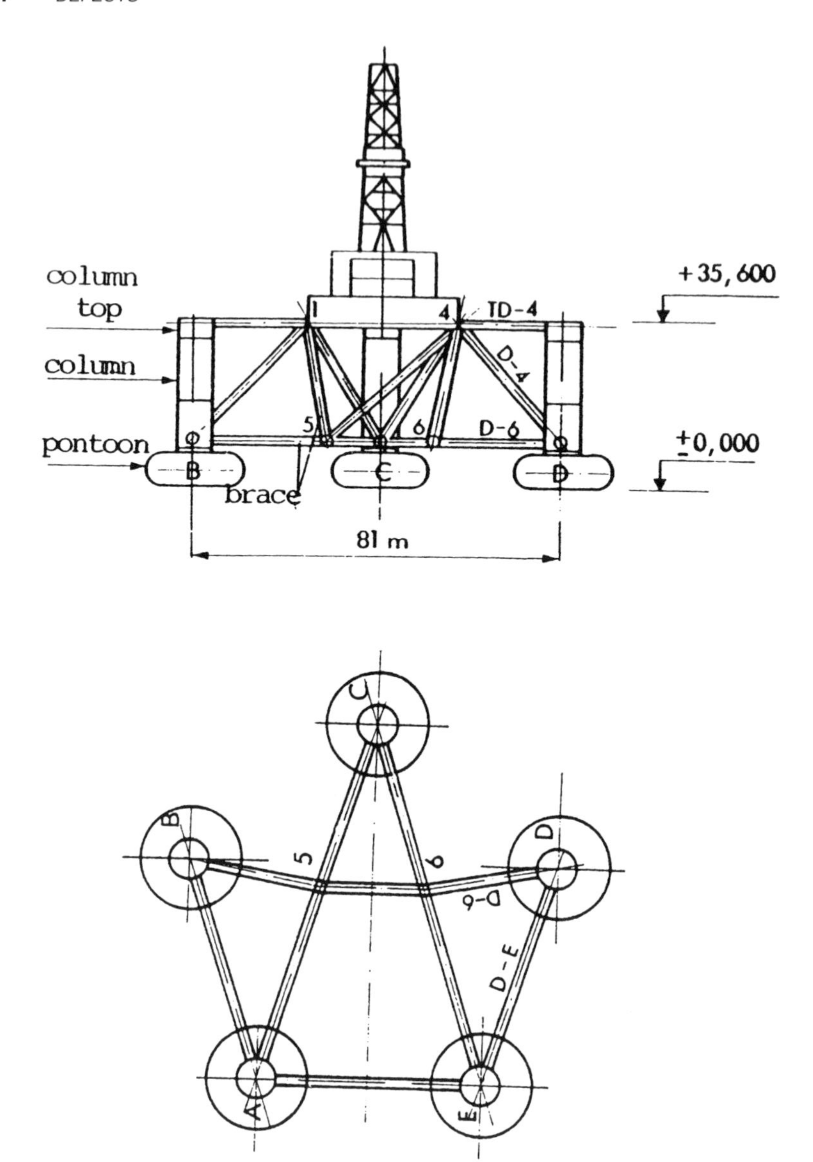

Fig. 11-4. The AK pentagon structure. (From The Alexander Kielland Accident, 3.)

35° until it capsized. From Edda it appeared as if only the wire from column B had kept the platform from capsizing, but when this wire broke, the platform overturned at 1853 hours and floated upside down in the sea.

As mentioned, horizontal brace D-6 broke first; see Fig.11-5. In this brace an opening had been cut out and a hydrophone for control of positioning had been

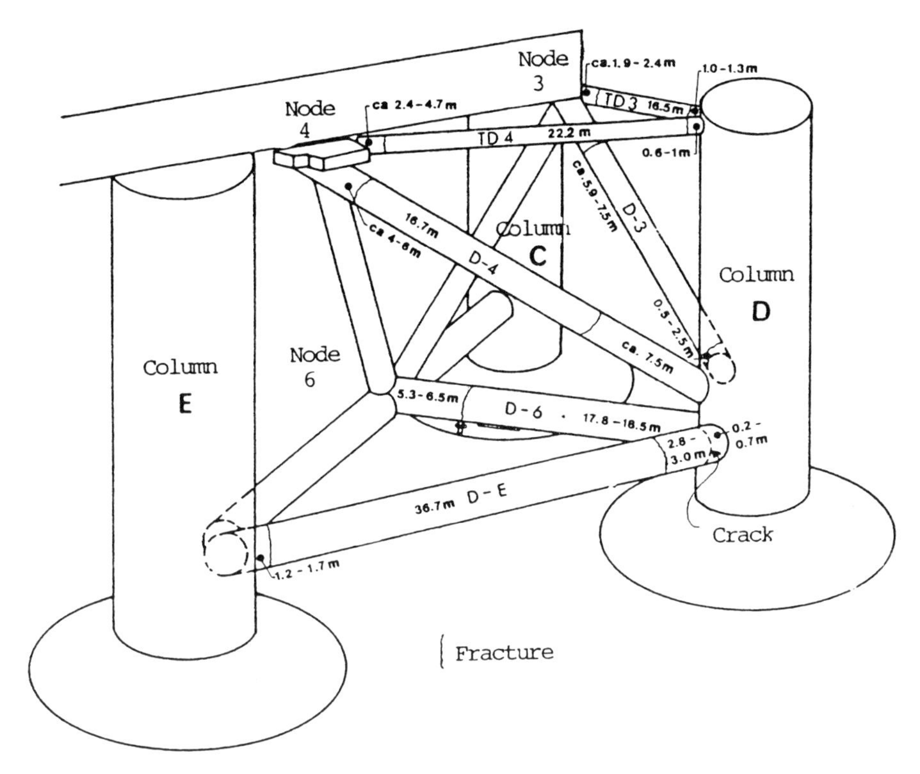

Fig. 11-5. Location of the fractures in the braces connected to column D. The fracture in brace D-6, the first to fail, was located at a hydrophone which had been welded to the brace. Approximate lengths of structural components are given in meters. (From The Alexander Kielland Accident, 3.)

welded in. The hydrophone was considered by the builder as equipment rather than as a load-bearing component of the structure. For this reason, no complete strength evaluation had been made of the welding-in of the hydrophone.

There were 212 men aboard the AK at the time of the accident. Most tried to reach the platform's highest point, the B column. Several were unable to don life vests as there was no time to go to the cabins to get them. The permanent crew on board, as well as the employees of some companies, had survivor suits. Eight were able to don these, but only four survived. There were seven covered lifeboats, each of which could take fifty men. An attempt to launch five lifeboats was made. Three of these lifeboats failed to release properly and were smashed against the platform and crushed. One lifeboat containing 26 men got away and the men were rescued by Norwegian helicopters. Another lifeboat containing 14 people failed to launch properly, but came to the surface when the platform capsized, and then picked up 19 survivors from the water. All 33 were rescued. A total of 16 men were saved when they swam to rafts that were self-launched at the time of capsizing or thrown from Edda. Seven survivors were picked up by supply boats in the area. A further seven survivors were brought on board Edda by a personnel basket. Of the 212 on board, 89 were saved and 123 men lost their lives.

B. Accident Investigation

The main structural support elements of the pentagon platform consisted of the five columns supported by pontoons, designated A through E. The existing rules required that the rig remain floating in a stable position after damage that resulted in flooding of two adjacent tanks in a column. No consideration was given to the possible loss of a column.

The pontoons were circular, 22 m in diameter and 8.5 m in height. Each pontoon supported a column with a diameter of 8.5 m. The columns were 27.1 m in height and were connected with a set of horizontal and angular braces made of high-strength steel. Ten anchor lines, two from each column, were used together with a hydrophone system to keep the platform properly positioned at the selected location.

Annual inspections of limited extent and a more thorough inspection every four years were called for. Three annual inspections in situ had been carried out, but the lower horizontal braces could not be inspected. The main inspection of the hull and machinery was scheduled to begin in April 1980.

Corrosion protection was afforded by a coat of paint consisting of a brown undercoating and a black covering paint ("brai epoxy"), as well as a red covering coat to protect against fouling on the external surfaces of the bracing. In addition, a cathodic protection system using an applied current was used. The system could supply a maximum current of 600 amps, and was controlled by adjusting the potential on various parts of the structure. Sacrificial anodes were used during the period of construction.

Welding electrodes of the type ESAB OK 48.30 were used for manual fillet welds at braces, and in particular, in the fillet weld between the hydrophone fitting and brace D-6. Welds made from these electrodes will have a yield strength 30% higher than that of the base metal, and with tensile strength and ductility properties similar to that of the base metal. Based upon the perceived criticality of the welds, the welds were classified into three groupings, which affected the qualification requirements for the welders and the inspections to which the welds were subjected. The welding of hydrophones to the braces fell into the lowest category. The specifications called for 100% inspection of all critical joints. Inspection methods included radiography, dye penetrant, and magnetic particle methods.

To attach the hydrophone to the tubular brace, a hole was burned into the brace (26 mm thick, 255 mm in diameter) of diameter 3–5 mm larger than that of the hydrophone (325 mm in diameter). The hydrophone was then welded into place with fillet welds by manual arc welding and 5-mm covered electrodes, with two passes being made, one on each side of the brace. There was no preheating. Examination of the hydrophone fitting on brace D-6 after the accident showed evidence of cold bending and welding from both sides of the brace (X-joint). The weld also showed a marked "root defect." The weld had poor adhesion to the base metal in some places. The form of the weld was also unfavorable, with contact angles of up to 90°

present. The fillet weld at the hydrophone fitting had been inspected for cracks with a dye penetrant.

C. Fracture in Bracing Member D-6

Two independent fractures were initiated: one from an outside fillet weld, one from an inside fillet weld; see Fig.11-6. The exact origins were difficult to pinpoint because fracture markings and striations were weak, and the details on the fracture surfaces had been ruined by the hammering together of the opposing fracture surfaces. The first 200–300 mm of the fracture surfaces show the typical markings of fatigue fracture. When the fractures had reached a length of about 300 mm, growth occurred more rapidly in leaps, giving intermittent fracture marks. The fracture marks here were coarse and fibrous, and shear lips occurred along the edges. Contraction was small, of the order of 2–4%. The final break constituted about one-third

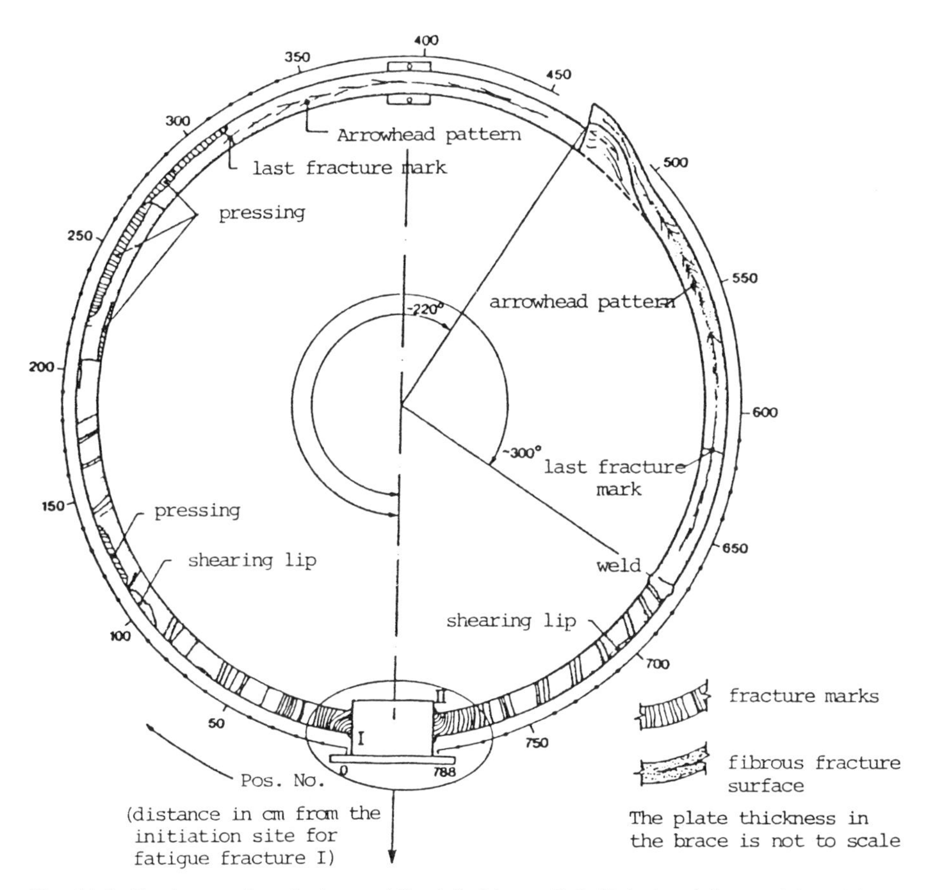

Fig. 11-6. Fracture surface features of the failed brace D-6. Note two fatigue origins at locations I and II at the welded in hydrophone. (From the Alexander Kielland Accident, 2.)

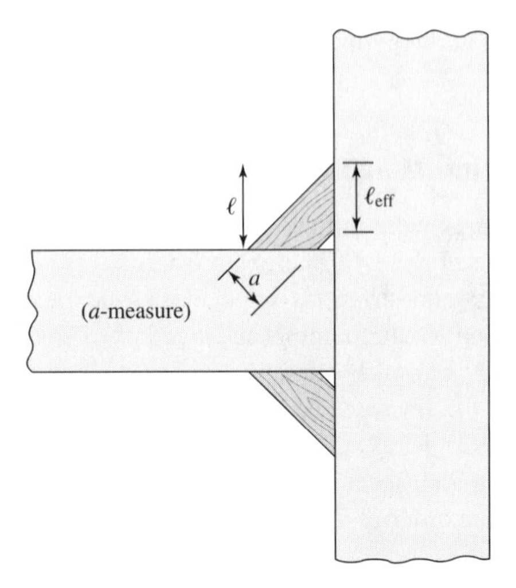

Fig. 11-7. Fillet weld *a*-measure.

of the circumference and was coarse with some chevronlike markings, and there was considerable contraction.

A portion of the fracture surface in the weld between the hydrophone and the brace contained the remains of a coat of paint, indicating that cracks were present at the time of fabrication. This paint was present on both the fracture surface of the inside weld and the bracing D-6 to an extent of 70 mm. A crack of this length must have existed at the time of paint application.

The fillet weld joints showed shallow, flat fractures in and near the fusion zone, typical tear-off fractures, and in part, fractures in the weld itself. The formation of the initial cracks could have been due to excessive thermal strains during welding, and/or excessively high external loads on the platform, or insufficient cracking resistance of the weld metal. In addition, the local strain level may have been too high. A load-bearing fillet weld will have as large a failure capacity as a butt weld when the fillet weld a-measure is of the order of 40–70% of the plate thickness; see Fig. 11-7. For a 26-mm plate, this indicates an a-measure of 10–18 mm, while the nominal a-measure of the weld in question was 6 mm. In addition, the weld angle was not optimum and fusion in the hydrophone holder was low. These factors reduced the effective a-measure of the fillet weld.

D. Conclusions

The Accident Report dealt with overall issues relating to the safety of mobile drilling platforms such as an improved analysis of potential problems in the design stage. A better system for exploring the consequences of all possible types of failures (in this case the loss of a column had not been anticipated), together with fault trees, was needed. Fault trees are used to explore in a systematic manner the consequences of the failure of individual components. An example of a fault tree for a gasoline tank in marine service is given in Fig. 11-8. Stability of the platform in the event

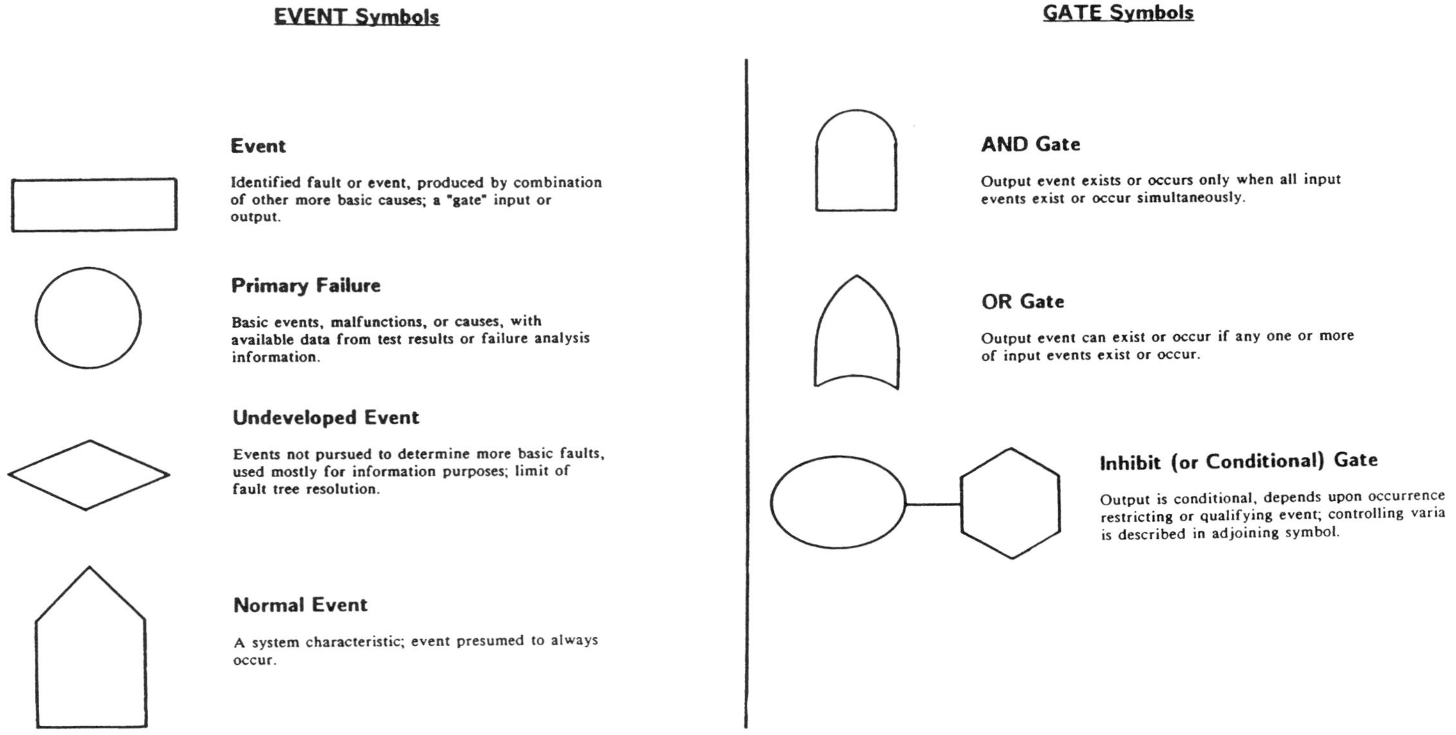

EVENT Symbols

Event

Identified fault or event, produced by combination of other more basic causes; a "gate" input or output.

Primary Failure

Basic events, malfunctions, or causes, with available data from test results or failure analysis information.

Undeveloped Event

Events not pursued to determine more basic faults, used mostly for information purposes; limit of fault tree resolution.

Normal Event

A system characteristic; event presumed to always occur.

GATE Symbols

AND Gate

Output event exists or occurs only when all input events exist or occur simultaneously.

OR Gate

Output event can exist or occur if any one or more of input events exist or occur.

Inhibit (or Conditional) Gate

Output is conditional, depends upon occurrence restricting or qualifying event; controlling varia is described in adjoining symbol.

(*a*)

Fig. 11-8. Fault trees. (*a*) Fault tree symbols. (*b*) An example of a fault tree. (From Witherell, 4.)

An Example of a Fault Tree

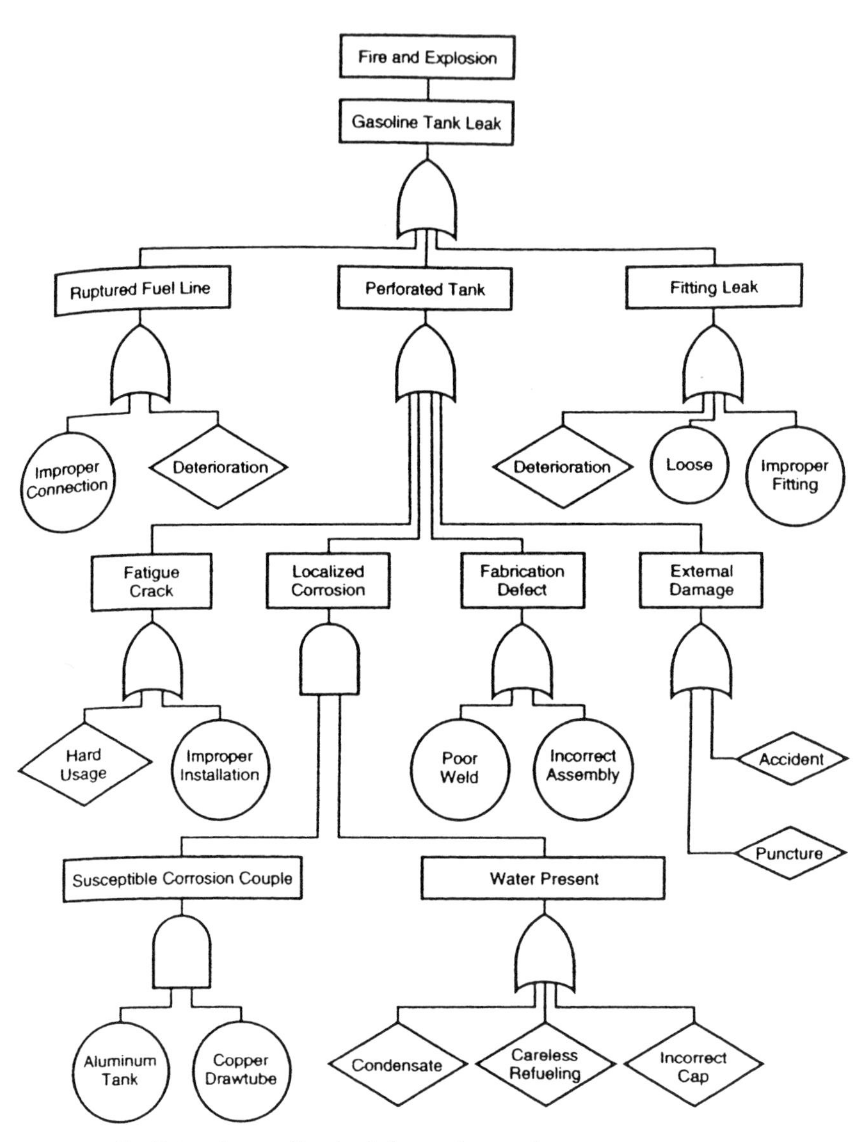

Fault tree for gasoline tank for marine service.

(*b*)

Fig. 11-8. *(Continued)*

of major damage was also a consideration. With respect to metallurgical matters, it can be concluded:

(a) We should beware of weld add-ons.
(b) When add-ons must be used on structural components, the welds should be stress relieved and carefully inspected.

IV. CASTING DEFECTS

There are a large number of types of potential defects associated with the casting process, any of which can lead to product failure. For example, the new nuclear-powered French aircraft carrier Charles de Gaulle was engaged in long-distance sea trials in November 2000 when it was forced to return to its home port at reduced speed because a blade on one of its two propellers broke off. Each propeller weighs 17,000 kg and is 5.8 m in diameter. They were specially designed to keep noise to a minimum. The cause of failure has not as yet been established, but it is suspected that a defect in the failed propeller may have been created during the casting process.

Among the principal casting defects there are:

(a) *Metallic Projections:* Fins, and so on can occur.
(b) *Cavities:* These might be porosity, blowholes, or pinholes. Porosity in cast single crystal turbine blades is a cause of concern with respect to fatigue crack initiation, for example.
(c) *Discontinuities:* One example of a discontinuity would be a cold shot, which is a portion of an ingot or casting showing premature solidification caused by a splash of metal during solidification. Another discontinuity is known as a cold shut, which is a discontinuity that appears on the surface of cast metal as a result of two streams of liquid meeting and failing to unite. (Cold shuts can also develop during forging, where they appear as laps on the surface of a forging or billet that was closed without fusion during deformation.)
(d) *Defective surface:* This can take the form of roughness, or sand adherence.
(e) *Incomplete Casting:* A portion of the desired shape may be missing.
(f) *Incorrect Dimensions or Shape:* Shrinkage, core shift, and so on, can occur. For example, in casting a hollow rectangular shape, a polymeric form (core) may be used to shape the inside cavity. If during the pouring operation this core should shift, the resultant casting will have walls of differing thicknesses, with the thinner wall being more prone to fatigue than the thicker wall.
(g) *Nonmetallic inclusions:* Possible inclusions include slag, flux, sand, oxides, and so on.
(h) *Centerline Piping:* A pipe is a central cavity formed by contraction of an ingot during solidification.

V. CASE STUDY: CORNER CRACKING DURING CONTINUOUS CASTING

Corner cracking in continuous square casting occurs due to shrinkage, phase transformation, and the lack of mold support. To avoid oxidation, the mold is sealed with argon, and oil is used as a lubricant, which is probably graphite at 3000°F. During the early stage of the continuous casting process, the exterior of the casting solidifies while the interior remains a liquid. Since the casting direction has a vertical component, there is an increase in hydrostatic pressure with increase in the vertical component. This increase can lead to cracking at the corners, which are prone to cracking due to trapped solute and an elongated grain structure. If such corner cracks are present after casting, the cracked corners have to be ground away, a costly operation. To avoid this problem, the mold should be double tapered to provide lateral support during the shrinkage and phase transformation processes.

VI. FORMING DEFECTS (3)

(a) *Over-Under Cambering:* In a rolling mill, in order to obtain a product of uniform thickness, the rolls may have a larger diameter (camber) at mid-width than at the edges to compensate for bending of the rolls during the rolling operation. Cracking in a rolled product can occur due to either the over-cambering or the under-cambering of the rolls. Over-cambering leads at mid-width to a thinning of the product and a greater tendency to extend in the rolling direction. This tendency is restrained by adjacent, less highly thinned material, with the result that the restraining material is placed in tension, which can lead to the development of transverse cracks. Under-cambering, on the other hand, results in the potential for cracking at mid-width.

(b) *Edge Cracking:* In rolled sheet metal, edge cracking is due to a state of plane stress (rather than plane strain) at the edges and a lack of edge support. Since the material at the edges can expand in the transverse direction, there is less material moving in the rolling direction than there is away from the edges. As a result, a tensile stress is set up at the edges in the rolling direction, and this stress can result in edge cracking.

(c) *Stretcher Strains:* Elongated markings that appear on the surface of a steel that has an upper and lower yield level are known as stretcher strains. These are objectionable in exterior sheet metal automobile parts as they detract from a smooth finish.

(d) *Alligatoring:* The longitudinal splitting of a rolled flat slab in a plane parallel to the rolled surface is called alligatoring.

(e) *Residual Stresses:* Such stresses can develop during most cold forming operations. They may have a deleterious effect on subsequent machining operations because of distortion when the remaining residual stresses are equilibrated.

(f) *Banded Microstructure:* A heterogeneous microstructure with the segregates aligned in a direction parallel to the rolling direction is known as a banded microstructure. In steels, bands of pearlite may be separated by bands of ferrite. The microhardness will vary across a band, with bands of higher hardness being more prone to hydrogen-induced cracking.

(g) *Seams and Laps (Folds):* These are defects created during extrusion, drawing and forging operations. A lap is a surface defect, appearing as a seam, caused by folding over hot metal, fins, or sharp corners, and then rolling or forging them into the surface, but not welding them.

(h) *Undesirable Flow patterns in Forging:* During a forging operation, the grains elongate in the main direction of metal flow, which can lead to an anisotropy in material properties. If in use the principal tensile stress is transverse to the flow direction, cracking may occur due to material weakness in the transverse direction. A similar situation existed with respect to the Titanic rivet discussed in Chapter 5, where the slag inclusions were oriented perpendicular to the tensile stresses due to metal flow during the rivet head forming process.

VII. CASE STUDIES: FORGING DEFECTS

A. F-111 Aircraft (4)

On the basis of its fatigue resistance, fracture toughness, and resistance to stress-corrosion cracking, a high-strength steel, D6ac, of 220–240 ksi ultimate tensile strength, was selected for transferring load from the wing to the fuselage in the F-111 swing-wing aircraft. Machined forgings were used in the fabrication of the wing pivot fitting and the carry-through box. During the development of the F-111, three full-scale fatigue tests were carried out on wing assemblies, and these tests demonstrated a fatigue capability of six times the design life.

In December 1969, an F-111A was involved in an accident during a training mission. During a pull-up from a pass over a target area, the left wing separated in flight and the aircraft crashed. In the investigation, the flaw shown in Fig. 11-9 was found in the wing pivot fitting. A small band of fatigue striations, approximately 0.5 mm (0.02 inches) in depth, was found on the periphery of the flaw, and was attributed to the104.6 hours of flight operations prior to the accident. It was determined that the flaw had been initiated during the hot-forging process and had extended during the subsequent cooldown. The flaw had escaped detection during inspection primarily because the flux fields used with magnetic particle inspection were inadequate for a tight flaw in a part of the unusual shape and size of the wing pivot fitting. Also, the sound wave transmission during ultrasonic inspection had been directed almost parallel to the flaw surfaces and the reflected signal was insufficient to detect the flaw.

As result of this accident, an unprecedented rigorous proof test and nondestructive examination of each F-111 was undertaken. This program was directed, not only at the detection of flaws, but also at establishing inspection intervals for use with the fleet of F-111 aircraft in service. The inspection interval, which included a confidence factor, was determined by the initial size of an assumed flaw based upon the proof test, the growth of the flaw as a function of time under service loading and environmental conditions, and the critical size of the flaw for service operations.

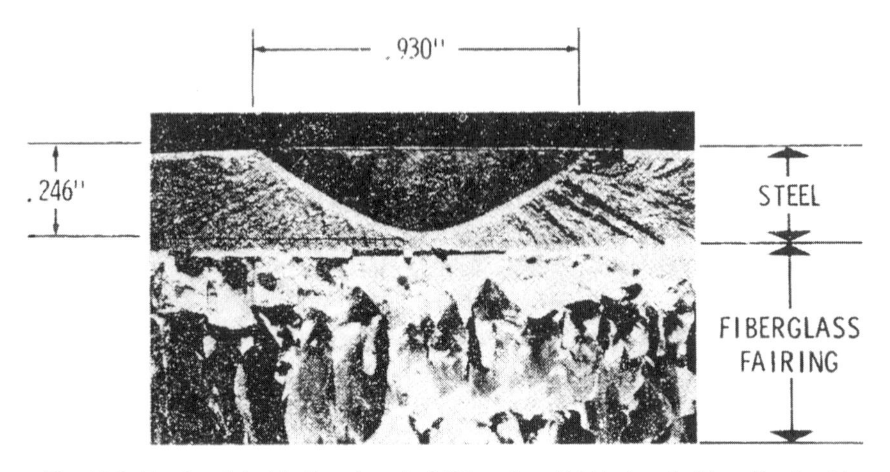

Fig. 11-9. Forging defect in the wing pivot fitting of an F-111 aircraft. (From Buntin, 4.)

B. Jet Engine Components

On January 2, 2001, a front-page article in the *Wall Street Journal* drew attention to problems with the General Electric Aircraft Engines (GEAE) CF-6 engines. These engines are used to power a broad range of Boeing and Airbus Industrie planes, such as the DC-10, the A300, and the Boeing 767. Almost 6000 of these engines are in service, and failures are quite rare. One problem first surfaced in 1991 when a type of cracking known as "dwell time fatigue" was found during inspection of a compressor component known as a spool, which is made of a forged titanium alloy. Since then, this type of cracking has led to two instances of engine failure on takeoff, and a third instance in which the takeoff was aborted. Fortunately, these failures did not result in deaths or injuries. The cause of this type of fatigue cracking is not well understood, but appears to be promoted in certain titanium alloys such as Ti-6Al-2Sn-4Zr-2Mo by large grains of susceptible orientation that are developed during a hot-forging process in which the billet is not heavily deformed. This type of fatigue is thought to be accelerated by hold periods at peak stress, hence the term dwell time fatigue. Since the service temperatures involved are well below the creep range, creep does not appear to be a factor, but environmental factors may be involved. GEAE has changed both the forging technique and the titanium alloy in an effort to eliminate this type of cracking in the future.

A second problem was discovered in April 2000. It was found that the locking mechanism holding in place the ring containing the engine vanes had cracked, allowing the ring to rotate and wear through the casing, thereby releasing a shower of debris. To remedy this situation, GEAE has developed new locking mechanisms to secure the vanes.

Until the replacement parts have been installed, these problems will require a more intensive engine inspection program, involving the use of ultrasonic inspection methods, and a reduction in the spool's service life from 15,000 takeoffs to 12,500. There are clearly important economic considerations as well. For example, since the spool inspections now require taking apart and reassembling the whole engine, a 60-day process, a shortage of inspected engines could result. Because of this, GEAE is developing computerized inspection tools in an effort to reduce the inspection time to the order of one week.

VIII. CASE STUDY: COUNTERFEIT PART (5)

The following case describes a failure in which a bogus part played an important role.

A four-engine, propeller-driven transport plane had just touched down when the left main wheel assembly separated. The aircraft slewed off the runway and a fire damaged the wing and engines on the left side. It was found that the failure of a trunnion (a pin or pivot on which something can be rotated or tilted) arm shaft in the landing gear, which had occurred during the previous takeoff, was responsible for the accident. Progressive beach markings were found on the fracture surface of the trunnion arm, which suggested that fatigue may have been involved. However, in a detailed examination, intergranular fracture was observed without evidence of fatigue striations. It was concluded that the trunnion arm had failed because of stress corrosion cracking.

The trunnion arm was made of a fine-grained, through hardened and tempered 4340 steel. It had been subjected to excessive wear in service, and during an overhaul it had been chrome plated to bring the dimensions of the worn trunnion arm back to the original level. Beneath the chrome plating and adjacent to the fracture, a regular pattern of intergranular cracks was found, some of which were not continuous through the chrome plating, an indication that they were not due to plating cracks. These cracks were about 0.5 mm in depth, and it was concluded that they were thermal checks, which had been created by overheating that occurred during a grinding operation at the time of an overhaul prior to hard chrome plating 18 months before the accident. A properly carried out magnetic particle inspection should have detected these cracks. Normally, local overheating due to coarse grinding leaves a shallow transformed untempered martensitic layer, but fine grinding following a coarse cut may erase this layer, as was the case in this instance.

In comparing the failed trunnion shaft with a nominally similar trunnion shaft from the failed landing gear, certain differences were noted. The failed shaft had been ground undersize and then brought up to the correct size by plating, whereas the unfailed shaft had been machined undersize on a lathe with a single point tool prior to plating. The failed shaft was grit-blasted but not shot-peened, whereas the unfailed shaft was shot-peened. A fine crack network in the plating on the failed shaft was typical of a high-speed, low-crack plating bath, whereas a large crack network in the plating of the unfailed shaft was typical of a chromic/sulfuric conventional plating bath. The unfailed trunnion conformed to the overhaul facility's work statement procedures, whereas the failed trunnion did not. It was concluded that the failed trunnion was a substandard, counterfeit part. All manufacturers and overhaul facilities need to be on the lookout for the substitution of cheap parts instead of parts specified by the original equipment manufacturer (OEM).

IX. THE USE OF THE WRONG ALLOYS; ERRORS IN HEAT TREATMENT, AND SO ON

Incorrect identification of alloys is another cause of failure (1). Examples include the misidentification of a monel alloy as a Type 304 stainless steel, and the welding of monel with stainless steel electrodes, resulting in a brittle joint because the copper content of the monel led to hot-short cracking of the weld. Other examples include the substitution of Type 430 stainless steel for Type 304.

In-service failures have occurred where carbon steels have been substituted for chromium-molybdenum steels in high-temperature high-pressure pipe lines. Because of such mix-ups, the entire superheater tube sections in major steam power plants have had to be replaced in at least two instances. Mix-ups have even occurred in submarine construction, where welded pipe has been substituted for seamless pipe. Undoubtedly, there are many more instances of this sort of problem. To minimize such problems, it is recommended that parts be properly identified by stamping or by etching with an electric pencil. However, even this procedure is not foolproof, for there are instances where properly identified but incorrect alloys have been substituted for specified alloys because of careless erection practice.

Problems can also arise from improper heat treatment. In one welded structure, a quenched and tempered HY-80 alloy steel was specified, but the same steel in a normalized condition was used instead. The low toughness of the normalized steel resulted in hydrogen-induced cracking.

A punched-in trademark on a metal component can also serve as a site for crack initiation, particularly in a part that is used repetitively, such as die. One is reminded of the wooden baseball bat with its burned in label and the warning "never hit the ball with the label facing forward." Too often, this sort of good advice is ignored by manufacturers.

X. SUMMARY

Defects are often the cause of premature failures. This chapter has provided a description of some of the more common types of defects. The avoidance of defects involves good quality control and inspection procedures.

REFERENCES

(1) H. Thielsch, Defects & Failures in Pressure Vessels and Piping, Reinhold Publishing Co., New York, 1965.

(2) The Alexander L. Kielland Accident, Norwegian Public Reports, NOU 1981:11, Oslo, Norway, 1981.

(3) W. D. Buntin, Concept and Conduct of Proof Test of F-111 Production Aircraft, Aeronautical J., vol. 76, no. 742, Oct. 1972.

(4) T. W. Heaslip, in Metallography in Failure Analysis, ed. by J. L. McCall and P. M. French, Plenum, New York, 1978, pp. 141–165.

(5) Fracture Surface Replicas, The Welding Institute, Abingdon, Cambridge, UK, 1973.

(6) C. E. Witherell, Mechanical Failure Avoidance, McGraw-Hill, New York, 1994.

12

Environmental Effects

I. INTRODUCTION

Corrosion combined with steady or alternating stresses can play a major role in several types of failure processes. The interaction of the environment, stress, and an alloy can be quite complex, and in some cases, a consensus may not as yet have been reached as to the exact mechanisms involved. Nevertheless, on a macroscopic level there is ample evidence of the deleterious effects that the environment can exert in promoting fracture. Corrosion is a time-dependent phenomenon, and its effects in service may not be realized until many years have passed. This is an important consideration, since many components are subjected only to short-time testing prior to being approved for in-service use, and consideration needs to be given to the time-dependent effects of corrosion in service, as in the case of aging aircraft. In this chapter, some basic aspects of corrosion processes are reviewed, and various type of corrosion-related fractures are discussed.

II. DEFINITIONS

(a) *Corrosion:* The deterioration of a metal by chemical or electrochemical reaction with its environment.

(b) *Stress-Corrosion Cracking (SCC):* Failure by cracking under combined action of corrosion and stress, either external (applied) or internal (residual). Cracking may be either intergranular or transgranular, depending on metal and corrosive medium.

(c) *Hydrogen Embrittlement:* A condition of low overall ductility in metals resulting from the absorption of hydrogen. Steel tensile specimens charged with hydrogen exhibit a reduction in percent of elongation and in the percent of reduction in area as compared to uncharged specimens. The failure mode may be cleavage, intergranular, or ductile, but in the latter case the dimples are more numerous and more shallow as compared to an uncharged specimen. Hydrogen embrittlement is most severe at room temperature, and can lead to a delayed or time-dependent form of cracking. It is also referred to as hydrogen-assisted fracture, which can denote either embrittlement on the atomic scale due to bond-weakening or the promotion of localized plastic deformation.

(d) *Hydrogen-Induced Cracking (HIC):* A form of hydrogen embrittlement. HIC can occur in the absence of applied or residual stress as the result of the combining of two hydrogen atoms to form an H_2 molecule inside the metal matrix. This reaction takes place at a convenient location such as an inclusion interface. The formation of a hydrogen molecule from two hydrogen atoms results in a local increase in pressure. As more and more molecules form at a given site, the pressure increases to the point where, if the site is near the surface, a blister may form. If the site is remote from a surface, the formation and growth of a crack may take place, and if a number of such cracks form, they can link up and greatly degrade the fracture resistance. In some cases when external stresses are present, it may be difficult to distinguish between hydrogen embrittlement due to atomic interaction with the lattice and HIC.

(e) *Hydrogen Attack:* An elevated temperature problem brought about by the diffusion of hydrogen into steel, wherein it combines with carbon to form methane gas, leading to porosity within the steel.

(f) *Liquid Metal Embrittlement:* Loss of ductility of a metal such as aluminum as the result of direct contact with a liquid metal such as mercury, or of austenitic steel in contact with molten zinc, or of molten lead in contact with ferritic steel.

(g) *Corrosion Fatigue:* The deleterious effect of an environment, including ambient air, on the purely mechanical fatigue process.

(h) *Localized Deformation:* Plastic deformation involving the motion of dislocations in microscopically narrow zones.

(i) K_{Iscc}: The threshold value of the Mode I stress intensity factor for stress corrosion cracking for a given alloy and environment.

III. FUNDAMENTALS OF CORROSION PROCESSES

A basic aspect of the electrochemical aqueous corrosion process is that the dissolution of metal takes place at the anode, and the removal of electrons takes place at the cathode. A typical dissolution reaction can be written as

$$M \rightarrow M^{++} + 2e^{-}. \qquad (12\text{-}1)$$

a reaction referred to as oxidation reaction in that electrons are liberated. A typical cathodic reaction can be written as

$$2H^+ + 2e^- \rightarrow H_2, \qquad (12\text{-}2a)$$

a reaction referred to a reduction reaction in that electrons are consumed. Another cathodic reaction that is important in neutral and alkaline solutions containing dissolved oxygen is

$$2H^+ + \tfrac{1}{2}O_2 + 2e \rightarrow H_2O. \qquad (12\text{-}2b)$$

In an electrochemical corrosion process, it is basic that the total anodic current, I_A, be equal to the total cathodic current, I_C. Figure 12-1 shows the reactions occurring at a corrosion pit in aluminum where the reduction process involves oxygen (Fig. 12-1*a*) and where the reduction process involves hydrogen (Fig. 12-1*b*).

Faraday's law gives the mass in grams going into solution from an anode in a given time period. In some models of stress corrosion cracking (SCC), anodic dissolution at a crack tip is envisioned. This law is derived as follows:

1. A metal ion going from the anode to the cathode carries z units of charge, where z is the valence. This is equivalent to $z \times 1.602 \times 10^{-19}$ coulombs.

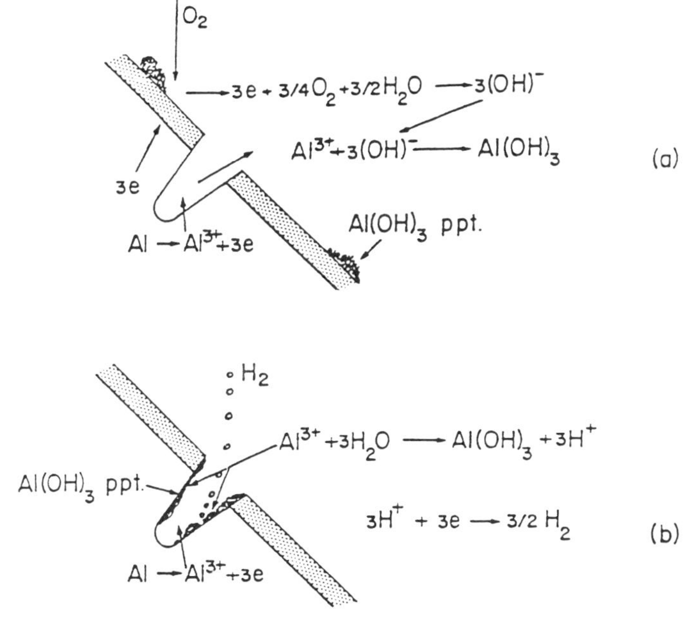

Fig. 12-1. Reaction at a corrosion pit in aluminum. (*a*) Reduction process involving oxygen. (*b*) Reduction process involving hydrogen.

2. If N mols go from the anode to the cathode, then the number of coulombs transported would be $N \times (A_v z 1.602 \times 10^{-19})$ coulombs, where A_v is Avogadro's number, 6.022×10^{23}.
3. This number of coulombs can be expressed as $N \times z \times 1.602 \times 10^{-19} \times 6.022 \times 10^{23} = N \times z \times 96{,}519$ coulombs $= N \times z \times F$, where F is Faraday's constant = 96,519 coulombs/mol.
4. The current is the number of coulombs transported per second, or $I = N \times z \times F/t$.
5. The above relation can be rewritten as $N = It/zF$.
6. To convert this expression to the metal weight loss in grams from the anode, multiply both sides of the above equation by the molecular weight in grams/mol, W, to obtain

$$NW = \text{total mass in grams lost} = M = \frac{WIt}{zF}. \tag{12-3}$$

This is Faraday's law. This expression can be rearranged to give the total anodic current, I:

$$I = \frac{MzF}{Wt}. \tag{12-4}$$

Faraday's law can also be written in terms of the current density i, where $i = I/A$, that is, $M = WiAt/zF$, where A is the area of the anode. For this reaction to proceed during SCC, it is necessary that any protective oxide film at the crack tip be ruptured. Continuing plastic deformation in the plastic zone at a crack tip may be required to expose the underlying base metal to the corrosive environment. In addition chloride ions, if present, may serve to weaken oxide films so that they are more easily ruptured by plastic deformation.

It is also noted that the cathodic reaction may involve hydrogen ions. If, instead of combining with electrons, these ions diffuse into the metal, the possibility for hydrogen embrittlement develops, especially since the cathode may be quite close to the anode.

When metal ions are created, and electrons are liberated at the anode, an electric potential develops. This potential is measured relative to a standard electrode when the following typical reversible reaction is established in a molar solution of metal ions:

$$Fe \leftrightarrow Fe^{++} + 2e^{-}. \tag{12-5}$$

By this means, a table of electrochemical potentials is established for the various metallic elements, which is similar to the Galvanic series. The potential for the reversible hydrogen reduction reaction,

$$2H^{+} + 2e = H_2, \tag{12-6}$$

is taken as zero, and it becomes the standard reference potential. The potential for the reaction in Eq. 12-5 is 0.440 volt below that of the standard hydrogen electrode. If the reduction process involves oxygen rather than hydrogen, the reduction reaction is

$$O_2 + 2H_2O + 4e^- = 4OH^-. \tag{12-7}$$

This reversible reaction occurs at potential of 0.82 volt above the standard hydrogen potential.

The actual potential in a corrosion process falls between that for the oxygen potential, +0.82 volt (if dissolved oxygen is more plentiful than hydrogen ions), and the metal potential, if iron, −0.44 volt. The realized potential is called the mixed corrosion potential, E_{corr}, with which is associated a corresponding corrosion current density, i_{corr}. The process whereby the potentials are altered is called mutual polarization, which occurs as the result of electron flow from the anode to the cathode. This information can be displayed on a potential-current diagram, as in Fig. 12-2.

Pourbaix diagrams, or potential-pH diagrams, indicate the thermodynamically stable phases that can form during aqueous corrosion. Figure 12-3 is a simplified Pourbaix diagram for iron in aqueous solution. In Fig. 12-3, the upper dotted line represents the reaction given by Eq. 12-7 and the lower dotted line represents the reaction given by Eq. 12-6. At potentials above the oxygen reaction line, oxygen is evolved. At potentials below the hydrogen reaction line, hydrogen is evolved. Between the two lines, water is stable. At one time, it was thought that if the potential-pH values fell between these two lines, the stress corrosion cracking pro-

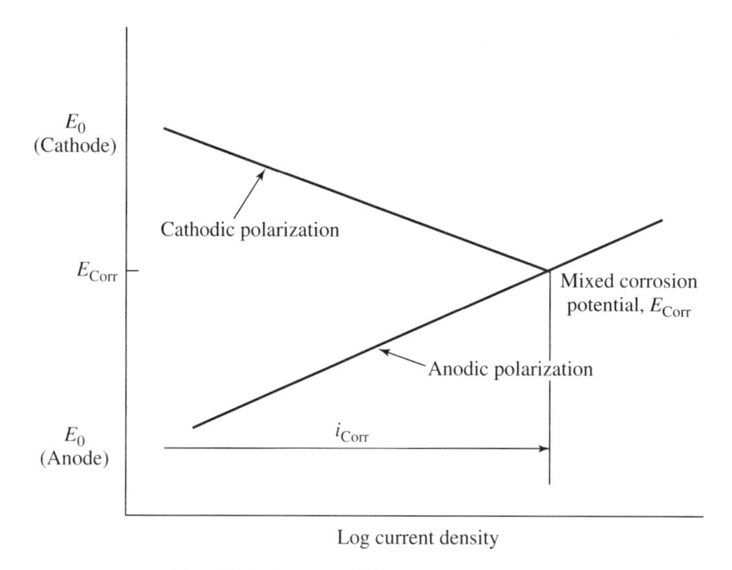

Fig. 12-2. A potential-log current diagram.

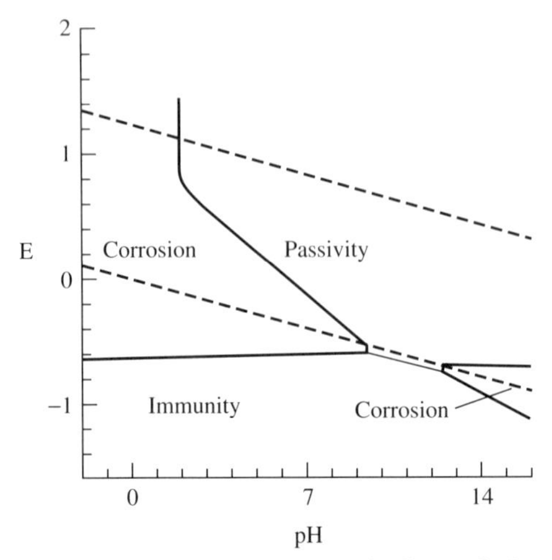

Fig. 12-3. A simplified Pourbaix diagram for iron.

cess could not involve hydrogen, and that anodic dissolution must control the cracking process. However, in more recent times, measurements of the pH at crack tips have shown that pH values can be much lower than the bulk values. Therefore, at a given potential, the crack tip pH value can lie below the line for the hydrogen reaction, meaning that hydrogen can be evolved and hydrogen embrittlement is possible. Figure 12-4 gives an example of this situation for a high-strength steel.

IV. ENVIRONMENTALLY ASSISTED CRACKING PROCESSES

There are currently four types of mechanisms that relate to the effects of the environment on cracking processes. One mechanism is based upon the repeated rupture and reformation of films at the tip of growing crack. A second mechanism is based upon the process of anodic dissolution at a crack tip. The third is liquid metal embrittlement, involving either the weakening of atomic bonds or the promotion of localized deformation. The fourth is hydrogen-assisted fracture, and since aqueous corrosion is the most common type of corrosion, the potential for hydrogen assisted cracking in many cases exists. There are two leading theories for hydrogen-assisted fracture. One is that hydrogen weakens atomic bonds, thereby resulting in embrittlement. The other is that hydrogen promotes very localized plastic deformation, which results in high shear strains and fracture within the localized shear bands. There may not in fact be a single mechanism, as one or the other may be operative under a given stet of circumstances. From the practical point of view, it is important to know of the situations in which hydrogen-assisted cracking can occur, and to take appropriate preventive steps.

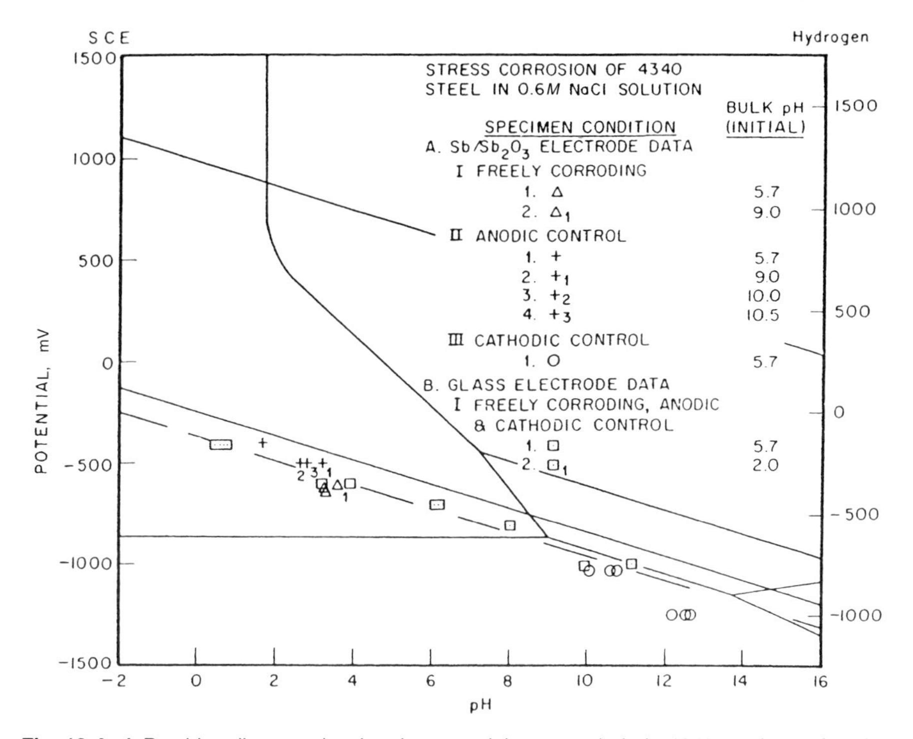

Fig. 12-4. A Pourbiax diagram showing the potential at a crack tip in 4340 steel as a function of the crack tip pH value. (From Brown, 13.)

Figure 12-5 is a schematic showing how stress corrosion cracks can develop from a free surface. Figure 12-6 shows how the SCC process may proceed by a film rupture model in cartridge brass.

SCC tests are carried out on either smooth, notched, or precracked specimens. Results for smooth stainless steel specimens in boiling magnesium chloride, a standard but severe environment, are shown in Figure 12-7. In SCC of stainless steel as well as other alloys, typical branching cracks are observed, which distinguish this form of cracking from fatigue cracking, where usually only a single major crack is

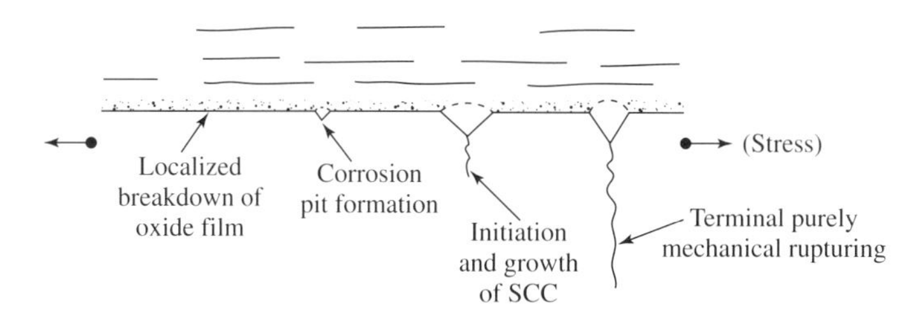

Fig. 12-5. Schematic of a stress corrosion crack developing at surface. (From Brown, 14.)

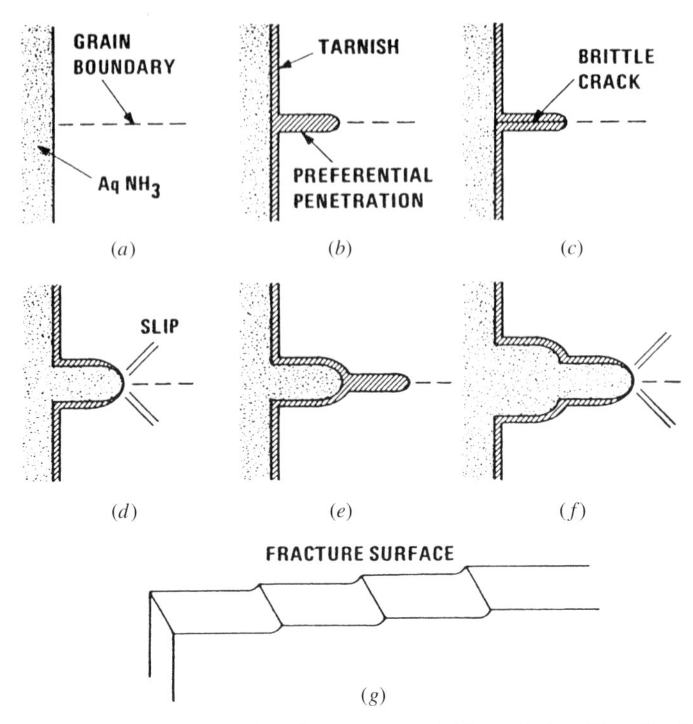

Fig. 12-6. A film rupture model for α-brass in a tarnishing solution. (From Pugh, 15.)

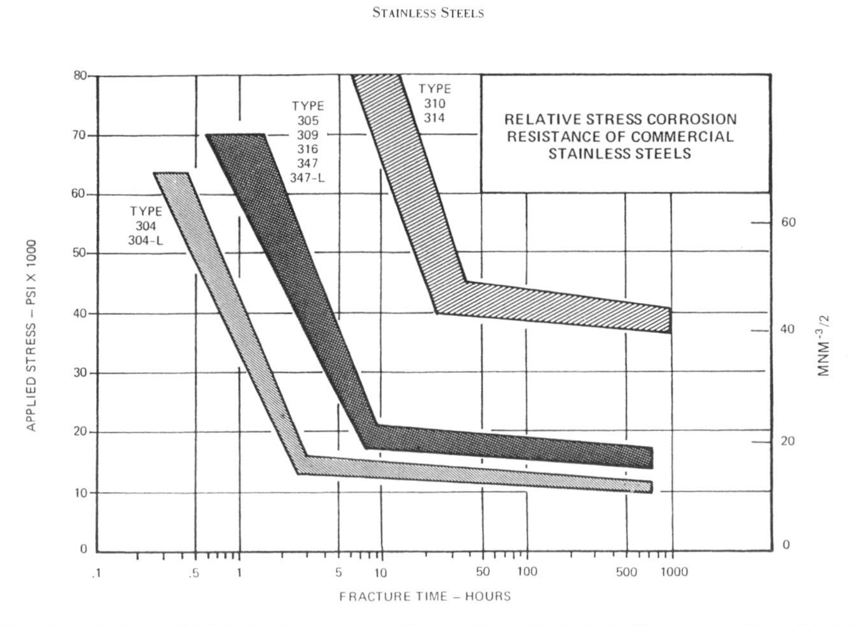

Fig. 12-7. Relative SCC behavior of austenitic stainless steels in boiling magnesium chloride solution. (From Denhard, 16.)

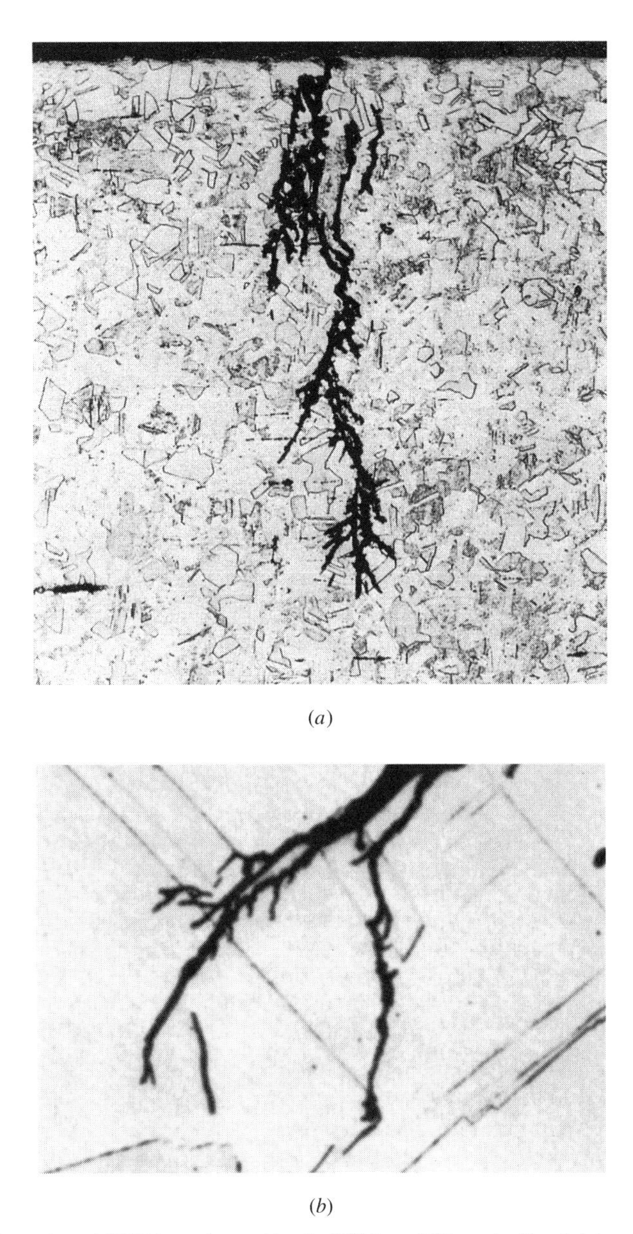

(a)

(b)

Fig. 12-8. Examples of SCC branch cracking in AISI type 304 austenitic stainless steels. (From Brown, 14.)

observed, Fig.12-8. Figure 12-9 is a schematic representation of the rate of stress corrosion crack growth as a function of K. Figure 12-10 shows results for precracked specimens of aluminum alloys. Note that a plateau exists where the rate of cracking is constant and independent of K. This plateau develops because the diffusion of the corroding species in the liquid at the crack tip becomes rate limiting. At the lowest rates of crack growth, the K_{ISCC} value is approached. The effect of flaw size

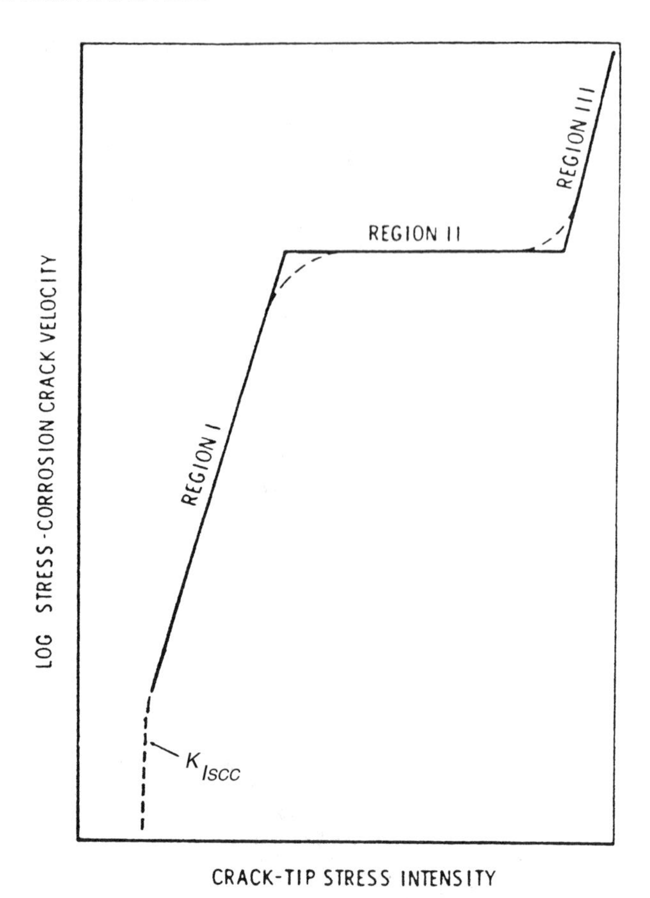

Fig. 12-9. Schematic of the rate of SCC as a function of the stress intensity factor, *K*. Region III is of little interest and is often missing. Region I is missing in some systems. Regions I and II are not always straight lines in such plots, but may be strongly curved. (From Brown, 14.)

on establishing the safe limits for the avoidance of SCC is shown in Fig. 12-11. In this case, the threshold found under smooth bar conditions sets the limit up to a flaw size of 1.5 mm where the stress intensity factor takes over. This behavior is analogous to a similar situation encountered in mechanical fatigue.

The stress intensity factor for a long, shallow surface flaw is given as

$$K = \sqrt{\frac{1.2\pi\sigma^2 a}{1 - 0.212(\sigma/\sigma_Y)^2}}. \tag{12-8}$$

If, as an upper limit, σ is set equal to σ_Y, then a value for a_{cr}, where a_{cr} is the critical crack length at K_{ISCC}, can be expressed as

$$a_{cr} = 0.2\left(\frac{K_{ISCC}}{\sigma_Y}\right)^2, \tag{12-9}$$

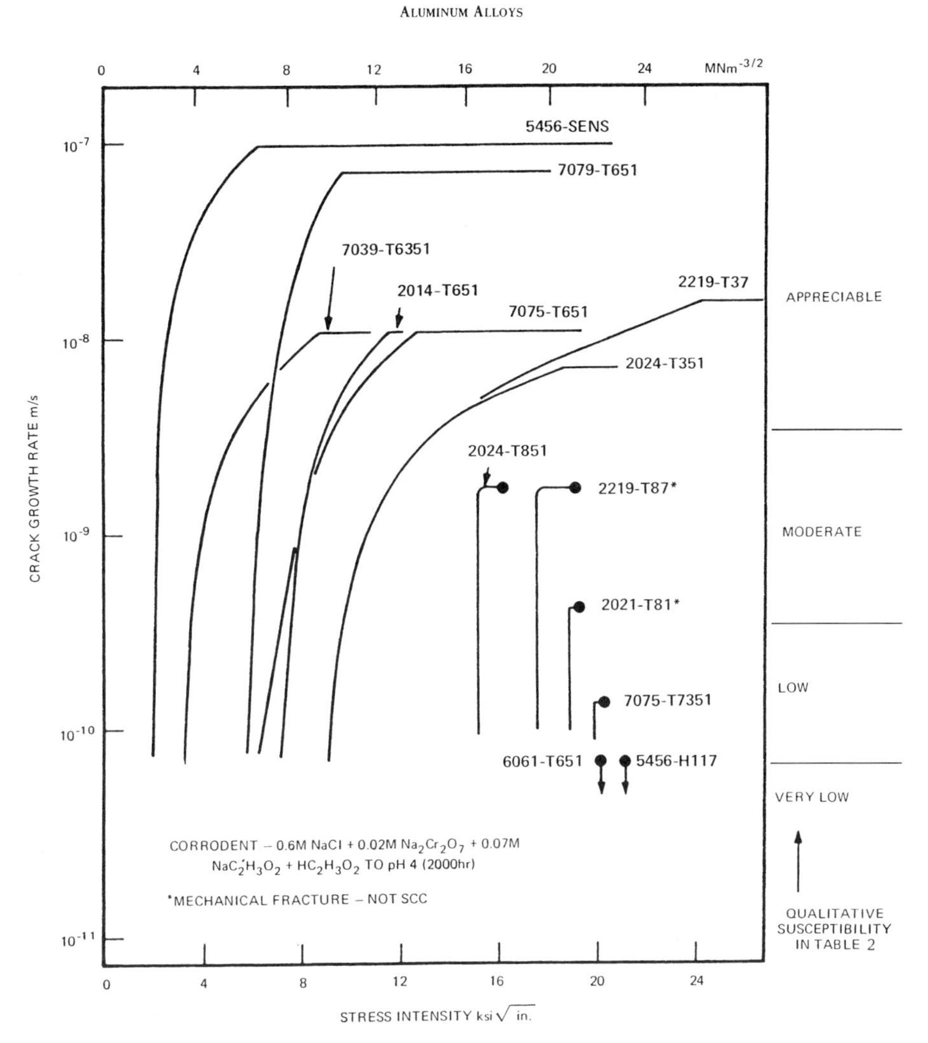

Fig. 12-10. The rate of SCC growth, *da/dt*, as a function of the stress intensity factor for aluminum alloys showing plateau levels. (From Brown, 14, and Sprowls et al., 18.)

and $K_{Iscc} = \sqrt{5a_{cr}}\sigma_Y$. This last relationship plots as a straight line of slope $\sqrt{5a_{cr}}$ on a plot of K_{Iscc} versus σ_Y. The relationship gives the value of K_{Iscc} required for a given a_{cr} and σ_Y to avoid stress-corrosion crack growth. Figure 12-12 shows this level of K_{ISCC} for several crack lengths as a function of σ_Y, and includes experimental $K_{ISCC} - \sigma_Y$ data points for a number of titanium alloys. Note that the K_{ISCC} datum point for Ti-8Al-1Mo-1V at a σ_Y just over 120 ksi is quite low, as is the corresponding a_{cr}. Because of its relatively high strength, at one time the U.S. Navy was considering this alloy for submarine hulls. The alloy performed well in smooth-

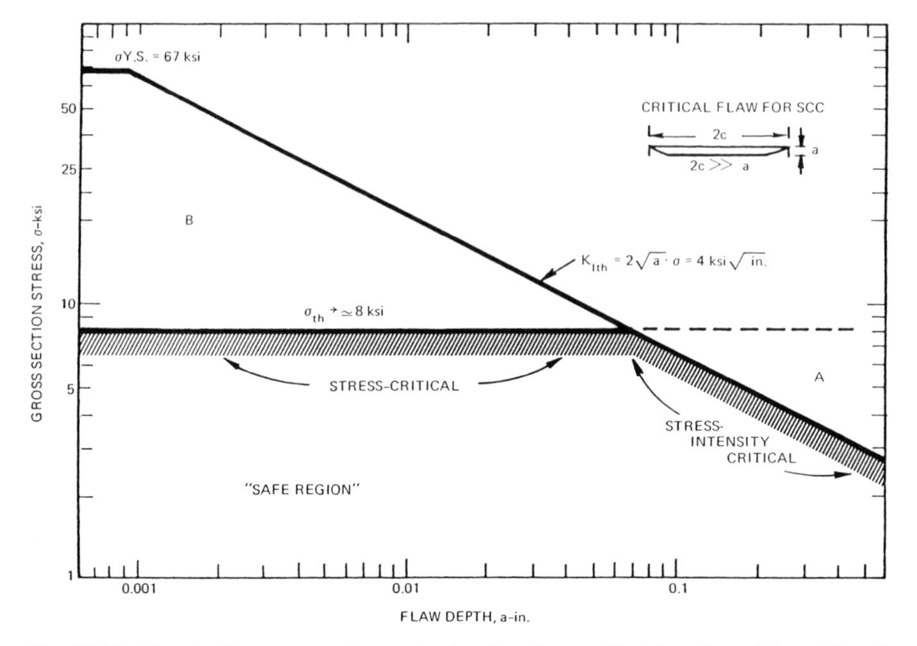

Fig. 12-11. Flaw depth *a* versus stress, showing the stress critical-(small cracks) and threshold stress critical (large cracks) boundaries for the aluminum alloy 7079-T651 plate. (From Brown, 14.)

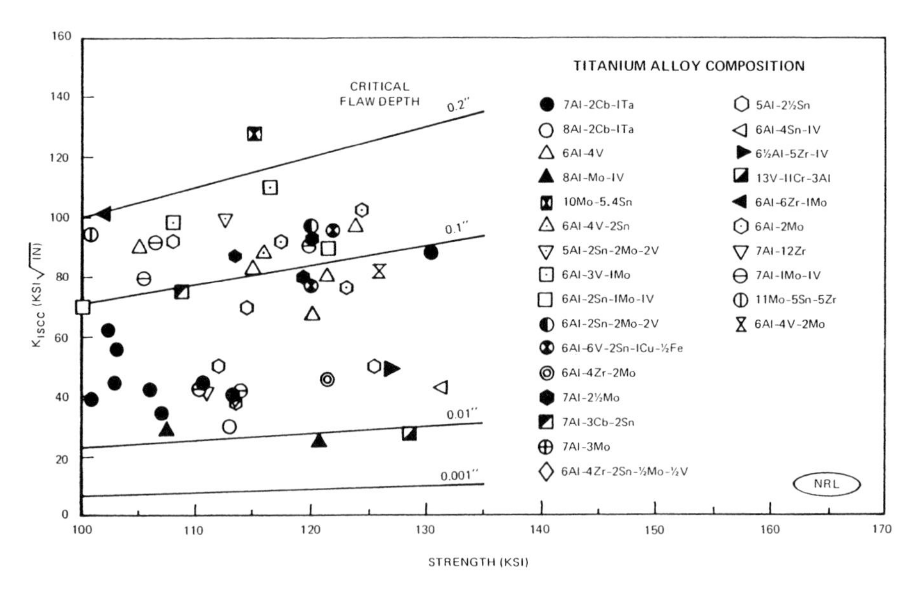

Fig. 12-12. K_{ISCC} data on titanium alloys in sea water. The critical flaw size is a function of K_{ISCC} and yield strength level is indicated (NRL = Naval Research Laboratories). (From Brown, 14.)

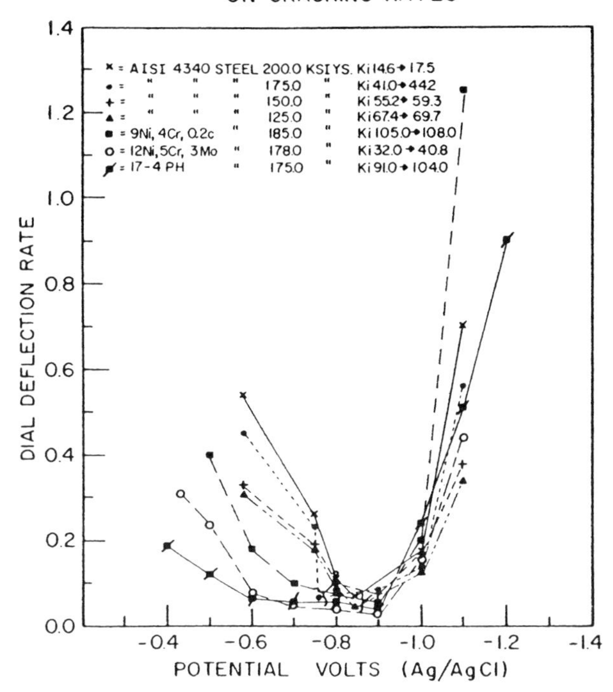

Fig. 12-13. Effect of varying degrees of cathodic protection on the stress corrosion cracking rates of several steels. The rate of cracking was taken to be proportional to the dial deflection rate. (From Brown, 14.)

bar stress corrosion tests, but its low K_{ISCC} value precluded further consideration for such an application.

Cathodic protection, either by the use of an impressed current or sacrificial anodes, can be used to lower the rate of SCC, as shown in Fig. 12-13. However, too much of a reduction in the corrosion potential can result in hydrogen embrittlement due to larger amounts of hydrogen being required at the cathode to maintain a higher i_c, as shown in Fig. 12-14. The effect of a reduction in potential, therefore, is to reduce the anodic current but to increase the cathodic current.

V. CASE STUDIES

A. Spring Failures

Failures associated with hydrogen-assisted cracking are delayed failures that involve discontinuous crack growth. They often occur without warning and with severe or potentially disastrous consequences. For example, consider an encased, precom-

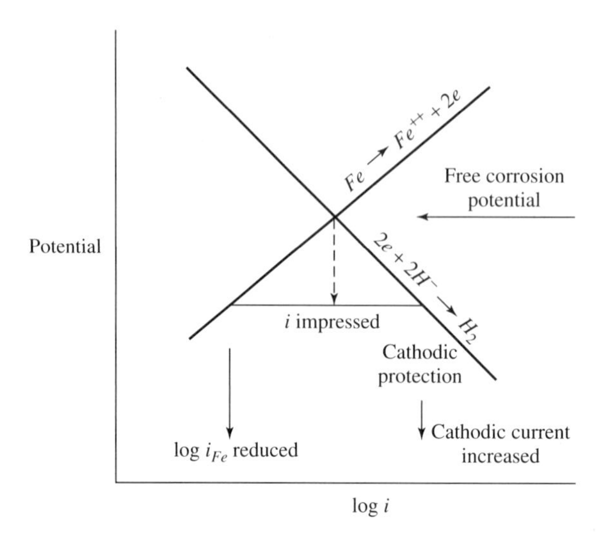

Fig. 12-14. Effect of an impressed current on reducing the corrosion rate of iron through cathodic protection.

pressed spring in a fire-protection system. In the event of a fire, a low-melting-point component melts, thereby releasing the spring and activating a water sprinkler system. However, if the spring contains hydrogen because of a plating operation, or if it corrodes with the uptake of hydrogen, it may fail prior to its time of need. Such a failure can remain undetected, which would negate the protective scheme with obvious potential for disaster. The extent of corrosion may depend upon the surface, for a highly polished surface often shows less corrosive attack than a rough surface for the same service conditions.

Spring failures can also occur because the spring material was inappropriate for the intended use (1). For example, a pressure relief valve was provided to vent gas from a high-pressure unit in the event that the pressure in the unit built up beyond the 7.6 MPa (1100 psi) operating conditions. The gas contained 1000 ppm by weight of H_2S at 20°C. The valve included a high-strength, compressed coil spring that was essential for the proper operation of the safety valve, but that was not resistant to the effects of the H_2S environment. Fortunately, the relief valve was never called upon to operate, for after nine months of service the spring was found to have broken.

The National Association of Corrosion Engineers (NACE) Standard MR-01-75 (2) provides information regarding the selection of materials to be resistant to sulfide stress cracking (SSC), a form of hydrogen-assisted cracking. The standard indicates that for the operating conditions, 1000 ppm of H_2S and 1000 psi total pressure (indicated by the V in Fig. 12-15), a material should have been selected that was resistant to SSC. MR-01-75 lists applicable materials for springs intended for such in-service conditions. The allowable hardness is normally restricted to below 22 HRC (Rockwell C Scale Harness), whereas the failed spring was a martensitic

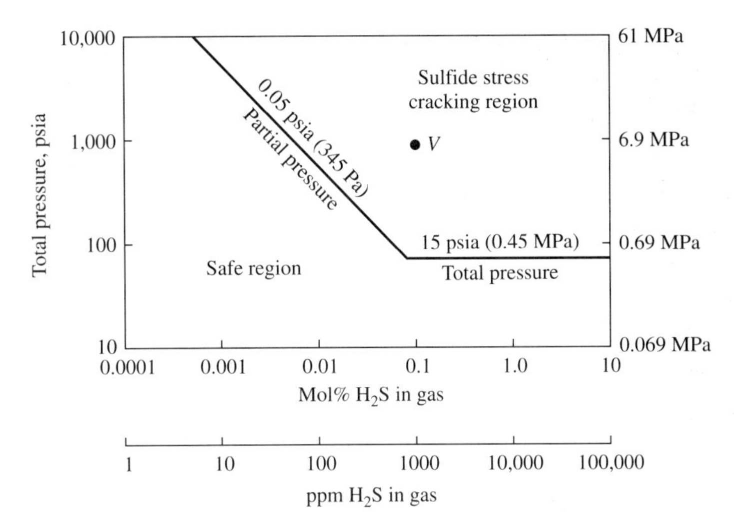

Fig. 12-15. Graph showing the limit of susceptibility of steels to sulfide stress cracking (SSC) in sour gas systems. Point V represents the conditions as seen by the pressure release valve spring. (After NACE Standard MR0175-2001.)

stainless steel with a hardness of 45–50 HRC. This failure was clearly due to a lack of awareness of the danger of hydrogen-assisted cracking in the presence of H_2S.

B. Failure of a Ladder Rung

Step-rungs are used to enable workers to mount the towers that support high-tension lines. At the lower levels of a tower, the rungs are removable to prevent children from climbing the towers. Workers install portable rungs in slots as they mount a tower and remove them as they come down. The rungs are made from a single rod of normalized 1018 steel which has been cold formed to a shape consisting of the horizontal rung and two vertical legs welded together, with the vertical legs fitting into the slots on the tower. On one occasion, as a rung was being tapped into a slot, the rung failed, and an investigation into the cause of failure followed. The appearance of the fracture surface is shown in Fig. 12-16 (1). After the cold forming operation, the rung had been cleaned and phosphated prior to painting in order to improve the adhesion of the paint. The dark region in the lower portion of Fig. 12-16 contains phosphorous, indicating that hydrogen-assisted cracking in this region occurred during the phosphating process. Fracture adjacent to the dark region occurred by a cleavage mechanism. Failure of the remaining ligament at the top of the figure occurred by a ductile tearing process.

Hydrogen is produced in the phosphating process, but generally does not lead to cracking in low-strength steels such as 1018. However, in this case, the properties of the steel had been modified by the cold forming operation. The local hardness at the inside of a bend was equivalent to that of steel of 120 ksi tensile strength, as

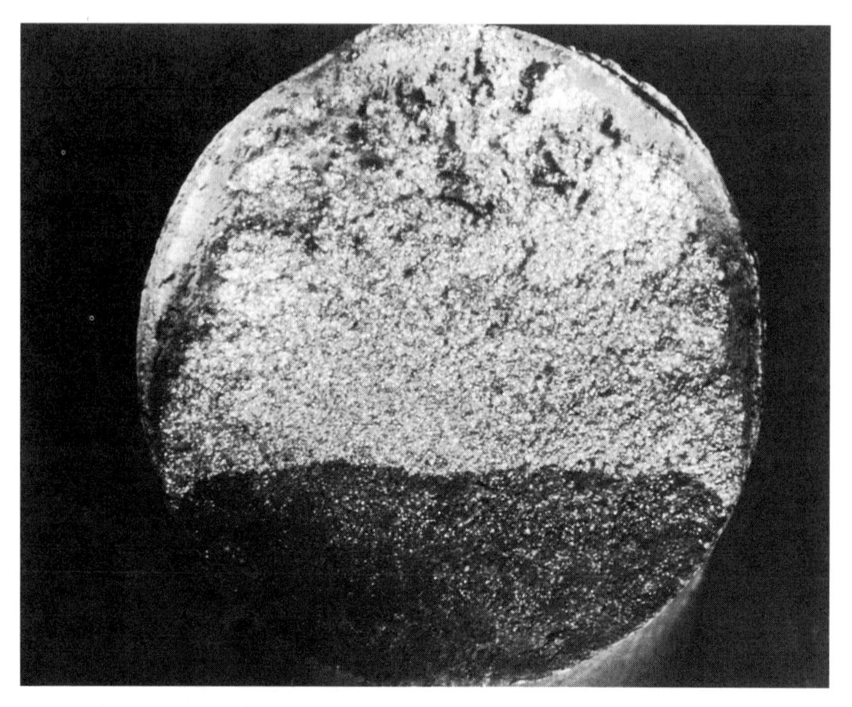

Fig. 12-16. Ladder rung fracture surface appearance. (Reprinted from Materials Characterization, Vol. 26, A. J. McEvily and I. LeMay, Hydrogen assisted cracking, pages 253–258, Copyright 1991, with permission from Elsevier Science, 1.)

compared to the original strength of 72 ksi, and in addition, the bending operation had resulted in a state of residual tensile stress at the inside of the bend. The combination of high hardness, residual tensile stress, and hydrogen ions led to crack formation. The solution to this cracking problem was relatively simple. The flame from the welding torch that was used weld the legs together after forming was directed at the bends to bring them to a cherry red color and then the rung was allowed to cool in air. This procedure softened the steel and removed the residual tensile stresses, thereby rendering the material's resistance to hydrogen-assisted cracking.

VI. CRACKING IN OIL AND GAS PIPELINES

The prevention of oil or gas pipeline failures is a matter of prime concern in the petroleum industry, and ASME and API codes are widely used in specifying, fabricating, and inspecting the steels used for pipelines. Nevertheless, on average, two failures per day due to corrosion-induced leaks or ruptures have occurred in Alberta, Canada, alone (3). Pipelines are often coated with a bituminous layer to protect against corrosion, and, if laid on a seabed, they may be encased in concrete to protect against external damage. To further guard against corrosion, the pipeline may be cathodically protected with an impressed current or by sacrificial anodes of aluminum or magnesium. For internal corrosion control, the most widely used method is the use of inhibitors. If a pipeline containing oil were to fail, there would not only be the loss of product, but also the danger of damage to the environment. When a pressurized gas pipeline fails, not only may there be an explosion endangering lives, but also a kilometer or more of pipeline may be destroyed due to a running crack that moves faster than the gas pressure can be relaxed.

These pipelines can be examined by internal electromagnetic or ultrasonic inspection devices known as "smart pigs" (pipe internal gauging system), which move along inside the pipeline with the oil or gas, measuring wall thicknesses, metal loss, and internal diameter, and detecting dents and flaws. In addition, external examinations using nondestructive examination techniques, such as radiographic and ultrasonic inspection, are used to check the welds for defects and to determine wall thickness. Because of economic considerations, the control of risk may take a higher priority than the control of corrosion per se (4). Risk R is defined as the probability of failure P multiplied by the consequence of failure C:

$$R = P \times C. \tag{12-10}$$

For example, a higher probability of failure may be tolerated in a remote area than in a densely populated area because the consequence of a failure may be much lower in the remote region.

Much of the concern with respect to fracture of pipelines in addition to general corrosion has to do with hydrogen-assisted cracking, particularly when hydrogen sulfide is present in the moisture contained within the gas or oil. A gas that has a hydrogen sulfide concentration of less than 20 ppm is known as sweet gas, and problems due to H_2S are not severe. When the concentration of H_2S exceeds 20 ppm, however, the deleterious effects of H_2S become pronounced, and for this reason the gas is referred to as a sour gas. The reason why H_2S is of concern is that, when it dissociates in water, the sulfide anion acts as a "poison" and promotes the penetration of hydrogen ions into a metal. Hydrogen ions are produced by the corrosion reaction between iron and moist sulfide (5). In pipeline, steels cracks that form because of hydrogen embrittlement (HIC) often develop at and follow manganese sulfide stringers because the interface is weak and hydrogen is trapped at such inclusions. Planar cracks can form near mid-thickness of the pipe, where the concentration of manganese sulfide particles is greatest, but they do not reduce the structural integrity of the pipe per se. However, several noncoplanar cracks may be linked up by short radial cracks, forming a "step wise crack," and thereby a critical crack length for fracture under the influence of the hoop stress may be reached.

Cracking in a buried pipeline can occur because of potential differences that develop between the soil and the pipe. For example, a 34-inch diameter, spirally welded Grade X-60 spheroidized steel pipe with a wall thickness of 7.8 mm (0.305 inch) was used in the fabrication of a pipeline in accord with API Specification 5L. The maximum allowable pressure in the line was 52 bar (775 psi), but leakage occurred at an operating pressure of 42 bar (626 psi). The line was cathodically protected by an impressed current as well as by sacrificial magnesium anodes. At the failure location, it was noted that the bituminous asphalt coating had been stripped off, and that the supporting ground had subsided, thereby inducing additional stresses at a low strain rate into the steel. Cracks had initiated on the outside of the pipe, and tended to follow the manganese sulfide inclusions. SEM examination revealed evidence of transgranular, quasi-cleavage-like brittle fracture. A jagged crack appearance with a tendency for stepwise cracking suggested that hydrogen was involved in the cracking process.

Cathodic protection is widely used to protect gas pipelines from corrosion. However, hydrogen evolution can occur during cathodic protection when the pipe-soil potential is more negative than −850 mV with reference to a Cu/CuSO4 reference electrode. In one case, the actual pipe-soil potential was −1100 mV, which was sufficiently negative for hydrogen evolution. In addition, the sacrificial anodes had come in direct contact with the steel, further lowering the potential. It was considered that slow plastic deformation due to the subsidence of the soil had occurred, which had facilitated the entry of hydrogen into the steel. To check on this possibility, slow strain rate tests (strain rate = 10^{-6}/sec) were carried out in artificial brackish groundwater with steel specimens held at a potential of −1100 mV. The reduction in area at fracture in these tests dropped from a normal value of over 60% to less than 30%, and it was concluded that the leakage had occurred because of hydrogen-assisted cracking (6).

In another instance (1), a steel pipeline had been heat-treated to have a spheroidized microstructure, and because of a relatively low yield strength (60 ksi), it was not considered to be susceptible to hydrogen-assisted cracking. The pipeline was laid in winter, and soil and snow were trapped within the line. No cleaning was done until the summer months, by which time the snow in the line had melted. The ground through which the line had been laid had an acid topsoil and an alkaline lower layer, and when they were mixed together with water, a visible reaction occurred. When the line was pressure tested in summer, numerous intergranular failures occurred at bends in the line where laying stresses occurred. At the failures, there was an absence of corrosion products, further evidence that hydrogen-assisted cracking had occurred because of poor housekeeping procedures in laying the line.

Stress corrosion cracking has been observed in gas-transmission pipelines downstream of a compressor station (7) where the pipe temperatures are the highest. The cracks originated on the external surface, and a black oxide layer formed on the fracture surfaces. Iron carbonate or bicarbonate was detected on the fracture surfaces, having been produced at a pore in the pipe coating as the hydroxide created by the cathodic potential was converted by the CO2 in the soil to a carbonate-bicarbonate environment. The fractures were intergranular and typical of SCC, the cracks were branched.

VII. CRACK ARRESTERS AND PIPELINE REINFORCEMENT

Steel rings are used as crack arrester devices in steel gas-transmission pipelines to reduce the possibility of long running longitudinal cracks. These arresters have the effect of reducing the pipe opening as the crack propagates. This decreases the available crack-driving force and, as a result, crack arrest can take place. This essentially is a second line of defense against catastrophic failure in the event that crack initiation cannot always be prevented (8).

For economic reasons, it is often undesirable to shut down a pipeline to replace sections in which defects or loss of wall thickness have been discovered during inspection. Instead a "sleeve" is used to reinforce the original section. These sleeves are sections of steel that snugly encase the original pipeline and are welded in place. A modified method uses an epoxy layer between the original pipe and the outer steel sleeve. When the epoxy sets, it expands and can create a radial pressure equal to that in the pipeline, thereby taking load off the original pipe and transferring it to the reinforcing sleeve.

VIII. PLATING PROBLEMS

During the electroplating of steel with protective coatings of chromium, cadmium, or zinc, large amounts of hydrogen are generated at the metal surface, some of which diffuses into the steel. In steels of high hardness, this hydrogen can lead to embrittlement. For example, it was found that zinc-plated steel screws of 28–36 HRC failed after being in service only two weeks. This particular problem was remedied when the fasteners were replaced by zinc-coated ones having a lower hardness, 100 Rockwell B Hardness (HRB).

In cases where plated, high-strength steels must be used, it is common practice to bake out the components after plating. The solubility of hydrogen at room temperature is extremely low, of the order of 1 ppm. A plating operation results in a supersaturation of hydrogen, and some of this hydrogen will diffuse out of the steel after the plating operation. However, critical amounts of hydrogen will be retained in the steel, particularly at so-called trapping sites. To reduce the hydrogen content, it is common practice to bake out the component after plating. It has been shown (9) that, after baking a hydrogen-charged 4130 steel at 150°C (302°F) for 24 hours, embrittlement is almost completely eliminated.

IX. CASE STUDIES

A. Welding Electrodes (10)

A highway bridge in Melbourne, Australia, was supported by girders made of a high-strength low-alloy steel (yield strength 550 MPa, 80 ksi). The girders were reinforced at mid-span by welding cover plates on to the lower flange at the erection site. The cover plates were narrower than the flange and were machine-welded to the flange by a pair of single-pass fillet welds. In order to smooth out the load transfer between the reinforced and nonreinforced portions of the flange, the width of the coverplate was reduced gradually over the last 460 mm of its length from its full width of 325 mm down to 76 mm. It is important to note that the tapered regions, as well as the transverse ends of the coverplates, were welded by hand using three-pass fillet welds. After a year of service, a partial collapse of the span occurred as the result of progressive crack formation at the ends of the coverplates. Some of these cracks had existed for some time, as indicated by paint primer and rust on the crack surfaces. All fractures had initiated in the heat-affected zone (HAZ) at the toe of the cover plates, and showed no evidence of corrosion products.

The HAZ cracks were most likely due to hydrogen-assisted cracking, with the source of the hydrogen being the moisture in the flux coatings on the electrodes. The specifications called for the flux coatings to be of a low-hydrogen type, but the flux coatings could still have absorbed moisture from the atmosphere. It was therefore decided to bake out the electrodes immediately prior to use at a temperature of 150°C for thirty minutes. After the accident, however, it was determined that a higher temperature should have been used. Further, there was evidence that the electrodes used in the hand-welding operations had been lying about and exposed to the atmosphere prior to use.

Concern about moisture absorbed in the flux coatings arises because during welding, the moisture provides hydrogen that is dissolved in the weld pool and diffuses into the HAZ. As the HAZ cools, the steel becomes increasingly saturated with hydrogen, and upon transformation from austenite to ferrite, bainite, or martensite, the solubility of hydrogen in the steel is markedly decreased. There is a strong tendency for the hydrogen either to escape from the metal or to be trapped at sites such as dislocations, grain boundaries, and particle-matrix interfaces. The shrinkage of the weld metal during cooling leads to the development of tensile residual stresses in the HAZ, which increase with the hardness of the weld metal. The tensile residual stresses are locally further increased by the stress concentrations associated with the irregular geometry of the weld toes. Rapid cooling can also result in the presence of brittle, untempered martensite within the HAZ, which is susceptible to hydrogen-assisted cracking.

In the case of the Melbourne highway bridge, it was concluded that the combination of hydrogen, tensile residual stresses, and a susceptible HAZ resulted in the observed delayed cracking.

B. Stack Corrosion (11)

A 123-m high flue gas stack of 3-m diameter was built from 7-m tubular sections of a self-weathering copper-bearing steel. The wall thickness varied from 38 mm in the lower sections to 19 mm in the upper sections. Flanges were welded at each end of the sections and provided the means for bolting sections together. After 18 months of service, severe corrosion was observed on the upper two-thirds of the interior surface of the stack, with metal losses approaching 1 mm. In addition, bolt failures due to fatigue were found in both the upper and lower sections. The severe interior corrosion was attributed the formation of H_2SO_4 as the result of the condensation of sulfur-containing flue gases on the upper 100 m of the stack. The corrosive fluid also ran into the bolted joints, and caused additional damage. Examination of the broken bolts also showed evidence of fatigue due to the cyclic nature of the wind loading of the stack. The fatigue problem had been made more severe because of the mode of fabrication of the flanges. The flanges were fillet welded to the tubular sections, and then triangular reinforcing gusset plates were added. These welding operations were performed without first bolting the flanges together. As a result, severe distortions were introduced when the flanges were finally bolted together, which increased the mean stress on the bolts and decreased their resistance to fatigue under the fluctuating wind loads.

Subsequent repairs were extensive, and included lining the upper tube section with stainless steel, welding stainless steel cover plates over each flange location, completly insulating the exterior of the stack so that the dew point temperature, 150°C, would be above the rim of the stack, and adding a tuned mass damper to reduce wind-induced deflections.

C. Backing Rings

In the 1960s, it was common practice to locate a backing ring in the interior of a steam pipe at the site of a butt weld in order to insure that full penetration of the weld would be achieved. However, it was found that branched stress corrosion cracks often initiated in areas of pitting at such sites due to (1) a stress concentration at the weld root, and (2) the interface between the backing ring and the pipe providing an ideal place for ions such as Cl^-, which emanated from the water phase during shutdown periods, to concentrate. Thermal cycling may have assisted crack growth when units were used over a long periods for peak load operations only (11).

X. PITTING CORROSION OF HOUSEHOLD COPPER TUBING

The pH level in the home water supply carried by copper tubing can be a critical factor affecting corrosion and pitting. In some municipalities, the pH is maintained at a level of 7.5. A pH value below 7 is not desirable because of concerns about lead from piping in the distribution system entering the water. A pH above 8 is usually not desirable because the water becomes hard and less suitable for washing purposes. However, in some communities, the water may contain large amounts of carbonates, and it is necessary to maintain a high pH, above 8.5, to prevent the carbonate from precipitating onto the walls of the copper tubing to form a layer of cupric carbonate hydroxide [$CuCO_3*Cu(OH)_2$] a greenish-white substance whose color is a direct indication that copper is corroding to form the compound. At susceptible sites, this corrosion of copper may occur as pitting, and as the pits grow, the wall of the copper piping may be penetrated and leaks develop. The solders used in joining sections of copper tubing are also of concern. The appropriate solder fluxes minimize the content of constituents that are known to promote pitting corrosion on copper, such as zinc and chlorides. It is also noted that the depth of the maximum size corrosion pit developed as a function of time has been studied using the statistics of the extreme (17).

XI. PROBLEMS WITH HYDROGEN AT ELEVATED TEMPERATURES

The above examples relate to the effect of hydrogen at room temperature. At elevated temperatures, there can be a different hydrogen-related problem. For example, a carbon steel that was used in a catalytic cracking unit was found to have lost its strength because of hydrogen diffusing into the metal. The hydrogen combined

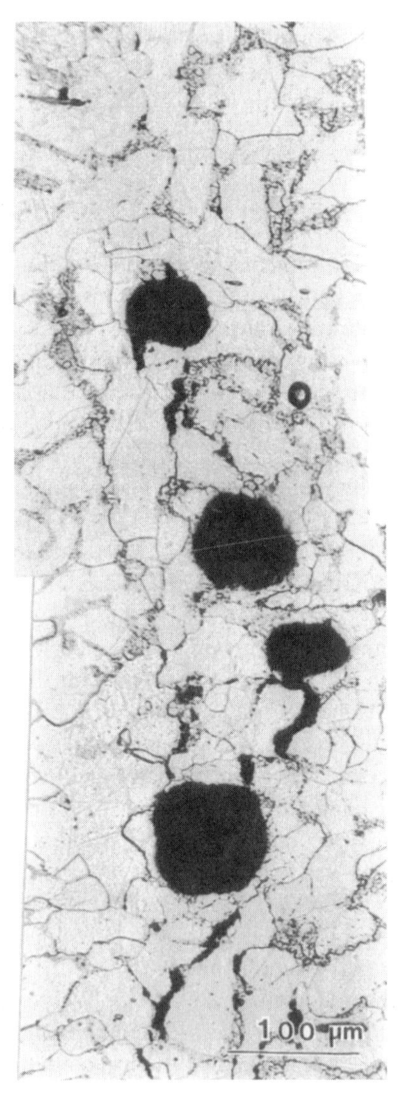

Fig. 12-17. Example of hydrogen attack. (Reprinted from Materials Characterization, Vol. 26, A. J. McEvily and I. Le May, Hydrogen assisted cracking, pages 253–258, Copyright 1991, with permission from Elsevier Science, 1. Photo courtesy of Dr. Tito Luiz da Silveira, Rio de Janiero, Brazil.)

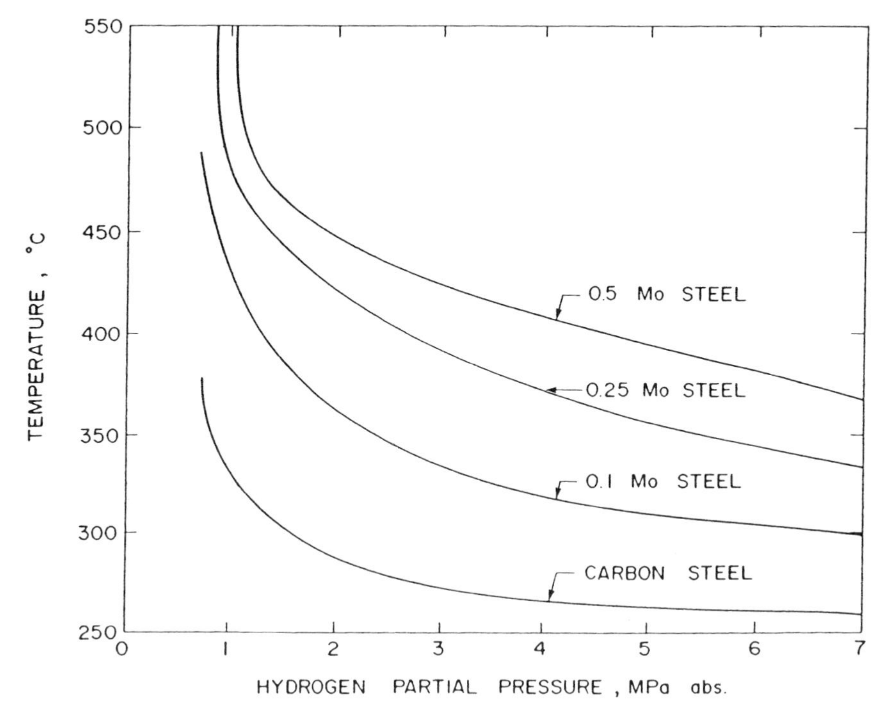

Fig. 12-18. Nelson Diagram. (From API, 12.)

with carbon in the Fe_3C to form methane and eliminated the pearlite phase. Figure 12-17 (1) shows the microstructure with voids caused by the methane gas produced in the hydrogen-methane reaction. Such problems can be avoided by the use of low-alloyed steels, which contain carbide-stabilizing elements. The Nelson diagram (12), Fig. 12-18, is an important guide in the selection of materials for service in hydrogen environments at elevated temperatures.

XII. HOT CORROSION (SULFIDATION)

The nickel-base and cobalt-base components of a jet aircraft engine are subject to a form of high-temperature corrosion known as sulfidation, which is related to the amount of sulfur in the fuel. Sodium sulfate in the environment can lead to a depletion of the alloying elements required to form a protective film on the parts. When nickel combines with sulfur, the melting point drops from 1455°C for pure nickel to 635°C for Ni-22% sulfur. Unless the alloy is protected by a coating, excessive creep and intergranular fracture may occur. The high velocity of flow through the engine can also result in loss of material through a process referred to as erosion-corrosion.

XIII. SUMMARY

The corrosion process can degrade resistance to fatigue and fracture by causing loss of cross-section, etch pits, pack-out, stress-corrosion cracking, and accelerating fatigue crack growth rates. In addition, the corrosion process can lead to the introduction of hydrogen into materials with consequent embrittlement. Where aggressive environments are anticipated, it is particularly important that consideration be given to the selection of the material best suited to withstand the effects of the environment.

REFERENCES

(1) A. J. McEvily and I. Le May, Hydrogen Assisted Cracking, Materials Characterization, vol. 26, 1991, pp. 253–258.
(2) NACE Standard MR-01-75 (1980 rev.), Sulfide Stress Cracking Resistant Material for Oil Field Equipment, NACE, Houston, TX ,1980.
(3) Pipeline Performance in Alberta 1980–1997, Report 98-G, Alberta Energy and Utilities Board, Calgary, Alberta, Canada, December 1998.
(4) R. W. Revie, Trends in Corrosion R&D, with a Focus on the Pipeline Industry, http://www.nrcan.gc.ca/picon/conference2/winston_revie2.htm, Jan. 1, 2000.
(5) D. A. Jones, Principles and Prevention of Corrosion, Macmillan, New York, 1992.
(6) A. Punter, A. T. Fikkers, and G. Vanstaen, Hydrogen-Induced Stress Corrosion Cracking on a Pipeline, Materials Performance, vol. 31, June 1992.
(7) R. J. Eifner and J. F. Kiefner, ASM Metals Handbook, 9th ed., ASM, Materials Park, OH, 1986, pp. 695–706.
(8) P. E. O'Donoghue and Z. Zhuang, A Finite Element Model for Crack Arrestor Design in Gas Pipelines, Fatigue and Fract. Eng. Mats. and Structs., vol. 22, 1999, pp. 59–66.
(9) J. O. Morlett, H. Johnson, and A. Trioano, J. Iron Steel Inst., vol. 189, 1958, p. 37.
(10) D. R. H. Jones, Materials Failure Analysis, Engineering Materials 3, Pergamon Press, 1993.
(11) C. Bagnall, H. C. Furtado, and I. Le May, in Lifetime Management and Evaluation of Plant, Structures and Components, ed. by J. H. Edwards et al., EMAS, UK, West Midlands, 1999, pp. 295–302.
(12) Steels for Hydrogen Service at Elevated Temperatures and Pressures in Petroleum Refineries and Petrochemical Plants, API Publication 941, American Petroleum Institute, Washington, DC, June 1977.
(13) B. F. Brown, in The Theory of Stress Corrosion Cracking in Alloys, ed. by J. C. Scully, NATO Scientific Affairs Division, Brussels, 1971, pp. 186–204.
(14) B. F. Brown, Stress Corrosion Cracking Control Measures, NBS Monograph 156, National Bureau of Standards, Washington, DC, 1977.
(15) E. N. Pugh, in The Theory of Stress Corrosion Cracking in Alloys, ed. by J. C. Scully, NATO Scientific Affairs Division, Brussels, 1971, pp. 418–441.
(16) E. Denhard, Corrosion, vol. 16, no. 7, 1960, p. 131.
(17) Application of Statistics of the Extreme to Corrosion (in Japanese), ed. by M. Kowaka, J. Soc. Prevention of Corrosion, Maruzen, Tokyo, 1984.
(18) D. O. Sprowls, M. B. Shumaker, and J. D. Walsh, Marshall Space Flight Center Contract No. NAS 8-21487, Final Report, Part I, May 31, 1973.

13

Flaw Detection

I. INTRODUCTION

A combination of a fracture mechanics analysis and flaw-detection procedures provides the most reliable method for insuring the integrity of critical structures. The most common nondestructive examination (NDE) methods for the detection of flaws are visual inspection, dye penetrant inspection, eddy current inspection, ultrasonic inspection, and radiography. Acoustic emission is also used but to a lesser extent. This chapter discusses the principal aspects of these inspection methods. Also included are case studies to illustrate the dangers that can be involved when nondestructive examinations are not carried out properly.

II. VISUAL EXAMINATION (VE)

Visual examination for surface flaws is widely used, for it is obviously a simple, fast, and inexpensive method. However, VE lacks good resolution, and eye fatigue, boredom, and distractions can reduce the quality of this type of inspection. Good illumination is needed, and the observations can be aided by such items as magnifiers, dental mirrors, tubular borescopes, and glass-fiberscopes (1).

III. PENETRANT TESTING (PT)

Penetrant testing is used for the detection of surface flaws. It is a low-cost, easy-to-use, rapid, and portable method. The procedure in making such a PT inspection is as follows (1):

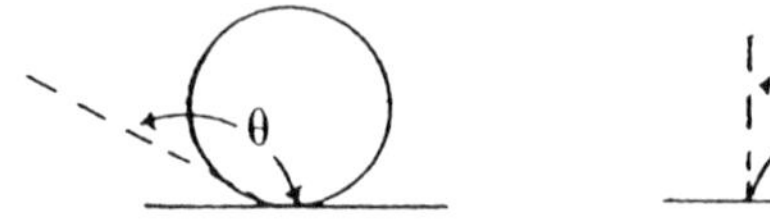

Fig. 13-1. The effect of the contact angle θ on wettability. (From Survey of NDT, 1.)

1. Carefully clean the surface.
2. Apply the penetrant, a liquid of low viscosity and high surface tension containing dyes, suspensions of colored particles, or a radioactive gas. Allow time for penetration ("dwell time").
3. Remove the excess penetrant (special agents can be used).
4. Apply a developer to the surface. The developer acts as a blotter to draw the penetrant to the surface from the flaw, and provides a contrasting background.
5. Observe with good illumination or, if fluorescent penetrants are used, with UV in a darkened area.

For the penetrant to flow over the surface and migrate into flaws open to the surface, the liquid must wet the surface, that is, the angle θ in Fig. 13-1 must be less than 90°. A low viscosity is also desired. The depth of penetration increases with the surface tension T, as shown in the following example.

A. Example

Determine the depth d_L that the liquid penetrant will penetrate down into a flaw of width w, depth d_0, and surface extent e. The resistance to penetration is provided by the buildup of pressure as the entrapped air is compressed.

The volume of the flaw is $d_0 we$, and $PV = C$ (ideal gas law at constant T). From equilibrium of forces: $Pwe = 2T \cos \theta e$,

$$\frac{C}{V} w = 2T \cos \theta, \qquad V = we(d_0 - d_L) \tag{13-1}$$

$$\frac{C}{we(d_0 - d_L)} w = 2T \cos \theta, \qquad \frac{C}{2eT \cos \theta} = d_0 - d_L \tag{13-2}$$

$$d_L = d_0 - \frac{C}{2eT\cos\theta} = d_0 - \frac{P_0V_0}{2eT\cos\theta} = d_0 - \frac{P_0wd_0}{2T\cos\theta}$$
$$= d_0\left(1 - \frac{P_0w}{2T\cos\theta}\right). \tag{13-3}$$

Therefore, the smaller the width of the flaw, and the higher the surface tension and the smaller θ, the greater will be the depth of penetration.

As indicated in Figures 13-1 and 13-2, in order for a liquid to wet a solid, the angle θ must be less than 90°.

$$\gamma_{sa} = \gamma_{sl} + \gamma_{la}\cos\theta = \gamma_{sl} + T\cos\theta, \tag{13-4}$$
$$T\cos\theta = \gamma_{sa} - \gamma_{sl} = \Delta\gamma_{sa-sl}, \tag{13-5}$$

where γ and T are surface tensions. The subscripts a, s, and l refer to air, solid, and liquid, respectively.

The surface tension of a liquid can be determined from a capillary rise experiment, Fig. 13-2,

$$T = \frac{h(\rho_l - \rho_g)gr}{2\cos\theta},$$

where h is the capillary rise, ρ_l is the density of the liquid, ρ_g is the density of the gas in the capillary, g is the acceleration due to gravity, r is the radius of the capillary, and θ is the contact angle of the liquid in the capillary). If T is assumed to be fixed, cos θ then depends on the difference in the surface energies of the solid and

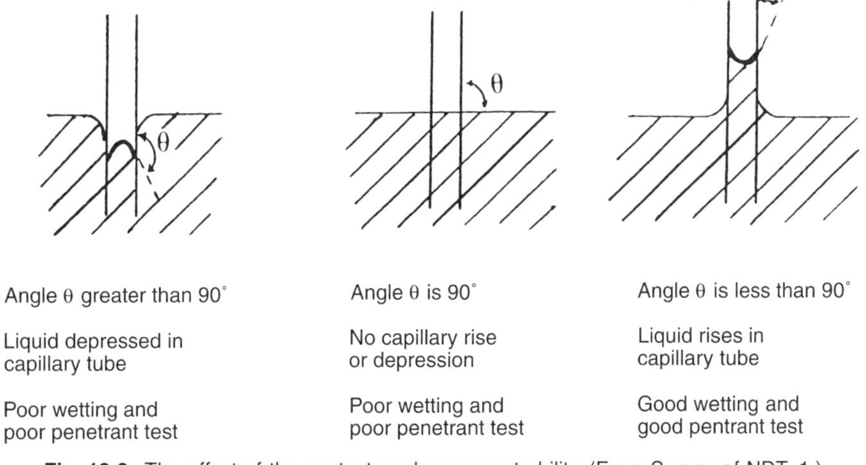

Fig. 13-2. The effect of the contact angle on penetrability. (From Survey of NDT, 1.)

the solid-liquid interface. The smaller γ_{sl}, for a given T, the larger will be $\cos \theta$, and hence the smaller θ will be, and the liquid is therefore considered to be more wettable. Making the liquid more wettable increases $\cos \theta$ and therefore also increases the depth of penetration.

The surface energy of water is 72 dynes/cm. The surface energy for solids is much higher, thousands of dynes per centimeter. The surface energy of glass is 1.75×10^3 dynes/cm^2 = 1.75 J m^{-2}.

IV. CASE STUDY: SIOUX CITY DC-10 AIRCRAFT (2)

A. Summary

On July 19, 1989, at 1516, a DC-10-10, tail number N1819U, operated by United Airlines as Flight 232, while en route from Denver to Chicago, experienced a catastrophic failure of the tail-mounted, No. 2 engine during cruise flight at Mach 0.83. The separation, fragmentation, and forceful discharge of stage 1 fan rotor assembly parts from the No. 2 engine led to the loss of three hydraulic systems that powered the airplane's flight controls. The flight crew experienced severe difficulties controlling the airplane, which subsequently crashed during an attempted landing at Sioux Gateway Airport, IA. There were 285 passengers and 11 crew members onboard. One flight attendant and 110 passengers were fatally injured.

In aircraft gas turbine engines, the engine shroud is designed to contain small pieces of hardware, such as turbine blades, should they break loose during engine operation. However, since the disk is much more massive, its kinetic energy is so high that it would not be feasible to design the engine shroud to contain the disk fragments should the disk fracture. Instead, reliance is placed upon the detection of fatigue cracks long before they reach a critical size. However, in this case this procedure did not work, for the NTSB determined that the probable cause of this accident was the inadequate consideration given to human factor limitations in the inspection and quality control procedures used by the United Airlines engine overhaul facility. This resulted in the failure to detect a fatigue crack originating from a previously undetected metallurgical defect located in a critical area of the stage 1 fan disk that was manufactured by General Electric Aircraft Engines (GEAC). The subsequent catastrophic disintegration of the disk resulted in the liberation of debris in a pattern of distribution and with energy levels that exceeded the level of protection provided by design features of the hydraulic systems that operate the DC-10's flight controls.

B. Factual Information

About one hour and seven minutes after takeoff, the flight crew heard a loud bang, followed by vibration and shuddering of the airframe. The No. 2 engine had failed, and the airplane's main hydraulic pressure and quantity gauges indicated zero. The flight crew deployed the air driven generator (ADG), which powers an auxiliary hydraulic pump, and the hydraulic pump was selected "on." However, this action did not restore the hydraulic power. The captain reduced the thrust on the wing-mounted, left-side engine (No. 1 engine), and the airplane, which had entered a right descending turn, began to roll to the wings-level attitude. Fuel was jettisoned to the level of the automatic system cutoff, leaving 33,500 lb. About eleven minutes before landing, the landing gear was extended by means of an alternate gear-extension procedure. The airplane touched down slightly to the left of runway at 1600 (44 minutes after the engine failure). First ground contact was made by the right wing tip, followed by the right main landing gear. Witnesses saw the plane ignite and cartwheel. The airplane was destroyed by impact and fire.

The airplane, valued at $21,000,000, had been delivered to United in 1971. It had acquired 43,401 flight hours and 16,997 cycles. It was powered by GEAE CF6-6D high-bypass-fraction turbine engines. The CF6-6 engine had been certified by the FAA in 1970. The total time on the No. 2 engine was 42,436 hours, and the number of cycles it had undergone was 16,899. 760 cycles had elapsed since the last maintenance, and the engine had been installed on October 15, 1988. The engine could be installed in either the wing or tail position.

C. Stage 1 Fan Disk Historical Data

The stage 1 fan disk had been processed in the GEAE Evandale, OH, factory from September 3 to December 11, 1971. It was a new part in an engine that was shipped to Douglas Aircraft on January 22, 1972, where it was installed on a new DC-10-10. During the next 17 years the engine was removed six times for inspection, the last being in February 1988, 760 cycles prior to the accident. This disk was accepted after each of six fluorescent penetrant inspections (FPI). (The second inspection was at the GEAE Airline Service Department in Ontario, CA. The other five inspections were at the UAL CF6 Overhaul Shop in San Francisco.) FPI is the accepted industry inspection technique for interrogating nonferrous (nonmagnetic) component surfaces for discontinuities or cracks. The technique relies upon the ability of the penetrant (a low-viscosity penetrating oil containing fluorescent dyes) to penetrate by capillary action into surface discontinuities of the component being inspected. The penetrant fluid is applied to the surface and allowed to penetrate into any surface discontinuities. Excess penetrant is then removed from the component surface. A developer is then applied to the component surface to act as a blotter and draw the penetrant back out of the surface discontinuity, producing indications that fluoresce under ultraviolet (black) lighting.

About three months after the accident, parts of the No. 2 stage 1 engine fan disk were found in farm fields near Alta, IA. There were two sections that constituted nearly the entire disk, each with fan blade segments attached. These parts were initially taken to the GEAE facility in Evendale, OH, for examination under the direction of the NTSB. The smaller of the two segments was later taken to the NTSB laboratory in Washington, DC, for further evaluation.

The stage 1 fan disk weighs about 370 pounds (168 kg) and is a machined titanium alloy forging (Ti-6Al -4V) about 32 inches (81 cm) in diameter. The various portions of the disk are the rim, the bore, the web, and the disk arm. These are shown in Fig. 13-3. The rim is about 5 inches (12.7 cm) thick and is the outboard portion

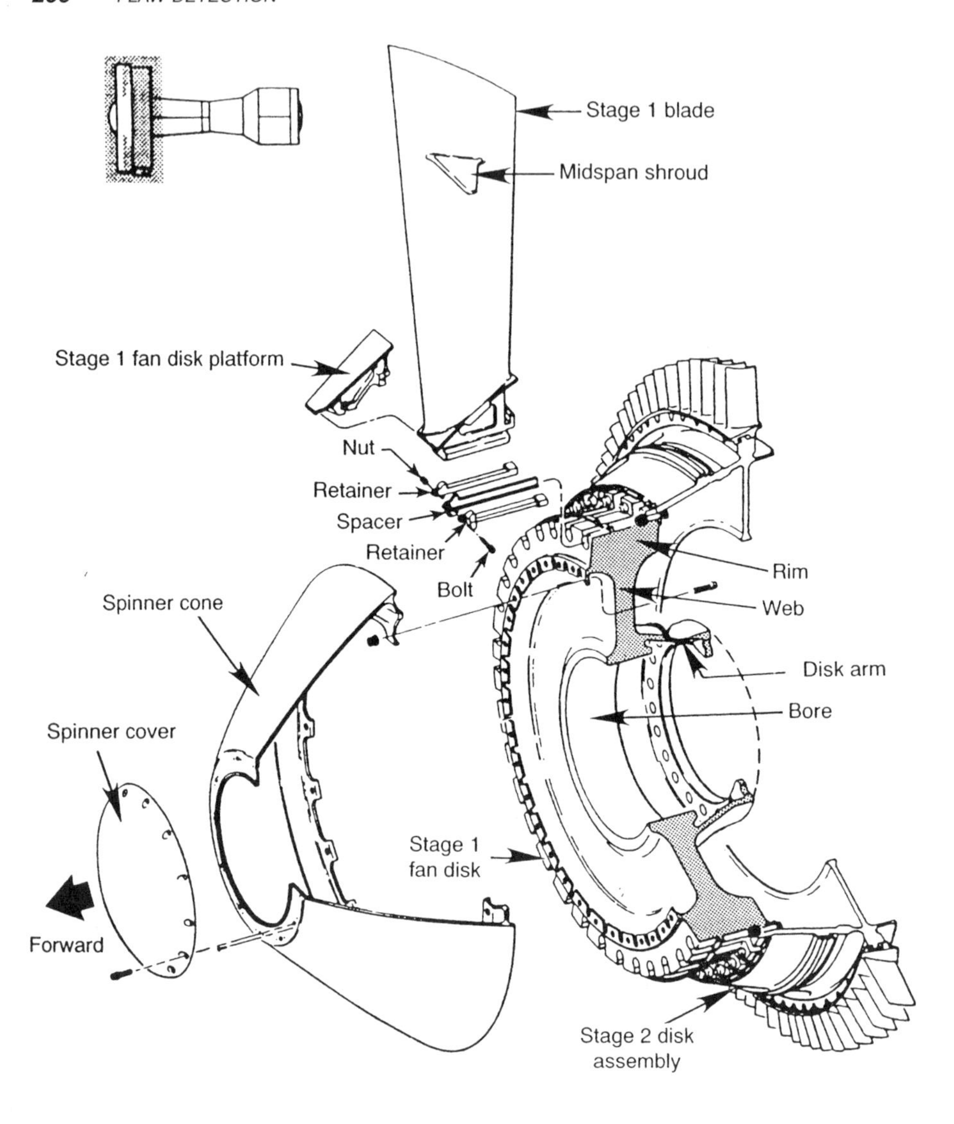

(*a*)

Fig. 13-3. The stage 1 fan disk. (From NTSB, 2.)

of the disk. The rim contains the axial "dovetail" slots, which retain the fan blades. The stage 2 fan disk is bolted to the aft surface of the rim. The bore is about 3 inches (7.6 cm) thick, and is the enlarged portion of the disk adjacent to the 11-inch (27.9-cm) center hole. Extending between the rim and the bore is the disk web, which is about 0.75 inch (1.9 cm) thick. The conical disk arm extends aft from the

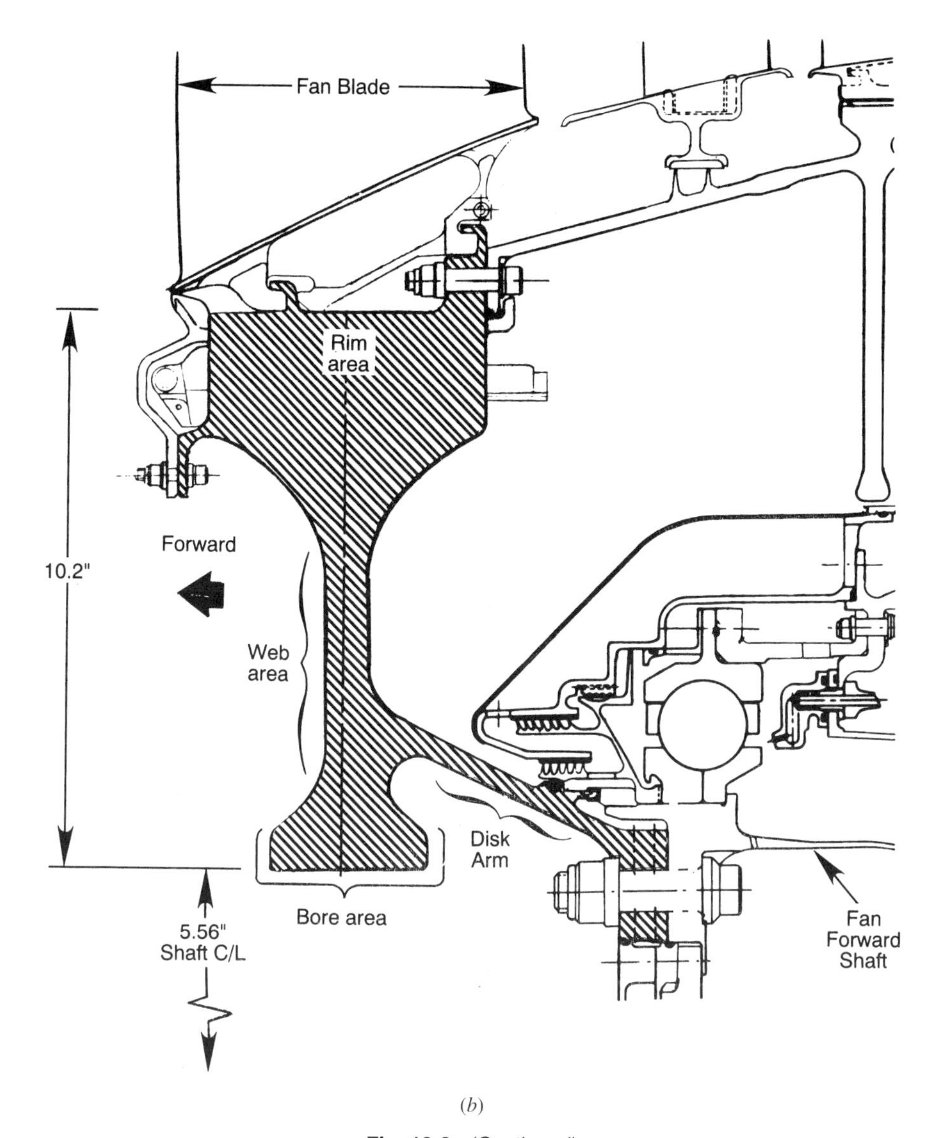

(b)

Fig. 13-3. *(Continued)*

web at a diameter of about 16 inches (40.6 cm). The conical arm diameter decreases in the aft direction to about 10 inches (25.4 cm) at the disk arm flange where the disk bolts to the fan forward shaft.

The primary loads imposed on the stage 1 fan disk are radially outboard loads in the dovetail slots, which arise as the disk holds the fan blades against centrifugal force during rotation of the assembly. These loads result in radial stresses in the disk rim, which decrease toward the disk bore, where they are supplanted by circumferential (hoop) stresses. These hoop stresses are at a maximum along the inside diameter of the bore. Because the disk arm acts to strengthen the aft face of the disk, the forward corner of the bore is the area on the disk that experiences the maximum hoop stress.

D. Examination of the No. 2 Engine Stage 1 Fan Disk

The two recovered pieces of the No. 2 engine fan disk comprised the entire separated disk, with the exception of an unrecovered dovetail post. Figure 13-4 shows the reconstructed pieces of the disk after the larger disk piece had been cut during metallurgical examination. The gap between the smaller and larger pieces does not represent missing material, but is the result of mechanical deformation that occurred during disk separation. The disk contained two principal fracture areas, resulting in about one-third of the rim separating from the remainder of the disk. One of the fracture areas progressed largely circumferentially through the web and the rim. The other was on a near-radial plane, progressing through the bore, the web, the disk arm, and the rim. Features on the circumferential fracture were typical of overstress separation stemming from multiple origin areas in the radius between the disk arm and the web. The near-radial fracture surface also contained overstress features over

Fig. 13-4. The reconstructed stage 1 fan disk of the No. 2 engine. (From NTSB, 2.)

Fig. 13-5. Fatigue crack fracture area of the stage 1 fan disk. (*a*) The fatigue crack extends from the cavity (arrow "C") to the dashed line position. The discolored portion of the fatigue crack is between the cavity and the dotted line. Magnification: 2.26×. (*b*) An enlarged view of the discolored area of the fatigue crack. Arrowhead indicates cracking direction away from the cavity. (From NTSB, 2.)

most of its surface. However, on this break the overstress features stemmed from a preexisting radial/axial fatigue crack region in the bore of the disk. Figure 13-5 shows the fatigue region at the bore.

The fatigue crack initiated near a small cavity on the surface of the disk bore, about 0.86 inch (2.2 cm) aft of the forward face of the bore. A portion of the fatigue crack around the origin was slightly discolored. The topography of the main fracture surface in the fatigue zone was the same outside the discolored area as it was inside the discolored area. The overall sizes of the fatigue crack, the discolored area, and the cavity were as follows:

	Length	Radial Depth
Fatigue zone	1.24 inch (3.1 cm)	0.56 inch (1.4 cm)**
Discolored area	0.48 inch (1.2 cm)	0.18 inch (0.46 cm)
Cavity	0.055 inch (0.14 cm)*	0.015 inch (0.04 cm)

*The width of the cavity across the mating fracture surfaces was also 0.055 inch (0.14 cm).

**GEAE estimated that the fatigue crack was 0.5 inch (1.3 cm) deep at the time of the last inspection.

Fractographic, metallographic, and chemical analyses of the fatigue region revealed the presence of an abnormal nitrogen-stabilized hard-α phase around the cavity. This phase extended slightly outboard of the cavity to a maximum radial depth of 0.018 inch (0.046 cm) and an overall length of 0.44 inch (1.12 cm). The altered microstructure associated with this hard phase extended significantly beyond the area containing only stabilized α-structure, gradually blending into the normal microstructure, which consisted of a mixture of approximately equal amounts of primary α-structure and transformed β-structure. The stabilized α-region contained microcracks that were generally oriented parallel to the cavity surface, and some microporosity was also found.

SEM examination revealed that fatigue striations existed just outboard of the stabilized alpha region. Close to the cavity, areas of brittle fracture intermixed with ductile appearing bands were observed. The fatigue striation spacing generally increased as the distance from the origin increased. At a radial distance of 0.145 inch (0.37 cm), areas with more closely spaced striations were also found. The more closely spaced striations were referred to as minor striations, and the more widely spaced striations were referred to as major striations. The total number of major striations along a radially outward direction from the origin area was estimated by graphically integrating a plot of striation density versus distance. The estimate correlated reasonably well with the total number of flight cycles on the disk, indicating that fatigue crack growth had been taking place since early in the life of the disk.

Analytical procedures were developed to determine if chemical residues from the United Airlines dye penetrant inspection were present on the fatigue fracture surface. The fracture surface was gently washed in deionized water, and then in an ultrasonic unit also using deionized water. Secondary ion mass spectroscopy (SIMS) measurements showed an ion fragmentation pattern that was consistent with chemical compounds used in the FPI fluid. Gas chromatograph (GC) mass spectroscopy (MS) of the water used in the ultrasonic cleaning provided further evidence for the presence of the dye penetrant on the fracture surface.

E. Fan Disk Manufacturing Process and Hard Alpha

The three primary steps in the manufacturing of titanium alloy fan disks are: material processing, forging, and final machining. In the first step, metals are combined in a heat, which is numbered, and then processed into a titanium alloy ingot in a vacuum furnace melting operation. The ingot is then mechanically formed into a billet. The billet is then cut into smaller pieces, which are then forged. The next step is the machining of the forged shape into the final configuration and the shot peening of fatigue-critical regions.

Hard α-inclusions constitute one of the three main anomalies in titanium alloys, the other two being high-density inclusions and alpha segregates or β-flecks. Most of the hard α-inclusions result from a local excess of nitrogen and/or oxygen introduced through atmospheric reactions with titanium in the molten state. A typical hard α-inclusion contains an enriched α-zone in the α-plus β-matrix, and voids and cracks are frequently associated with these zones. Hard α-inclusions have a melting point significantly greater than the normal structure, and to promote their melting or dissolution, it is desirable to increase the temperature of the molten pool in the furnace or to increase the time during which the material is in the molten state. Successive melting operations, such as double or triple vacuum remelting, provide additional opportunities for dissolution of hard inclusions but do not guarantee their complete dissolution, and all GEAC fan disks manufactured after January 1972 have been triple-vacuum remelted. The double-melted 16-inch (40.6-cm) billet for the failed disk was produced in 1971, and the NTSB concluded that, at the time of the manufacture of the disk, the cavity at the fatigue origin was filled with hard alpha material, making the defect difficult to detect by ultrasonic inspection.

During manufacture of the failed disk, GEAE inspected the forged shape using a macroetch technique to bring out any anomalies on the surface, and a final FPI procedure was carried out in December of 1971 with no anomalies found. The United Airlines FPI procedure warned inspectors that titanium parts resist the capillary action of the penetrant and that "complete penetrant coverage is required for these materials." Also, United Airlines inspectors were cautioned not to overwash the parts or the penetrant might be flushed out of true indications. Along with certain other areas, the disk bore was mentioned as a critical area for inspection.

F. Initiation and Propagation of the Fatigue Crack

Fracture mechanics calculations by GEAE were consistent with the fatigue crack reaching critical size at the time of separation. The analysis was also consistent with fatigue crack initiation on the first application of stress from a defect slightly larger than the cavity found at the fatigue origin. The NTSB concluded that the hard alpha defect area cracked upon the first application of stress during the disk's initial exposure to full-thrust engine power conditions and that the crack continued to grow and entered material unaffected by the hard alpha defect. From that point, the crack followed established fracture mechanics predictions for theTi-6Al-4V alloy until it reached the critical size for fracture. It seems a bit odd that fracture occurred under cruise conditions rather than under the full-thrust conditions of the last takeoff.

V. MAGNETIC PARTICLE TESTING (MT)

Magnetic particle testing is used for the detection of surface and near surface flaws, and is low cost, fast, and portable. However, the material must be ferromagnetic, and the surface must be clean. The method is more sensitive than dye penetrant inspection for the detection of tightly closed cracks, which may also contain corrosion products (3).

The MT procedure involves (1):

1. The establishment of a magnetic field in the part, usually by an electric current.

2. The application of magnetic particles in the form of liquid suspension or dry powder.
3. An examination and evaluation.
4. A repeat test with the magnetic field at 90° to the original direction of magnetization.

If there is a flaw, a leakage flux of the magnetic field in the part will develop, as shown in Fig. 13-6. This leakage flux will attract the magnetic particles and indicate the location of the discontinuity. For an indication to form, it is necessary that the flaw be at a large angle (90°) to the lines of flux. If the flaw is parallel to the lines of flux, no leakage current will occur, and the flaw will be undetected. At angles between 0° and 90°, the leakage flux will be proportional to the sine of the angle between the discontinuity and the lines of flux.

When an electric current passes through a conductor, a magnetic field is formed around the conductor. The direction of the lines of flux are determined by the right-hand rule. If the current flows along a bar, the lines of flux will form as a circular field along the conductor, and the MT method will be sensitive to longitudinal flaws. If the conductor is a coil around the specimen, then the lines of flux will form a longitudinal field in the specimen, Fig. 13-7, and the MT method will be sensitive to circumferential flaws. In the magnetic particle inspection of an aircraft crankshaft, a "head shot" using alternating current or rectified AC, which develops greater penetrating ability, is used to establish a circular magnetic field for the detection of longitudinal flaws (Fig. 13-7*a*). A coil is then used to establish a longitudinal magnetic field for the detection of circumferential flaws (Fig. 13-7*b*). In both cases, the applied currents can range up to 3000 amps. After the examination, the part may be demagnetized by applying an AC current whose magnitude gradually decreases in a controlled manner.

The current I in amps required to develop an adequate magnetic field with a coil is given as $I = 45{,}000D/LN$, where D is the specimen diameter, L is the specimen length, and N is the number of turns. L/D should be at least 2 to avoid end effects, but not greater than 15.

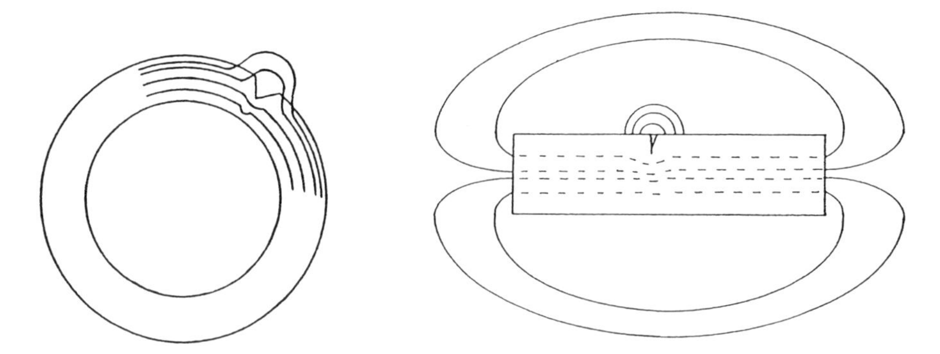

Fig. 13-6. Leakage of magnetic flux lines due to surface discontinuities. (From Survey of NDT, 1.)

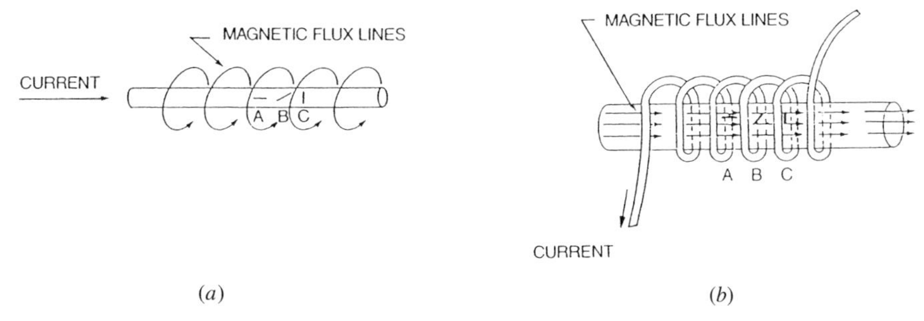

Fig. 13-7. The magnetic fields set up by (*a*) a "head shot" and (*b*) by a coil. (From Survey of NDT, 1.)

A standard practice is to use fluorescent particles that are suspended in a liquid. The liquid is sprayed over the part being examined, and those particles attracted to a flaw will remain while the rest are washed away. The observation is then made under ultraviolet (UV) light.

VI. CASE STUDY: FAILURE OF AN AIRCRAFT CRANKSHAFT

A single-engine, piston/propeller, four-passenger private plane had taken off with a pilot and two passengers when, shortly after the takeoff, power was lost. The pilot attempted an emergency landing, but unfortunately the plane struck power lines and all perished. It was subsequently determined that the cause of the accident was the fatigue failure of the crankshaft. Since the crankshaft had been examined by the magnetic particle method only 80 flight hours prior to the crash without detecting a defect, the organization that carried out the inspection was accused of negligence.

The crankshaft was a 4340 low-alloy steel forging. At the time of manufacture, it had been nitrided to improve both its wear and fatigue resistance. After 2000 hours in service, the crankshaft had been worn enough to warrant surface refinishing by grinding and renitriding. After an additional 1000 hours of service, the engine was overhauled. No work was done on the crankshaft, but because it had been removed from the engine during the overhaul, it was mag particle inspected in accord with regulations. Eighty flight hours later, the accident occurred. Examination of the failed crankshaft revealed that in addition to the fatal fatigue crack, a number of additional small surface cracks were present, as shown in Fig. 13-8. It is thought that these cracks are grinding cracks that had been introduced at the time of 2000-hour overhaul. It is noted that the cracks are filled with a white substance, which is related to the renitriding process. This material is Fe_4N and is a by-product of nitriding. Where it forms on the smooth surface of the crankshaft, it is easily removed by polishing prior to putting the crankshaft back into service.

Figure 13-9 is the phase equilibrium diagram for the Fe-N system. Of particular interest is the fact that the Fe_4N phase is ferromagnetic at room temperature. Not only that, but as shown in Fig. 13-10, the magnetic properties of this phase are

Fig. 13-8. An example of tight grinding cracks completely filled with nitride. Magnification: 50×.

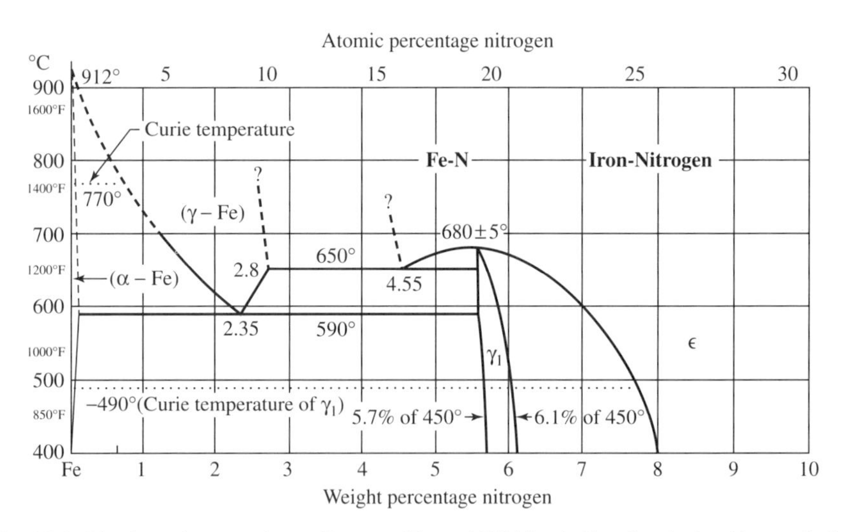

Fig. 13-9. The iron-nitrogen phase diagram. (From ASM Metals Handbook, 4, with permission of ASM International®.)

$\sigma_{sat} = 203$ emu/gm
$\sigma_{res} = 0.32$ emu/gm
$H_R = 6.0$ Oe

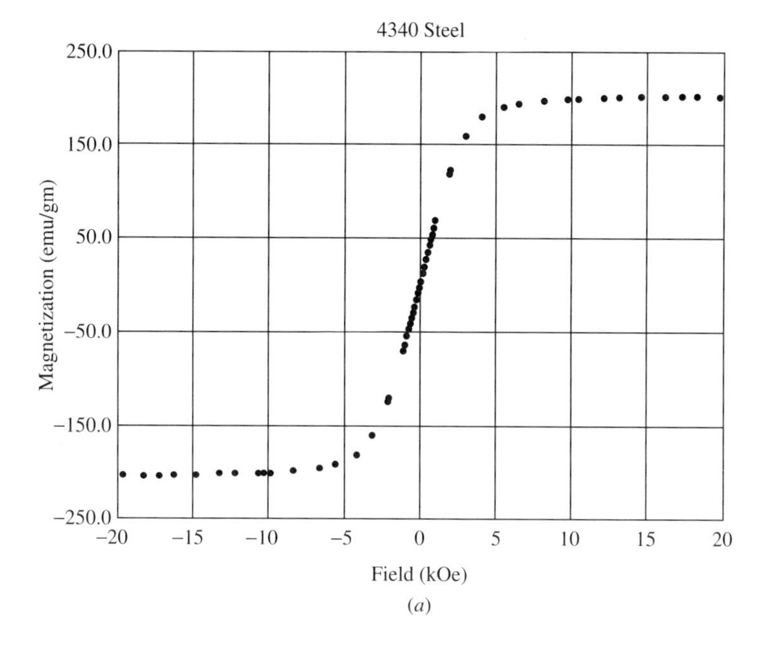

(*a*)

$\sigma_{sat} = 187$ emu/gm
$\sigma_{res} = 1.0$ emu/gm
$H_R = 12.5$ Oe

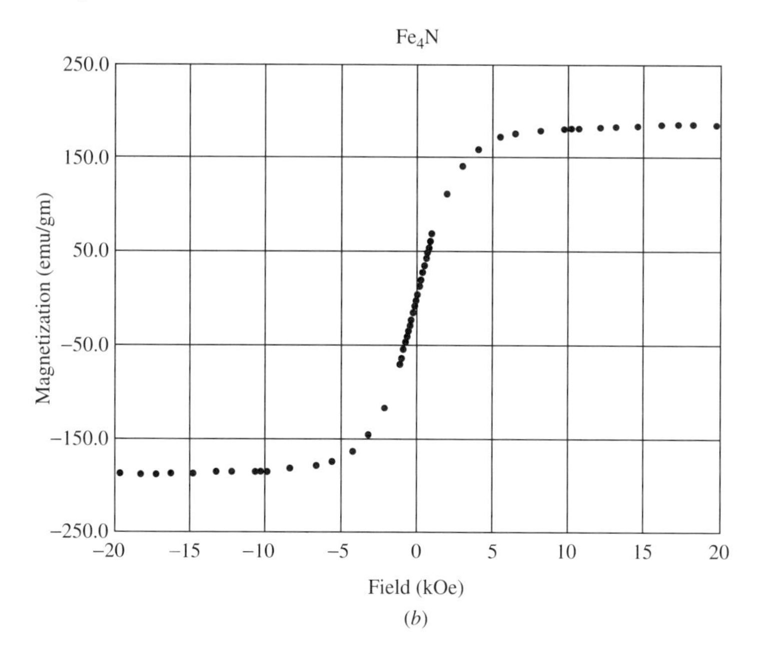

(*b*)

Fig. 13-10. Comparison of the magnetization behavior of 4340 steel with Fe_4N (γ'). (Courtesy of W. A. Hines, University of Connecticut.)

remarkably similar to those of 4340 steel. The implication is that, if a tight crack such as a grinding crack is completely full of this phase, then in a magnetic field there will be no disruption of the flux lines and the crack will go undetected if examined by the magnetic particle method. To check on this possibility, a fatigue crack was grown in a 4340 compact specimen, and the specimen was examined by the mag particle method. Figure 13-11*a* clearly shows the distribution of magnetic particles along the fatigue crack. The specimen was then nitrided and reexamined by the magnetic particle method. As shown in Fig. 13-11*b*, the fatigue crack could no longer be detected.

The conclusion drawn is that the organization that performed the last mag particle inspection had not been negligent. It was not possible to detect the grinding cracks after nitriding. It was the organization that had performed the 2000-hour overhaul that was at fault, for they should have inspected the crankshaft by the mag particle method prior to renitriding. Based upon an analysis of the fatigue striations, it appears that the grinding cracks did not propagate until after the second overhaul. An increase in the compression ratio at that time may have raised the stress intensity factor at the critical grinding crack to a level above the propagation threshold.

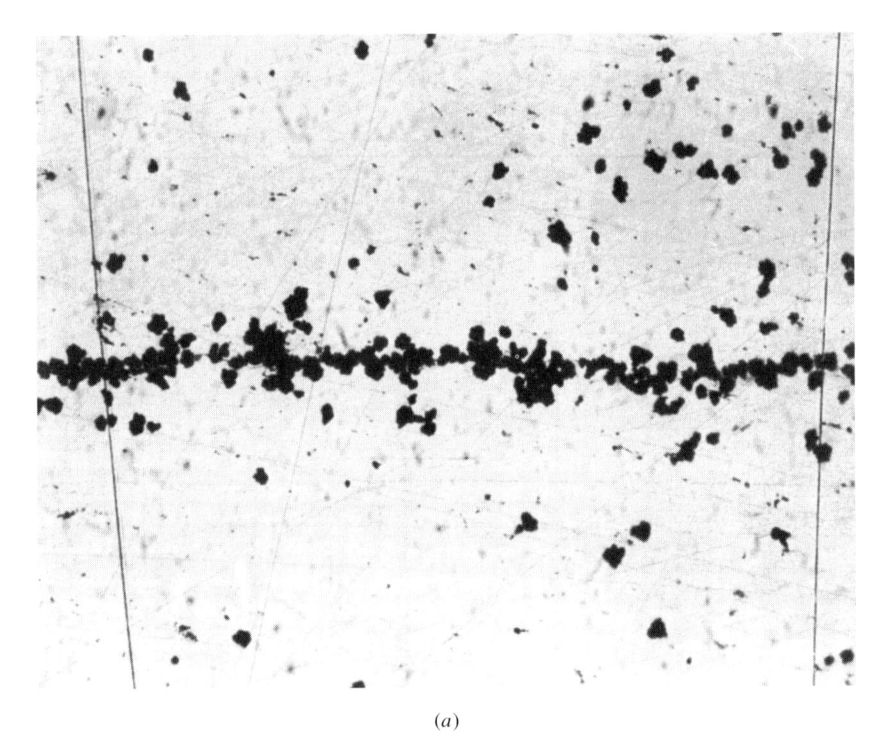

(*a*)

Fig. 13-11. Magnetic particle distribution at a fatigue crack. Magnification: 200×. (*a*) Prior to nitriding. (*b*) After nitriding with the white layer removed.

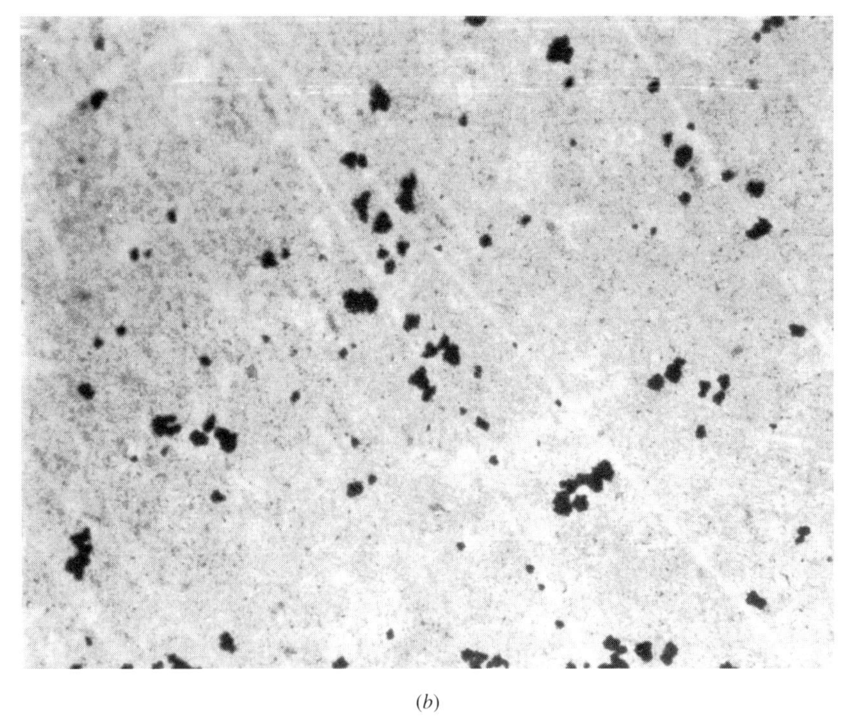

(*b*)

Fig. 13-11. *(Continued).*

VII. EDDY CURRENT TESTING (ET)

Eddy current testing is used for the detection of surface and near surface flaws (¼-inch maximum depth). It is rapid, can be automated, is sensitive, surface contact is not necessary, and it can provide a permanent record. The method is based on inducing small circular electrical currents (eddy currents) in metallic materials using a coil excited by electrical current, Fig. 13-12. Disruption of the eddy currents by a discontinuity is similar to the disruption of magnetic fields except that a wider variety of discontinuities and physical properties affect eddy currents. The depth of penetration S is frequency dependent, the higher the frequency, the less the penetration.

$$S(\text{inch}) = 1980\sqrt{r/\mu f}, \qquad S(\text{mm}) = 50{,}292\sqrt{r/\mu f},$$

where r is the resistivity in ohm-cm, μ is the magnetic permeability (taken to be 1 for nonmagnetic materials), and f is the frequency in hertz. The following table (1) lists the depths of penetration for several metals as a function of the frequency.

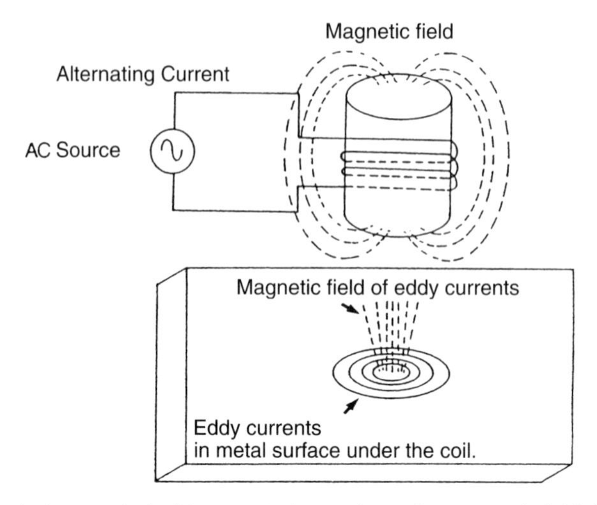

Fig. 13-12. The inducement of eddy currents by an alternating magnetic field. (From Survey of NDT, 1.)

Metal	Conductivity (% IACS)*	Standard Depth of Penetration** mils (mm) at a frequency of		
		1 kHz	100 kHz	10 MHz
Copper	100	80 (2.0)	8 (0.2)	0.8 (0.02)
Aluminum	61	160 (4.0)	16 (0.4)	1.6 (0.04)
Titanium	3.1	800 (20)	80 (2.0)	8.0 (0.2)
Iron	10.7	14 (0.36)	1.4 (0.04)	0.1 (0.004)
304 SS	2.5	550 (14.0)	55 (1.4)	

* International Annealed Copper Standard.
**Depth at which signal intensity is $1/e$ that at the surface.

A flaw in the specimen affects the eddy currents and produces a change in the impedance of the coil. Impedance changes in either primary or secondary (sensing) coils are sensed as changes in current through the coil or as phase changes in the voltages or currents. The readout of these changes may be the deflection of a meter, an oscilloscope presentation, a strip chart recording, lights or alarm activation, digital readouts, or operational control of manufacturing processes.

Specimen to probe effects include:

(a) Edge effects resulting from a distortion of the magnetic field near the edge of a specimen. Inspecting within $\frac{1}{8}$-inch from the edge of a nonmagnetic material or within 6 inches of the edge of magnetic material is likely to produce signal distortions.

(2) The gap between a circular specimen and the encircling coil can greatly affect readings. The closer the specimen comes to filing the hole in the center of the coil, the greater will be the sensitivity (fill factor = 1).

(c) A gap between specimen and probe decreases sensitivity because the field of the coil is strongest close to the coil. The "lift-off" effect can be used to measure the thickness of paint or nonmagnetic plating on a magnetic substrate. Probes often use spring loading to maintain a constant gap.

Eddy current inspections are also used to determine the thickness of coatings and other dimensional characteristics, but a degree of sophistication in the proper use of the technique is needed. In one case, the method was specified as a check on the positioning of cooling passages in turbine blades. However, none of the technicians was properly trained to carry out the inspection, and as a result, turbine blades with improperly positioned cooling passages entered service. Fortunately, no harm resulted from this oversight.

VIII. CASE STUDY: ALOHA AIRLINES

This case was discussed in Chapter 1. The eddy current method had been used prior to the accident in the inspection of the 737 Aloha Airlines fleet. However, readily detectable flaws such as the disbonding of skins and fatigue cracks emanating from multiple rivet holes were not discovered. Inadequate training and monitoring of the inspectors in the use of this technique were the main factors contributing to the failure of the method in this case.

IX. ULTRASONIC TESTING (UT)

Ultrasonic testing is used for the detection of surface and subsurface flaws (1). It can give the location and size of a flaw, is portable but slow, and a couplant is required. The orientation of the flaw is important, the results are operator dependent, and good standards are needed. Sound waves with frequencies of from 0.5 to 10 MHz are often used in inspections, much higher than the 20,000 Hz associated with the limit of human hearing. The most common ultrasonic inspection technique is called "pulse-echo," in which a pulse of sound waves is introduced into the specimen by a piezoelectric transducer (an anisotropic crystal capable of converting an applied voltage into vibration of the crystal or converting deformation of the crystal into a voltage. The pulse of sound waves travels through the specimen until reflected at the back surface or some discontinuity, Fig. 13-13. The oscilloscope presentation relates distance (time) traveled by the pulse of sound versus signal detected. The echo indication of the discontinuity will appear in the oscilloscope presentation in the same position relative to the front and back reflections as the discontinuity occupies in the specimen.

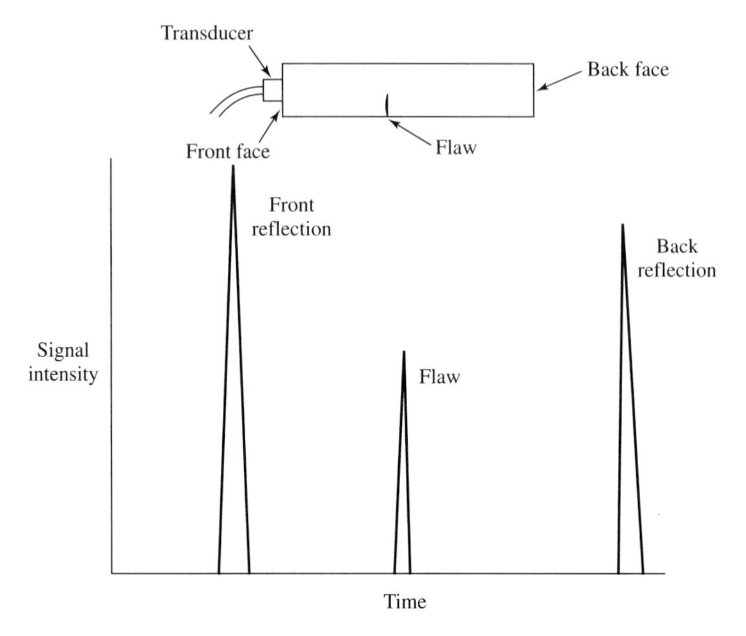

Fig. 13-13. Signals developed in the pulse-echo ultrasonic technique.

There are three scan modes, A, B, and C. In the A-scan mode, the one most widely used, a quantitative display of signal intensity (size of flaw) and time of flight (depth of flaw) is obtained at one point on the surface of the test piece. The method uses a fixed transducer and provides information on the flaw type, its depth, and location. In the B-scan mode, a quantitative display of time of flight along a line on the surface is obtained. In a B-scan, the transducer and the test piece move with respect to each other. The method can provide information on the size (in one direction), position, and depth, and to a certain degree, the shape and orientation of internal flaws. The C-span mode results in a two-dimensional, semiquantitative display of the echo intensity of the test piece. It displays a plane section of the test piece but provides no depth or orientation information. In a C-scan, the entire surface is traversed or rastered.

A problem with UT is the transmission of the ultrasonic energy from the transducer to the test piece. Placing a transducer in direct contact with a surface usually results in a very small amount of energy transfer because of the presence of air in the interface. The acoustic impedances Z of the air and the test piece are very different. ($Z = \rho V$, where ρ is the density and V is the wave velocity.) A "couplant" is used to couple the transducer ultrasonically with the test piece to ensure efficient sound energy transmission. The couplant acts to smooth out surface irregularities and to exclude all air between the surfaces. Ideally, the couplant should have an acoustic impedance between the transducer and the test piece. Two methods of coupling are used, immersion and contact.

In immersion testing, clean, deaerated water with a wetting agent is used as a couplant. In contact testing, transducers are held directly on the test surface with a thin liquid film for a couplant. The couplant is an oil- or greaselike material that fully wets the surface of the transducer and the test piece. Contact testing consists of three techniques, which are determined by the sound wave mode: normal beam (longitudinal wave), angle beam I (shear waves), and angle beam II (surface waves).

A. Normal Beam Technique

In the normal beam technique, the ultrasonic pulse is projected into the test piece perpendicular to the surface. This technique is further subdivided into pulse-echo techniques and transmission techniques. The transmission technique uses a detector on the back face of the test piece. The pulse echo technique uses either a single transducer (to send and receive signals) or two transducers (one to send the signal and the other to receive the reflected signals).

B. Critical Angles

For normal incidence, transmission and reflection occur with no change in beam direction. For other angles of incidence, however, some interesting and useful changes in beam direction occur in accord with Snell's law:

$$\frac{\sin \theta_I}{V_I} = \frac{\sin \theta_R}{V_R}, \tag{13-6}$$

where θ_I is the angle of incidence, θ_R is the angle of reflection or refraction, V_I is the velocity of the incident wave, and V_R is the velocity of the reflected or refracted wave.

When an incident longitudinal sound wave passes through an interface between materials with different acoustic impedances (see Fig. 13-14), Snell's law can be written as:

$$\frac{\sin \theta_I}{V_{L_I}} = \frac{\sin \theta_L}{V_{L_I}} = \frac{\sin \theta_T}{V_{T_i}} = \frac{\sin \theta_L}{V_{L_2}} = \frac{\sin \theta_T}{V_{T_2}}. \tag{13-7}$$

A single incident, nonnormal longitudinal beam results in two refracted waves, one a longitudinal wave and one a transverse wave, and a similar pair of reflected waves. As angle of incidence is increased, the angle of the refracted longitudinal wave approaches 90°. At 90° the first critical angle of incidence is found, Fig. 13-15*a*.

$$\sin \theta_{I_1} = \frac{V_{L_1}}{V_{L_2}}. \tag{13-8}$$

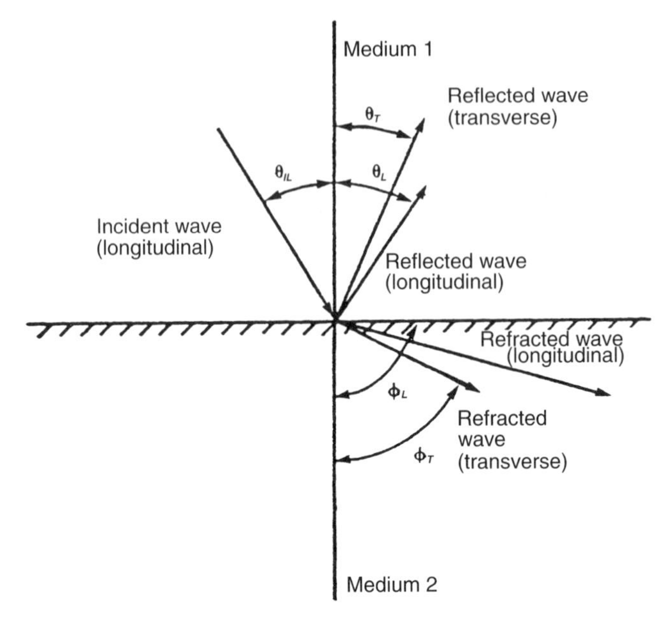

Fig. 13-14. Waves generated by the incidence of an oblique longitudinal sound wave on a medium of different acoustic properties. (From Survey of NDT, 1.)

As the angle of incidence is increased further, only shear waves are present in the second medium. The second critical angle is reached when the angle of the refracted transverse wave is increased to 90°, Fig. 13-15*b*:

$$\sin \theta_{I_2} = \frac{V_{L_1}}{V_{T_2}}. \tag{13-9}$$

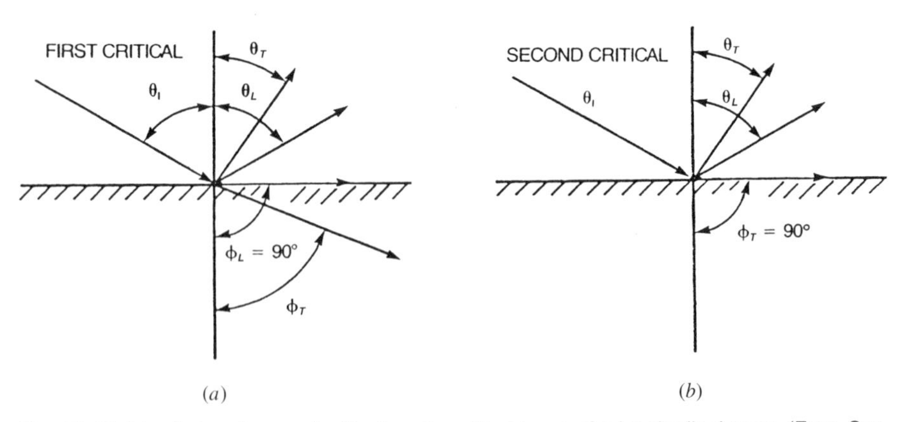

Fig. 13-15. The first and second critical angles of incidence of a longitudinal wave. (From Survey of NDT, 1.)

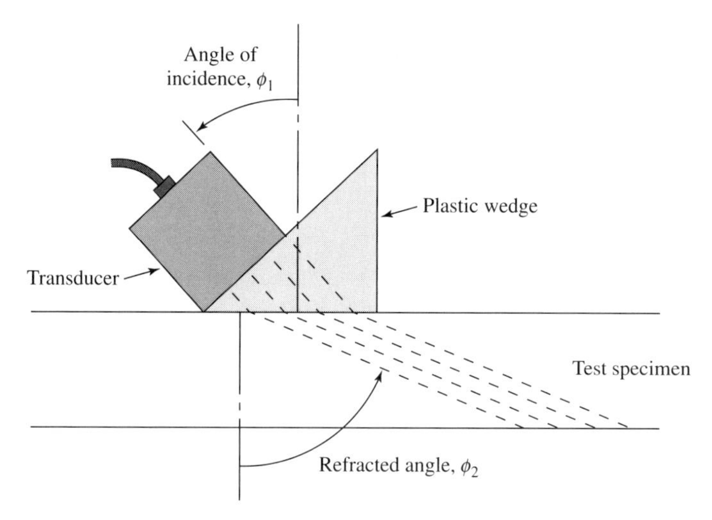

Fig. 13-16. Illustration of the setup for the angle beam technique with transducer mounted on a plastic wedge. (From Survey of NDT, 1.)

At the second critical angle, the shear wave has become a surface wave. If the angle of incidence is increased further, no energy will be transmitted to the second medium.

C. Angle Beam Techniques

The angle beam technique is used to transmit sound into the test piece at a predetermined angle to the surface. Figure 13-16 shows a typical angle beam search unit with the transducer mounted on a plastic wedge. Depending upon the angle of incidence, there may be mixed longitudinal and shear modes, shear modes only, or surface mode only produced in the test specimen as indicated in Fig. 13-14. Having more than one mode present in the test piece is unacceptable because interpretation becomes difficult. Therefore, angle beam transducers are always used at angles of incidence greater than the first critical angle. The use of the angle beam technique is illustrated in Fig 13-17. This method requires a signal detector. The method is also used with a single transducer-detetector unit.

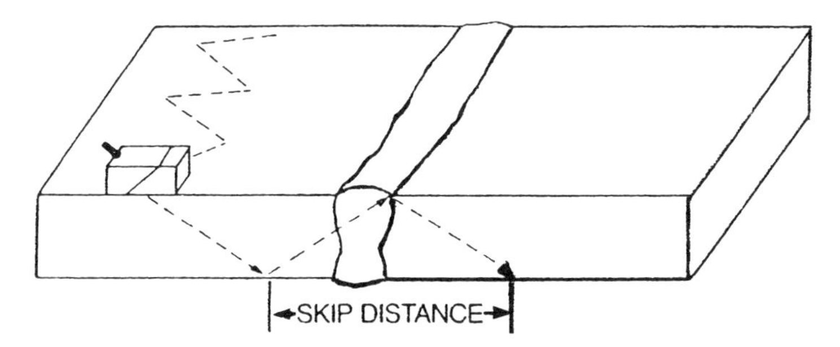

Fig. 13-17. Illustration of the use of the angle beam technique. (From Survey of NDT, 1.)

D. Wave Velocities and Acoustic Properties

The velocity of a plane longitudinal wave is given as

$$V_L = \left(\frac{E(1 - \nu)}{\rho(1 + \nu)(1 - 2\nu)} \right)^{1/2}. \tag{13-10}$$

The velocity of shear wave is given as

$$V_T = \left(\frac{G}{\rho} \right)^{1/2}. \tag{13-11}$$

The acoustic properties of some common materials are given below (1).

Material	Density, g/cm^3	V_L, cm/μsec	V_T, cm/μsec	Z_L, g/cm^2-μsec
Carbon steel	7.85	0.594	0.324	4
Al 2117-T4	2.80	0.625	0.310	1.75
304 SS	7.9	0.564	0.307	4.46
Air	0.000129	0.0331	——	0.00004
Glass	2.5	0.577	0.343	1.44
Lucite	1.18	0.267	0.112	0.32
Water	1.0	0.149	——	0.149
Copper	8.89	0.470	0.226	4.18

E. Other Considerations

(a) *The Reduction of Signal Intensity as a Function of the Distance Traveled (Attenuation):* $I = I_0 e^{-kd}$, where I is the intensity at d, I_0 is the incident intensity, k is the absorption coefficient, and d is the distance into the specimen.

(b) *The Near Field:* Close to the transducer emitting sound, the intensity of sound is very irregular, due to interference between the sound waves emitted from different parts of the transducer. The length of the near field L_{nf} is given as $L_{nf} = D^2/4\lambda$, where D is the diameter of the transducer and λ is the wavelength of the ultrasonic beam. Interpretation of data obtained from the near field may be difficult.

X. CASE STUDY: B747 AIRCRAFT

In Chapter 8, the fatigue failure of fuse pins on a B747 freighter was discussed. A related matter involves the ultrasonic inspection of the fuse pins. No cracks were detected at the time of the last inspection, and yet it appears that a fatigue crack probably was present on the inboard side of the inboard fuse pin. Why was this crack not found? The ultrasonic inspection procedure was based upon the assumption that any fatigue crack would grow normal to the axis of the fuse pin. However, the recovered portion of the outboard fuse pin indicated that the crack grew at an angle to the radial direction of the fuse pin. In the ultrasonic inspection technique a signal is sent from one end of the pin which is reflected from a radial crack as shown in Fig. 13-18*a* and appears on the CRT before the signal from the inclined portion of the fuse pin is received, Fig. 13-18*b*. However, if the crack were oriented as in Fig. 13-18*c*, the path length of the reflected signal would be longer and might not appear on the CRT in the expected location for a crack. Further, if the crack

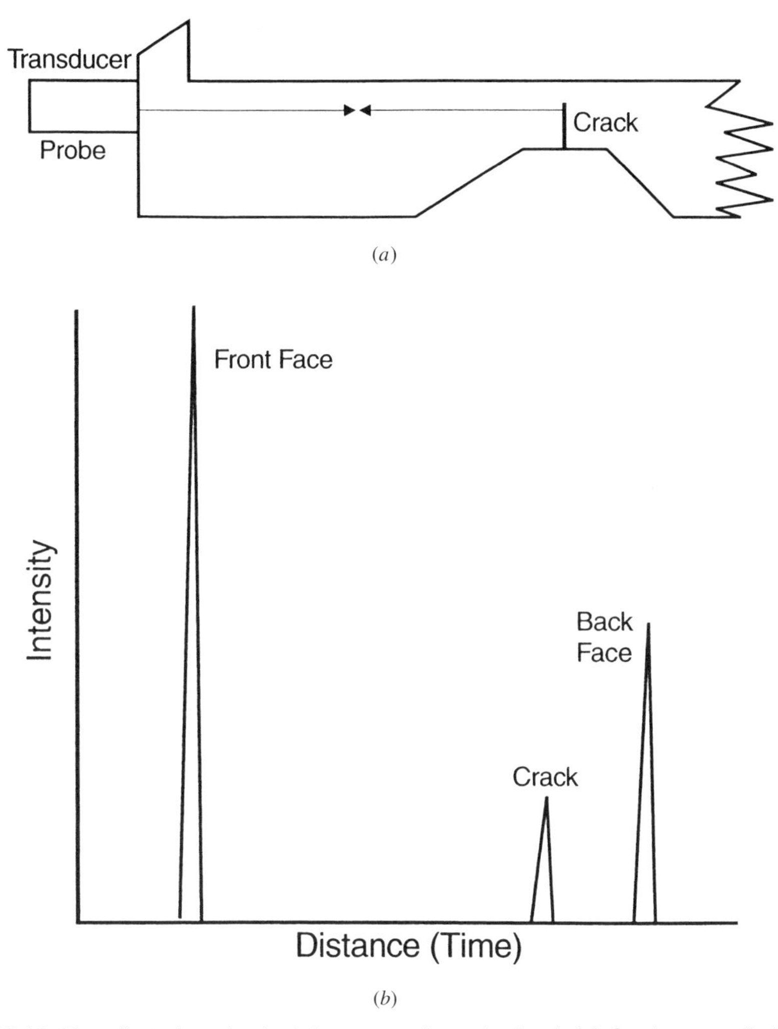

Fig. 13-18. The effect of crack orientation on an ultrasonic signal. (*a*) Crack perpendicular to ultrasonic beam. (*b*) Oscilloscope display corresponding to (*a*). (*c*) and (*d*) Ultrasonic signal paths when crack is not perpendicular to ultrasonic beam.

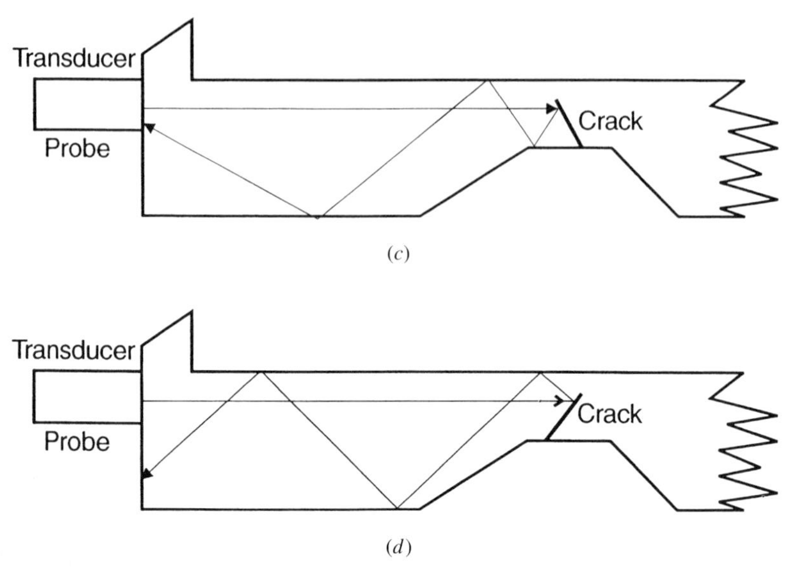

Fig. 13-18. *(Continued)*

were oriented as in Fig 13-18*d*, the return signal might not be detected at all. In addition to these complications, it was found that the fuse pins in service did not simply bend, as had been assumed, but instead developed a "crankshaft" configuration, which may also have had an effect on the ability to detect fatigue cracks. It is noted that the Netherlands Aviation Safety Board recommended that a review be made of the nondestructive inspection techniques

XI. RADIOGRAPHIC TESTING (RT)

Radiographic testing is used for the detection of subsurface flaws, particularly in welded joints. It is low cost, provides a permanent record, and is portable, but cannot detect laminations, is a radiation hazard, and needs trained operators (1).

In radiography, the specimen is exposed to a beam of X-rays, gamma rays, or neutrons. If the beam encounters a flaw of lesser density than that of the basic material, then more than the normal amount of radiation will transit that region and reach the detector, usually film. This leads to a darker area on the film when processed, Fig. 13-19.

GAMMA RADIATION SOURCES (1)

Source	Half-Life	Energy, MeV	Steel Thickness, inches	mm
Iridium-192	74 days	0.31, 0.47, 0.60	0.25–3	6–75
Cesium-137	30.1 years	0.66	0.5–4	13–100
Cobalt-60	5.3 years	1.17, 1.33	0.75–9	19–230

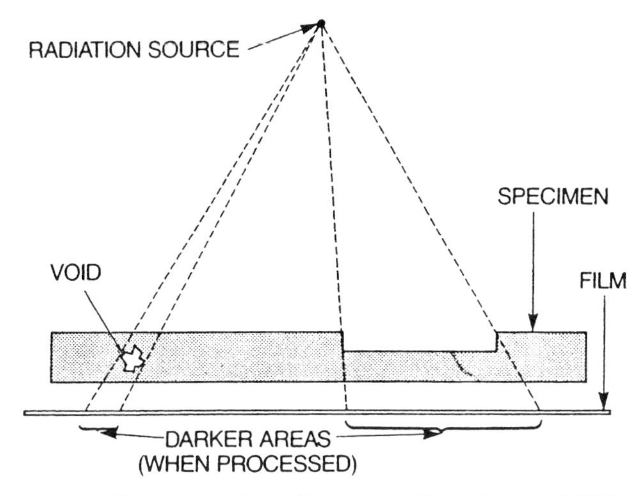

Fig. 13-19. The basic radiographic process. (From Survey of NDT, 1.)

Some X-ray machines can produce photons with energies in the 10–100 MeV range, which far exceeds the energies available from radioactive isotopes. Increasing the applied voltage used in generating the X-rays increases their penetrating power. The great advantage of isotopes is their portability for in situ inspection.

X-ray and gamma ray quality is usually expressed in terms of the thickness of a reference attenuator material (aluminum, copper, or iron) required to reduce the intensity of the beam to one-half of its original value. This thickness is referred to as the half-value layer (HVL).

A. Penetrameters

The two basic overall parameters that the interpreter must know in regard to a given radiograph are the radiographic sensitivity and definition that have been achieved. Radiographic sensitivity is measured in terms of the minimum percentage of the subject item that corresponds to the least discernible change in photographic density of the final radiograph. Definition refers to the smallest size (in lateral dimension) flaw of given (equivalent thickness). Both sensitivity and definition are established by the use of penetrameters.

A penetrameter is a device placed on the film side of a specimen whose image in a radiograph is used to determine radiographic quality. The standard penetrameter is a rectangle of metal with three drilled holes of set diameter, which is composed of a metal similar to that being radiographed. Normally the thickness T of the penetrameter is selected to be 2% of the thickness of the specimen. The penetrameter contains holes that may be $1T$, $2T$, $4T$, and $8T$ in diameter. A standard 2% sensitivity requires the technique to image the penetrameter whose thickness is 2% of the test piece and the $2T$ hole of the penetrameter. (See Fig. 13-20.)

ASTM E-142

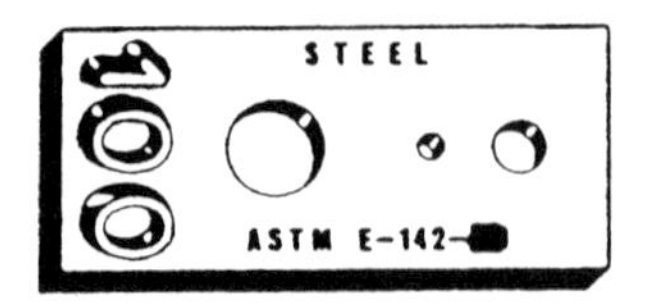

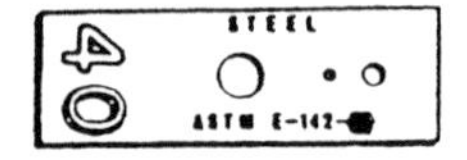

IDENTIFICATION:

The rectangular penetrameter is identified with a lead number attached to the penetrameter. The number indicates the thickness of the penetrameter in thousandths of an inch. The penetrameter thickness must be selected to indicate the proper quality level.

GENERAL DIMENSIONS:

2½" & smaller .... Length 1½" .... Width ½"
2⅝" to 8" incl. .... Length 2¼" .... Width 1"
9" & larger .... Diameter equal to 4 x thickness
(Number of holes — 2)

Thickness: 2% of the thickness, of the section, to be radiographed.
(To nearest standard fractional size)
Minimum thickness .... .005"

Hole sizes: Small hole dia. .... 1 x Thickness
Medium hole dia. .... 2 x Thickness
Large hole dia. .... 4 x Thickness
Minimum hole dia. .... .010"

ASTM E-142-72 & 74 are identical.

Fig. 13-20. Examples and dimensions of penetrameters. (From Survey of NDT, 1.)

The six methods discussed so far in this chapter are all passive methods in that the component being inspected need not be under stress or in service.

XII. ACOUSTIC EMISSION TESTING (AET)

Acoustic emission testing is used for the detection of surface and subsurface flaws (1). It can provide remote and continuous surveillance and a permanent record, but it may need many contact points and be expensive. This is an active method in that the component being inspected is loaded during an inspection or a crack grows under load during surveillance.

Acoustic emission is defined as the high-frequency (30 kHz–5 MHz) stress waves generated in a material by the rapid release of strain energy during crack growth, plastic deformation, or phase transformation (also involves plastic deformation). Using electronic sensing and analysis of data from a multiplicity of sensors, the location and relative severity of a flaw can be determined. AET can be combined with hydrostatic testing so that catastrophic failure from the growth of defects can be prevented. Nuclear reactor pressure vessels can be continuously monitored to guard against the growth of flaws. Careful interpretation of data is required to distinguish between plastic deformation and crack growth. As cracks increase in length, the strain energy released per increment of growth also increases, and so does the acoustic signal. Applications include checking on the safety of bridge structures, monitoring fatigue and stress corrosion crack growth, and the monitoring of the delayed cracking associated with hydrogen embrittlement.

XIII. COST OF INSPECTIONS

In carrying out an inspection program in order to ferret out defective parts, it is clear that the more parts that are examined in the case of a production run or the greater the frequency of the inspections in the case of an aircraft structure, the more likely it is that the defective parts will be discovered (1). However, each of these inspections has an associated cost. A balance between the cost of inspection and the benefit gained needs to be struck. The following example illustrates the considerations involved.

Assume that the losses due to substandard products decrease with the number of inspections N as

$$\text{total losses} = A - CN^{1/2}, \tag{13-12}$$

where A and C are constants. Assume further that the cost of inspections is given by

$$\text{cost of inspections} = BN, \tag{13-13}$$

where B is a constant. The total cost as a function of N is therefore equal to

$$\text{total overall cost} = A - CN^{1/2} + BN. \tag{13-14}$$

The optimum number of inspections N_{opt} is that which will minimize the total cost. N_{opt} is obtained by taking the derivative of the total cost with respect to N and setting it equal to zero, which leads to

$$N_{opt} = \left(\frac{C}{B}\right)^2. \tag{13-15}$$

Note that the constant A does not appear in this result. In a specific case where $A =$ \$100, $C =$ \$2.00, and $B =$ \$0.10, the optimum number of inspections would be 100. The total cost as a function of the number of inspections for this example is plotted in Fig. 13-21. If the cost per inspection were to increase, then N_{opt} would decrease. On the other hand, if C were to increase then N_{opt} would also increase.

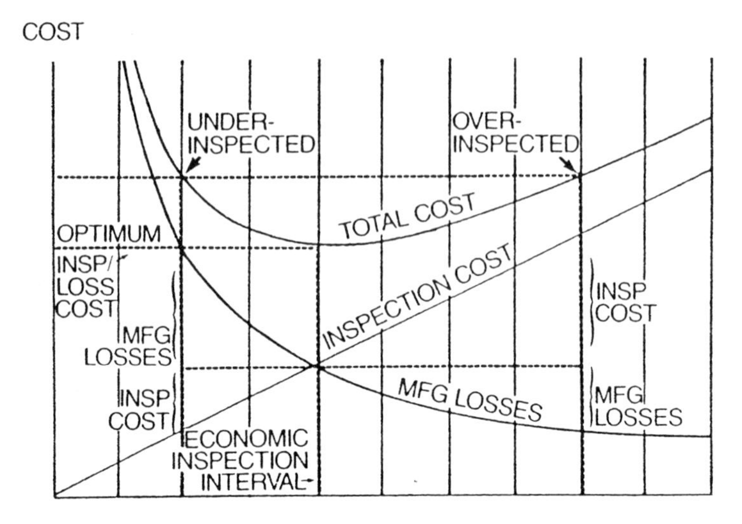

Fig. 13-21. Relationship of cost to inspection frequency. (From Survey of NDT, 1.)

When a new product is introduced, the term "bathtub curve" is used to describe the history of failures. The number of failures per unit time may be high initially, due to bugs in the manufacturing process or poor inspection procedures. Once these problems have been corrected, the number of failures per year remains fairly steady over a period of time before rising again as the useful life of the product is exceeded.

For further reading on nondestructive examination, see reference (1).

XIV. SUMMARY

The inspection for defects and cracks is as vital part of the procedures to insure the safety of structures. The use of fracture mechanics in design is coupled with a need for adequate inspection methods. This chapter has provided a review of the principal nondestructive techniques, and in case studies has pointed out their shortcomings.

REFERENCES

(1) Survey of NDE, EPRI Nondestructive Evaluation Center, Charlotte, NC, 1986.
(2) NTSB Aircraft Accident Report, NTSB/AAR-90/06,United Airlines Flight 232, Sioux City, Iowa, Washington, DC, 1989.
(3) C. Bagnall, H. C. Furtado, and I. Le May, in Lifetime Management and Evaluation of Plant, Structures and Components, ed. by J. H. Edwards et al., EMAS, West Midlands, UK, 1999, pp. 295–302.
(4) ASM Metals Handbook, 8th ed., vol. 8, ASM, Materials Park, OH, 1973, p. 303.

14

Crane Hooks, Coil Springs, Roller Bearings, Bushings, and Gears

I. INTRODUCTION

This chapter discusses the nature of the stresses developed in highly curved components such as crane hooks and coil springs. In addition, the factors contributing to the failure of roller bearings, bushings, and gears are considered.

II. CURVED BEAM THEORY (1)

The term "curved beam" is applied to a beam in which the neutral surface of the unloaded beam is a curved surface. Straight beam theory is not correct when applied to curved beams; however, if the inner radius is more than *four times the depth* of a curved beam, the maximum stress calculated as a straight beam will be too small by a factor of no more than 10%. If the ratio of inner radius to depth decreases, the error will increase rapidly.

The reason why straight beam theory is inapplicable to markedly curved beams is evident from Fig. 14-1*a*, which shows part of a curved beam subject to a bending moment M. It has been experimentally verified that every plane section of a curved beam remains plane after bending. Therefore, in a curved beam under bending, the total deformations are proportional to the distance from the neutral axis. However, since the lengths of fibers between two radial planes such as AB and CD are not all equal, the strains and the stresses do not vary as the total deformations. The inner fibers are shorter than the outer fibers, and consequently, the inner fiber stresses are higher in magnitude than the outer fiber stresses. The variation of stress through the beam is indicated in Fig. 14-1*b*.

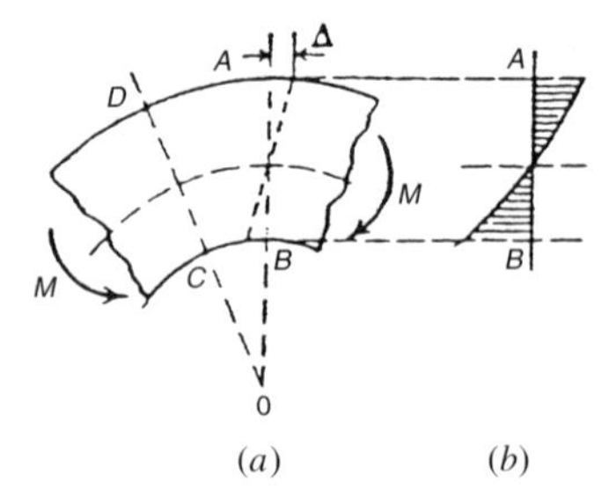

Fig. 14-1. (*a*) Curved beam. (*b*) Variation of stress through curved beam. (After Laurson and Cox, 1. With permission of John Wiley.)

Equilibrium requires that the total tensile force on one face of a radial segment be equal to the total compressive force. Since the stresses increase more rapidly from the neutral axis toward the center of curvature than from the neutral axis away from the center of curvature, the neutral axis is shifted away from the centroid of the section toward the center of curvature. This shift is designated by j, with the radius r to the neutral axis being given as

$$r = \overline{R} - j, \tag{14-1}$$

where $\overline{R}$ is the distance to the centroid from the center of curvature.

The stress at the inner surface of a curved beam of rectangular cross section of depth h is given by

$$\sigma_i = \frac{Mc_i}{R_i jA}, \tag{14-2}$$

where A is the cross-sectional area, R_i is the distance from the center of curvature to the inner fiber, and c_i is the distance from the neutral axis to the inner fiber. For a beam of rectangular cross-section, the distance r to the neutral axis is given by

$$r = \frac{h}{\ln(R'_0/R_i)}, \tag{14-3}$$

where R_0 is the distance from the center of curvature to the outer fiber. Hence

$$j = \overline{R} - r, \tag{14-4}$$

$$c_i = 0.5h - j. \tag{14-5}$$

As an example, if h is $2R_i$, $R_0/R_i = 3$, and $r = 0.91h$. Therefore, $j = h - 0.91h = 0.09\ h$.

The stress σ_i at the inner surface is

$$\sigma_i = \frac{M(0.5 - 0.09)h}{0.5h \times 0.09h \times bh} = \frac{9.14M}{bh^2}. \tag{14-6}$$

Straight beam theory would have given

$$\sigma = \frac{6M}{bh^2}. \tag{14-7}$$

The stress σ_o on the outer fiber is

$$\sigma_o = \frac{Mc_o}{R_o jA} = \frac{4.38M}{bh^2}. \tag{14-8}$$

Due to curvature, the stress at the inner fiber is 52% greater than in a straight beam, and the stress at the outer fiber is only 73% of the stress in a straight beam.

Under combined bending and axial loading, the stress at the inner fiber is given as

$$\sigma_i = \pm\frac{P}{A} \pm \frac{Mc_i}{R_i jA}, \tag{14-9}$$

and at the outer fiber,

$$\sigma_o = \pm\frac{P}{A} \pm \frac{Mc_o}{R_o jA}. \tag{14-10}$$

A question arises about the correct moment arm in calculating M. In curved beams, the moment arm is the distance from the line of action of the load to the centroid of the section and not to the neutral axis.

III. CRANE HOOKS (1)

A. Example

Figure 14-2 shows one of a pair of hooks used for lifting 125-ton ladles in a steel works. The width w of the hook is $16\frac{11}{16}$ inches, and the thickness of the metal is $5\frac{1}{4}$ inches. The inner radius is $6\frac{1}{16}$ inches. Calculate the maximum stress in the hook if the load on the hook is 135,000 lb, $P = 135{,}000$ lb:

$$w = h = 16.69 \text{ inches}, \qquad t = 5.25 \text{ inches}, \qquad R_i = 6.0625 \text{ inches},$$

$$\text{area} = 87.61 \text{ inches}^2, \qquad \frac{P}{A} = 1540 \text{ psi}. \tag{14-11}$$

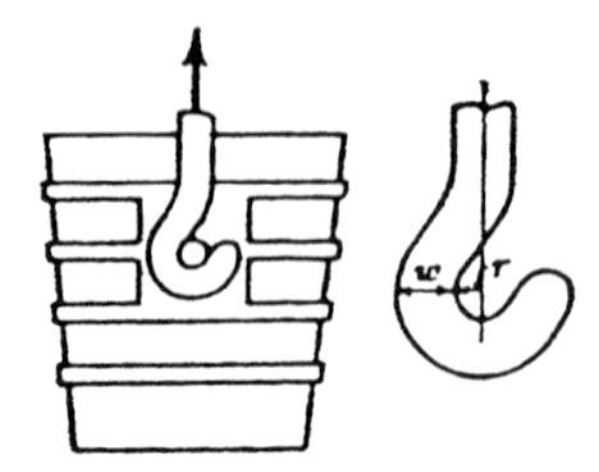

Fig. 14-2. A ladle hook. (After Laurson and Cox, 1. With permission of John Wiley.)

The cross section is rectangular:

$$r = \frac{h}{\ln(R_0/R_i)} = \frac{16.69}{\ln[(6.0625 + 16.69)/6.0625]} = 12.62 \text{ inches,} \quad (14\text{-}12)$$

$$\overline{R} = \frac{16.69}{2} + 6.0625 = 14.41, \quad (14\text{-}13)$$

$$j = 14.41 - 12.62 = 1.79 \text{ inches,} \quad (14\text{-}14)$$

$$c_i = \frac{16.69}{2} - j = 6.56 \text{ inches.} \quad (14\text{-}15)$$

The moment M is taken with respect to the centroid.

$$M = P\left(6.0625 + \frac{16.69}{2}\right) = 14.41P = 1{,}945{,}688 \text{ inch-lb} \quad (14\text{-}16)$$

$$\sigma_i = 1540 + \frac{1{,}945{,}688 \times 6.56}{6.0625 \times 1.79 \times 87.61} = 1540 + 13{,}425 = 14{,}965 \text{ psi.} \quad (14\text{-}17)$$

(Approximately 15,000 psi.)

IV. COIL SPRINGS (2)

Heavily coiled springs, especially those having a coil diameter less than four times the wire diameter, often fail in service under loads that are well below the safe load determined by the use of ordinary helical-spring formulas. A better stress analysis taking into account the effects of curvature is therefore needed.

Consider a heavy closely coiled helical spring of mean coil diameter $2r$ subjected to an axial load as shown in Fig. 14-3. The forces acting on any element $aa'b'b$ cut out by two neighboring radial planes may be resolved into a twisting moment Pr acting in a radial plane and a direct shearing force P acting in the direction of the axis of the spring.

The twisting moment Pr will cause rotation of two sections aa' and bb' with respect to each other through a small angle $d\alpha$. But since the length of the fiber $a'b'$ is much less than the length of the fiber ab, it is clear that the shearing strain, and hence

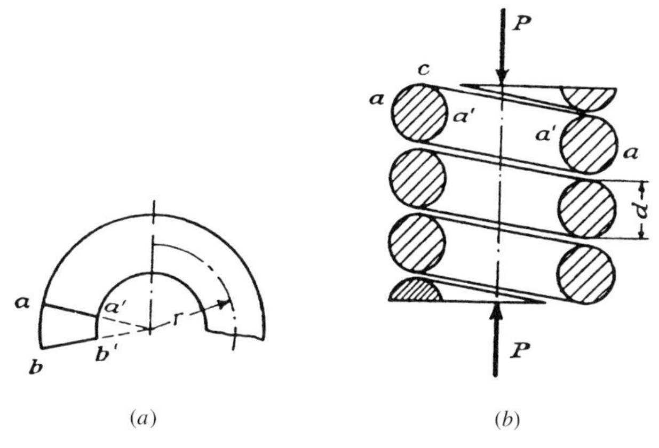

Fig. 14-3. A closely coiled helical spring, after Wahl, 11. (Reprinted with permission of the ASME.)

the shearing stress, for any given angular rotation of the two cross sections will be much greater in $a'b'$ than in ab. The stress at a is further increased by the direct shearing stress due to the axial load P. This stress corresponds to the shearing stress at the neutral axis of a cantilever beam of circular cross section loaded with a load P.

The shear stress distribution, due to the twisting moment Pr, along a transverse diameter of wire is as shown in the shaded area of Fig. 14-4, the stress at a' being much larger than at a. To this stress at a' must be added the direct shear stress due to the external load P.

In line with the above considerations, a more exact formula for determining the maximum shearing stress, $\tau_{\max}$, at a' for the case of a helical spring axially loaded was derived by Wahl (2).

$$\tau_{max} = \frac{16\mathrm{Pr}}{\pi d^3}\left(\frac{4c-1}{4c-4} + \frac{0.615}{c}\right), \tag{14-18}$$

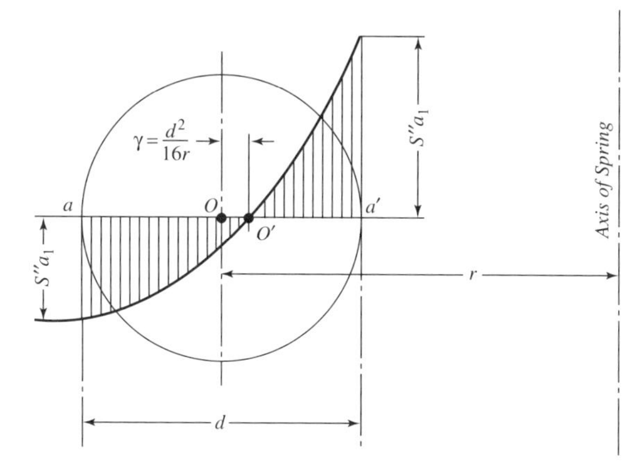

Fig. 14-4. Stress distribution in a coil spring along a transverse diameter, *a-a'*, assuming rotation about the point *O'*. (After Wahl, 11. Reprinted with permission of the ASME.)

where P is the axial load on the spring, d is the diameter of the wire, r is the mean radius of the coil, τ_{max} is the maximum shear stress in psi, and

$$c = \frac{2r}{d} = \frac{\text{mean coil diameter}}{\text{wire diameter}}. \tag{14-19}$$

The expression for τ_{max} consists of the ordinary formula for stress in helical springs multiplied by a factor k, which depends on the ratio c. Values of k, which can be considered to be a stress multiplication factor, for various values of c are given in the following table:

c	3	4	5	6	8	10
k	1.58	1.40	1.31	1.25	1.18	1.15

A. Example

Determine the maximum shear stress in a coil spring for an axial load of 25,000 lb for a wire diameter of 1 inch and an r of 4 inches.

$$c = 8, \qquad k = 1.18, \qquad \tau_{max} = \frac{25{,}000 \times 4}{\pi}(1.18) = 37{,}560 \text{ psi}. \tag{14-20}$$

Note that this stress occurs on the inside of the spring, the usual location for fatigue crack initiation.

V. ROLLER BEARINGS

A tapered roller bearing is shown in Fig. 14-5*a* (3). This type of bearing will be considered in some detail. The causes of failure of other bearing types, that is, spherical roller bearings, needle bearings, and so on, have many characteristics in common with this type of bearing.

A tapered roller bearing consists of four basic components, the inner race or cone, the outer race or cup, the tapered rollers, and the roller retainer or cage. The forces acting on a tapered roller bearing are shown in Fig. 14-5*b*. Under proper operating conditions, all components carry the load with the exception of the cage, whose primary function is to space the roller around the cone. The roller bearings as well as the races are tapered on the principle of a cone, and because of the tapered races, the bearing will handle a combination of radial and thrust loads. The rollers are crowned to assure uniform contact across the entire roller length. This improves fatigue behavior and extends the bearing life. The nominal contact stress between roller and race under rated loading is 200–250 ksi. The bearing cups and cones are machined from high-quality forged steel (52100, 58–60 HRC) of low nonmetallic

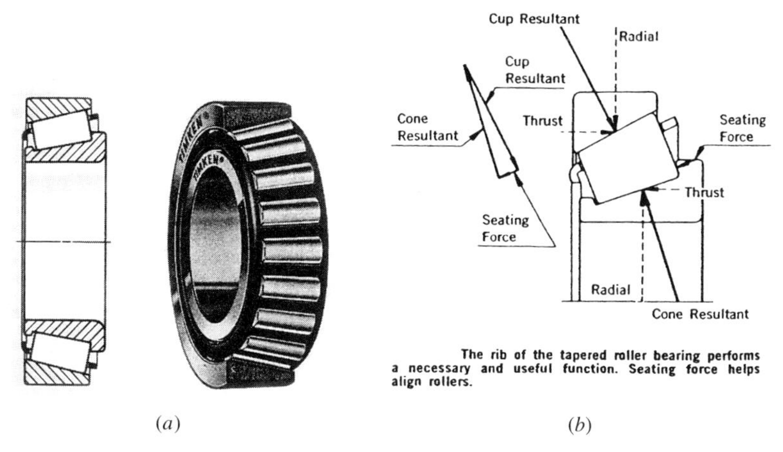

Fig. 14-5. (*a*) Example of a tapered roller bearing, type TS (pressed steel cage). (*b*) The forces acting on a tapered roller. (From Tapered Roller Beraings by The Timken Company. Copyright 1972, 3.)

inclusion content. Rollers are either cold formed from cold drawn wire or machined from hot rolled rods and bars. The steels are low-carbon carburizing grades, vacuum degassed bearing-quality alloy steels. After machining, carbon is added to the surfaces of the bearing components to a depth adequate to sustain bearing loads. This results in a hard, fatigue-resistant case and a tough, ductile core in the carburized and heat-treated part. An additional benefit from the carburizing is the compressive residual surface stress, which improves fatigue resistance. As cleaner steels have become available, the surface finish has become critical in controlling the fatigue life. Therefore, surface finishes are kept in the range 0.63–2 μm, and bearing noise levels are checked to ascertain that the noise levels meet specifications.

There are two fundamental load ratings for tapered roller bearings, a *basic dynamic* load rating and a *static* load rating. The basic dynamic load rating is used to establish the life expectancy of a rotating bearing. The static load rating is used to determine the maximum permissible load that could be applied to a bearing when it is not rotating without producing false Brinell markings. (False Brinelling has occurred when automobiles were shipped by rail over long distances. Under the vibratory loads experienced, the balls in ball bearings in the wheel assemblies wore indentations into the bearing races, causing the bearings to "run noisy" when the cars were subsequently driven. The indentations looked like Brinell hardness indentations, hence the term "false Brinelling".)

The basic dynamic load rating is subdivided into a basic dynamic radial load rating C(90), and a basic dynamic thrust load rating CA(90). The bearing *K*-factor indicates the ratio of the two, or

$$K = \frac{\mathrm{C}(90)}{\mathrm{CA}(90)}. \tag{14-21}$$

K can also be expressed in terms of one-half of the included cup angle α as

$$K = 0.389 \cot \alpha. \tag{14-22}$$

The life of a tapered roller bearing is considered to be completed after repeated stressing causes pitting or spalling over 1% of the contact surfaces. These defects result because of Hertzian contact of the bearing components, which leads to subsurface shear stresses, which alternate in sign as a roller passes a given spot; see Fig. 14-6. The repeated, alternating shear stresses lead to fatigue cracking and spalling in a process known as rolling contact fatigue. In this process, small subsurface fatigue cracks are formed that grow and break through to the surface, lead-

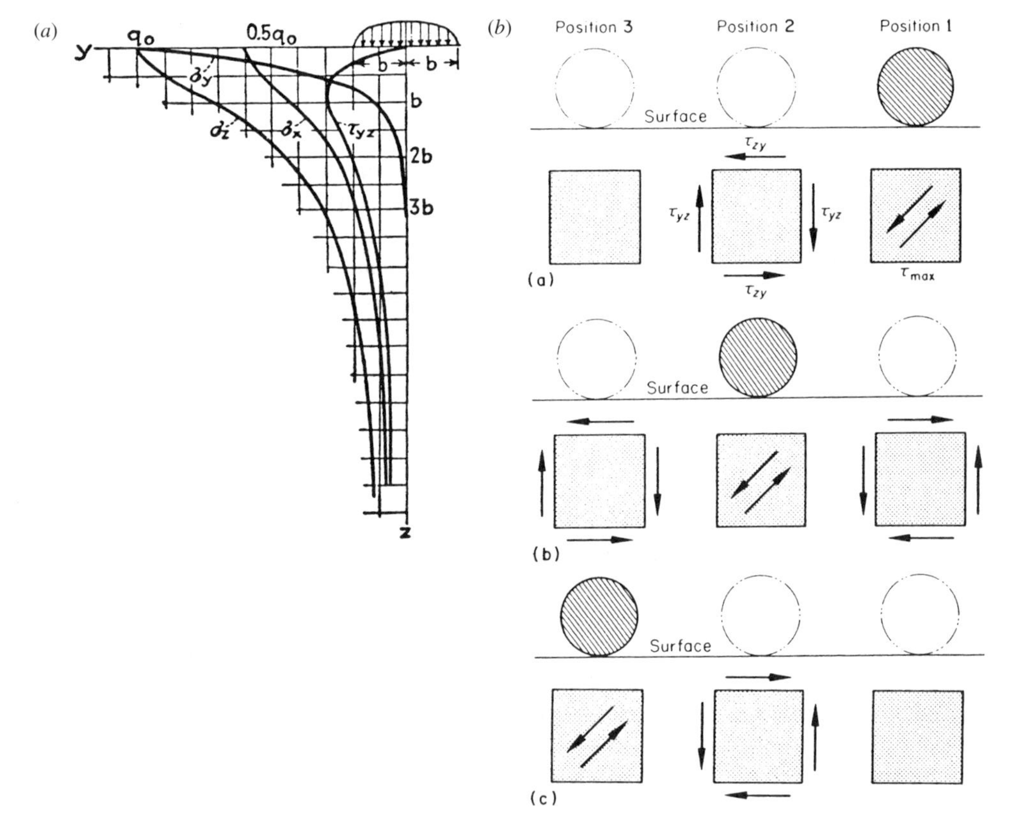

Fig. 14-6. (*a*) Subsurface stresses developed due to contact of a cylinder with a plane surface (From Timoshenko, 8. With permission of McGraw-Hill.) (*b*) Subsurface shear stresses developed at three fixed positions as a cylinder rolls from right to left. (From ASM Metals Handbook, 8th Edition, Vol. 10, Failure Analysis and Prevention, 1975. With permission of ASM International®.)

ing to pitting and spalling as fragments break away from the surface. This type of fatigue is found in many components of mechanical systems including railroad tracks and train wheels.

Since bearings exhibit considerable life scatter, a statistical method is used to evaluate the life.The Weibull extreme value distribution is used to determine the statistical life of bearings at any reliability level. The rated life for bearings is usually given in terms of the L_{10} level, where L_{10} is defined as the number of revolutions (cycles) that 90% of a group of identical bearings will exceed under standard loading conditions before the fatigue failure criterion is reached.

The basic dynamic load ratings for properly aligned and lubricated bearings are based on a "rated life" of 90×10^6 cycles (3000 hours at 50 rpm). The races should be less than 3–4 minutes out of alignment. Misalignment can occur due to shaft deflection, inaccuracies in machining of shaft or housing, and inaccuracies introduced during press fitting. The empirical relationship between "rated life" of a roller bearing and load is

$$L_{10}\,(P)^{10/3} = \text{constant}. \tag{14-23}$$

for $L_{10} = 90 \times 10^6$ cycles. For the corresponding load P is designated as $C(90)$; therefore, the L_{10} life for any other radial load P is given by

$$L_{10}\,P^{10/3} = 90 \times 10^6\,C(90)^{10/3}, \tag{14-24}$$

or

$$L_{10} = \left(\frac{C(90)}{P}\right)^{10/3} \times 90 \times 10^6 \text{ revolutions}. \tag{14-25}$$

(To account for thrust loads, P can be expressed as a P_{eq}.) The bearing catalog lists the $C(90)$ load for each bearing, and therefore the L_{10} life for other loads can be determined.

A. Bearing and System Reliability

The rated life of a bearing is an expression of reliability, that is, 90% reliability that a bearing will equal or exceed a given life. In some bearing applications, greater than 90% reliability is required. Figure 14-7 is an empirical Weibull plot that can be used to select the appropriate reduced life to give the increase in reliability.

Alternatively, to obtain the L_{10} life of a bearing at a level of reliability R other than 90%, the following relationship can be used:

$$a_i = 4.48\left(\ln \frac{100}{R}\right)^{2/3}, \tag{14-26}$$

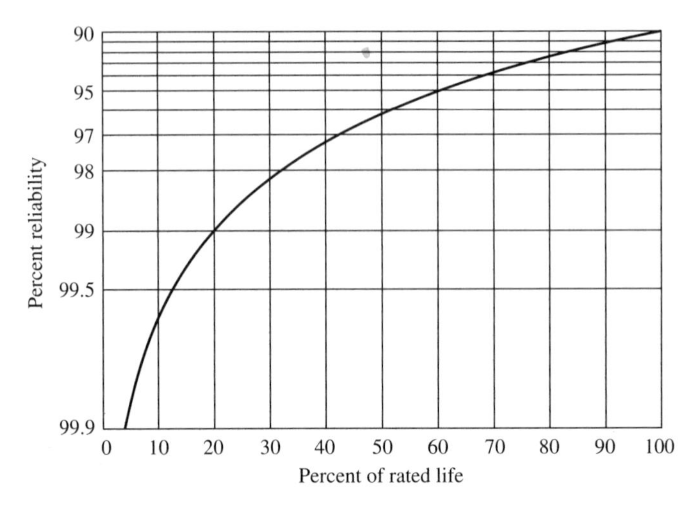

Fig. 14-7. Weibull plot indicating product reliability as a function of the percent of rated life. (From Tapered Roller Bearings, 3. Copyright 1972 by the Timken Company.)

where a_i is a life adjustment factor and R is the % reliability. (For 90% reliability, $a_i = 1.0$.)

$$R = 100 \times e^{(100a_i/448)^{-3/2}} \tag{14-27}$$

is the equation of the line in Fig 14-7, where $100a_i$ is the percent of the rated life. Multiply the L_{10} life by a_i to obtain the L_R life, which is the life for reliability of R percent. For a reliability greater than 90%, a_i will be less than unity, and for a reliability less than 90%, a_i will be greater than unity.

B. First Example

What is the adjusted rated life for an application that requires 98% reliability?

$$a_i = 4.48\left(\ln\frac{100}{98}\right)^{2/3} = 0.33. \tag{14-28}$$

Then $L_R = L_2 = 0.33L_{10}$. (Note that this factor can be obtained directly from the chart, Fig. 14-7.) If 10,000 hours of life are required at 98% reliability, then an L_{10} life of 30,300 hours would be required. For one company's bearings, the average or mean life is approximately 4 times L_{10}. Median life L_{50} is approximately 3.5 times L_{10}.

System reliability is encountered when it is desirable to consider the probability that all of several bearings will survive to a certain life, that is:

$$\text{reliability (system)} = (R_1)\,(R_2)\,(R_3)\,\ldots\,(R_n), \tag{14-29}$$

where R_1, and so on, represent the reliability of each bearing in a system to a required system life.

C. Second Example

What is the reliability of a system that requires 6000 hours of life, when the rated lives of the four bearings in the system are 12,000, 10,000, 9000, and 8000 hours?

Rated Life	System Life as % of Rated Life	% Reliabilty Bearing to System Life
12,000	50	96
10,000	60	95
9,000	66.7	94
8,000	75	93

$$\begin{aligned}\text{reliability (system)} &= 0.96 \times 0.95 \times 0.94 \times 0.93 \\ &= 0.8 \text{ or } 80\% \text{ for 6000 hours.}\end{aligned} \tag{14-30}$$

The L_{10} life of a system life for a number of bearings each having the same or a different L_{10} life is

$$L_{10system} = \left[\left(\frac{1}{L_{10A}}\right)^{3/2} + \left(\frac{1}{L_{10B}}\right)^{3/2} + \cdots + \left(\frac{1}{L_{10n}}\right)^{3/2}\right]^{-2/3}. \tag{14-31}$$

For example, if four bearings in a system each had an L_{10} life of 3000 hours at 50 rpm, the system L_{10} life would be approximately 1200 hours.

D. Bearing Failures (4)

Normally, the cause of bearing failure is rolling contact fatigue with subsurface crack initiation occurring at nonmetallic inclusions or carbides. The major factors that lead to premature failure are: incorrect fitting, excessive preloading during installation, insufficient or unsuitable lubrication, overloading, impact loading, vibration, excessive operating or environmental temperature, contamination by abrasive matter, ingress of harmful liquids, and stray electric currents. The deleterious effects resulting from the above are: flaking or pitting (fatigue), fluting (grooved pitting), cracks or fractures, rotational creep, smearing, wear, fretting, softening, indentation, case-crushing (heavy loads, thin case, soft core), and corrosion.

Proper lubrication of bearings is important for good performance. The intent of lubrication is to prevent direct metal-to-metal contact. There are two types of oil films: the reaction film and the elastohydrodynamic film. The reaction film is also known as a boundary lubricant and is produced by physical adsorption and/or chemical reaction to form a desired film that is soft and easily sheared but difficult to penetrate or remove from the surface. The elastohydrodynamic film forms dynamically on the wear surface as a function of surface speed. This film is very thin, has a very high shear strength, and is only slightly affected by compressive loads as long as constant temperature is maintained. Analysis of wear debris in a lubricant after a period of service can provide information as to the nature and extent of wear processes.

The basic life rating of a bearing assumes that the lubricant film thickness is at least equal to the composite roughness of the contacting surfaces σ_q, where

$$\sigma_q = (R_{q1}^2 + R_{q2}^2)^{1/2} \tag{14-32}$$

and R_{q1} and R_{q2} are the root mean square (RMS) roughness measurements of the two surfaces. The film thickness divided by the composite roughness is designated as λ, and the fatigue life increases with increase in λ. The mode of bearing fatigue failure is also influenced by λ. For values of λ in the range of 1.0–3.0, fatigue cracks initiate at subsurface inclusions and surface roughness is of relatively minor importance. When the value of λ is less than 1.0, then surface roughness is of much greater importance. A common type of bearing failure, known as burnup, is due to inadequate lubrication, too high a preload, or excessive speed. During burnup, the bearing temperature rises and the material flows plastically, thereby destroying the bearing geometry.

VI. CASE STUDY: FAILURE OF A RAILROAD CAR AXLE

During the late 1970s and early 1980s, weight saving was a strong driving force in the design of cars, trucks, buses, and trains because of anticipated fuel shortages. In one case involving passenger trains, hollow axles were substituted for solid axles to save weight. These axles were equipped with press-fitted tapered roller bearing assemblies that transferred the car loading to the axles, then to the wheels, and finally to the tracks. The cars had not been in service for long when an axle failed on one of them, and a wheel came off. Fortunately, there were no injuries. Examination of the failed axle revealed that it had been worn and overheated where the bearing had been seated. The wheel failure prompted inspection of the other axles in the fleet, and in many cases, excessive wear at the bearing seat was found. Excessively worn axles were taken out of service. To guard against any further failures of the axles while a fix was being developed, temperature-sensing devices were installed in the wheel hubs. These devices were examined for signs of overheating at selected stops along the routes of the trains. Unfortunately, these examinations were time consuming and resulted in considerable delay and expense. In addition, ultrasonic inspections of the bearing-axle interfaces were made to determine if a gap had developed due to excessive wear. These examinations were facilitated by the fact that the axles were hollow. When excessive wear was found, the axle was removed from service.

The use of hollow axles was not new. They had been used successfully before, but in prior usage the ratio of the inside to the outside diameter had been 0.5. In the case under consideration, the ratio was 0.6. In other words, the axles under discussion were more hollow than the older ones had been, and this turned out to have serious consequences. An analysis was carried out to determine how the axles deformed under load. It was found that the residual compressive stress on the bearing after pressing the bearing onto the axle was less in the case of the hollow axle than for a solid axle. Insufficient compressive stress on the bearing can encourage interfacial fretting, resulting in the reduction of the interference fit at the interface. It was also found that, in contrast to a solid axle, which remains essentially round under load, the cross section of these hollow axles "ovalized" into an elliptical shape under load. Because the interference fit was not sufficiently high, the ovalization caused cyclic relative motion of the bearing with respect to the axle. As a result, at a given location on the axle under the bearing, shear stresses were developed that oscillated in magnitude with each revolution of the axle. These shear stresses in turn led to the observed excessive wear at the bearing-axle interface.

The final fix was to replace the hollow axles with solid axles.

VII. GEAR FAILURES (5, 6)

The main factors controlling gear life are the accuracy of the tooth geometry, the gear-tooth contact conditions including lubrication, and the material. Figure 14-8

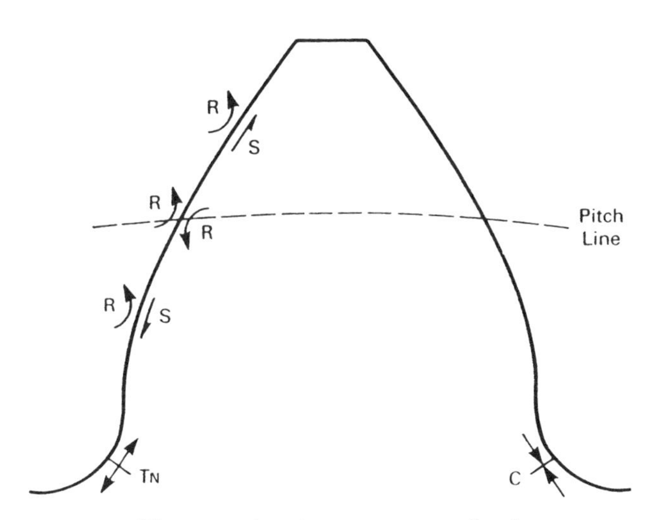

Diagramatic stress areas on basic spur gear tooth.

Fig. 14-8. The stresses developed in a spur gear tooth due to the rolling contact R of a mating gear. S is the shear stress, T_N is the tensile stress, and C is the compressive stress. (From Alban, 8. With permission of ASM International®.)

gives an indication of the stresses that are developed in a spur gear tooth as the result of rolling contact with a mating gear tooth. Gears can fail in a number of different modes, and as with bearings, an increase in the level of noise or vibration is an indication of impending failure. The leading cause of failure is due to either tooth bending, which leads to fatigue crack initiation at the root of a gear tooth, or to rolling-contact fatigue, which induces small-scale pitting, which leads to larger scale spallation. Additional causes of failure are impact, wear, and stress rupture. In at least one instance, the striking of a blade of a boat propeller has been known to cause enough of an impact to rupture gear teeth. In addition, other gear in the gear train contained cracks that were not detected, and a failure occurred shortly after the gear system was put back into service. Therefore, careful inspection of all gear teeth after an impact is warranted. Each part of a gear tooth surface is in action only for short periods of time. This continual shifting of the load to new areas of cool metal and cool oil makes it possible to load gear surfaces to stresses approaching the critical limit of gear metal without failure of the lubricating film. The maximum load that can be carried by gear teeth also depends upon the velocity of sliding between the surfaces, because the heat generated varies with the rate of sliding and the pressure. Too much frictional heat can cause scuffing and the destruction of tooth surfaces.

A. Gear Material Problems

A number of issues related to materials can contribute to fatigue failure in gears. These include the size and nature of nonmetallic inclusions, poor forging patterns, pitting, banding, segregation, forging laps, and improper material selection. Misidentification has led to the use of an incorrect steel in some instances, and the wrong grade of steel has sometimes been used in weld repairs.

Steel gears are usually case-hardened by carburization, nitriding, or induction hardening, and the properties of the case are important both at the mid-profile and at the root radius of the gear tooth. Case depth is usually understood to mean the "effective case depth," defined as the distance from the surface to the location where the hardness drops to specified level. For a carburized gear, this level is usually 50 HRC, whereas for an induction-hardened gear, the level depends upon carbon content, being 35 HRC for a steel containing 0.3 wt % carbon, and 50 HRC for a steel containing more than 0.53 wt % carbon. The shear stress due to rolling contact is at its maximum about 0.25 mm below the surface at the contact region of the mating teeth, and in order to avoid fatigue failure, the hardness in this region is typically specified to be at least 58 HRC at the surface with an effective case depth of 0.5 mm, and the case depth at the root radius, where the bending stress is at its maximum, is specified to be 0.3 mm.

The carburization process results in a biaxial compressive state of stress in the carburized layer, which is balanced by tensile stresses in the core of the gear tooth. This system of stresses can result in a time-delayed form of cracking of the gear tooth that initiates beneath the surface and that is probably hydrogen-related, as shown in Fig. 14-9. The tendency for this type of fracture to occur increases with

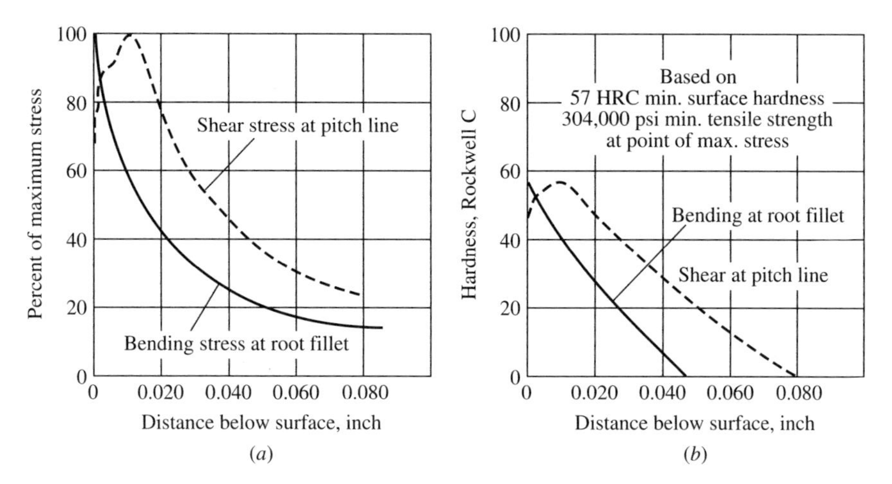

Fig. 14-9. (*a*) Subsurface stresses in a spur gear tooth. (*b*) The hardness gradients necessary to prevent subsurface failure. (From Alban, 8. With permission of ASM International.)

case depth, since a greater tensile residual stress will be needed to balance a larger residual compressive force.

Another type of subsurface-initiated failure as the result of cyclic loading has also been observed in gear teeth (7). If the yield strength of the substrate is low, and the forces developed along the pitch line are high, case crushing may occur, Fig. 14-10, particularly with a thin case. In cases where induction hardening is used to harden the entire gear, quench cracking can occur. Such a quench crack is distinguishable from a forging lap because the fracture surface of a quench crack is free of scale or oxides. The induction hardening process needs to be controlled to avoid irregular hardening patterns that can lead to easy fatigue failure.

B. Wear

Gears that have been in service for some time will show signs of wear. There are two types of wear: adhesive wear and abrasive wear.

Adhesive wear is related to friction between mating gear teeth and is pronounced under conditions of poor lubrication. This form of wear is associated with the contact of material asperities on each of the mating surfaces (9). It is considered that the actual contact area A_c associated with these asperities is a small fraction of the nominal contact area. Under a given interfacial load P_N, the asperities deform plastically until their cross-sectional area multiplied by the yield strength σ_y, is equal to the applied load. The contacting asperities are considered to be welded together, and under the influence of a tangential shear force P_T, they are sheared apart when the yield stress in shear τ_y is reached. The deformation and subsequent shearing of the asperities result in the following two equations:

$$P_N = A_c\sigma_y \qquad \text{and} \qquad P_T = A_c\tau_y, \tag{14-33}$$

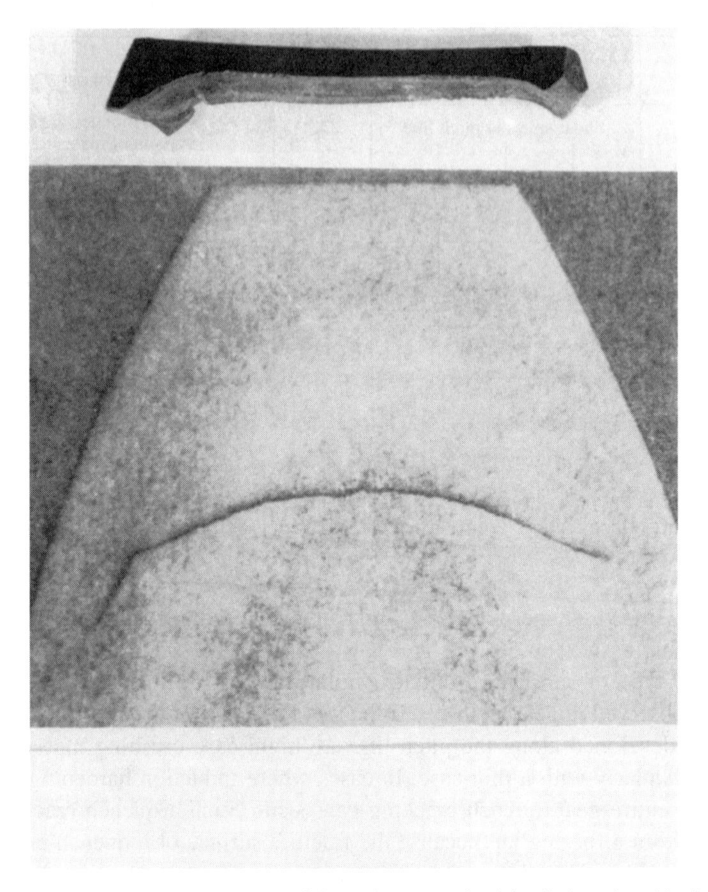

Fig. 14-10. An example of a gear tooth failure due to carburizing-induced residual stresses. (From Alban, 8. With permission of ASM International®.)

which lead to

$$P_T = \frac{\tau_y}{\sigma_y} P_N. \tag{14-34}$$

If τ_y is taken to be one-half of σ_y, then

$$P_T = 0.5P_N \tag{14-35}$$

This is Coulomb's law of friction with the coefficient of friction being 0.5, a not unreasonable value. The actual location of the shearing site is considered to be slightly above or below the junction where the material is not so heavily work hardened as at the junction itself. As two gear teeth slide past each other, material is transferred from one gear face to the other, and as a consequence, adhesive wear

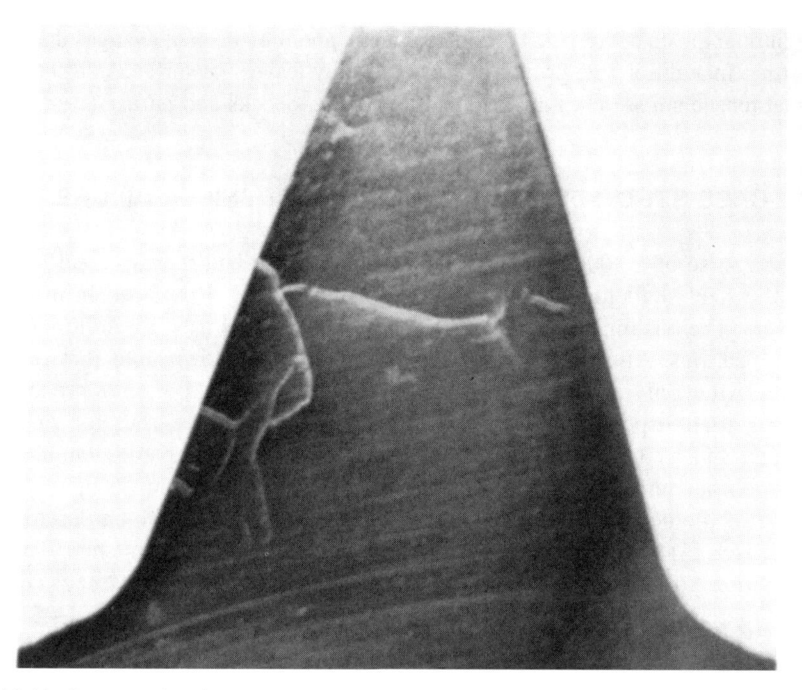

Fig. 14-11. An example of case crushing. (From Alban, 8. With permission of ASM International®.)

occurs. An increase in the surface hardness increases the yield strength and reduces A_c, thereby reducing this form of wear. Heat is generated during friction, sometimes with potentially disastrous results. For example, ignition of titanium has occurred in jet aircraft engines due to the rubbing of titanium alloy compressor blade tips against titanium alloy containment. As a result, a titanium alloy is no longer used for the containment.

Abrasive wear occurs as the result of plowing action of a hard particle across a softer surface, as in the case of the bushing to be discussed. A common example is the polishing action of diamond paste during metallographic preparation. Abrasive wear of gears can occur when hard particles become entrapped in the lubricant.

Alaska Airlines Flight 261 crashed on January 30, 2000, in the Pacific Ocean off Los Angeles, and 88 people perished. The crash appears to have been brought about by the excessive wear of a gimbal nut through which passed a jack screw that controlled the position of the horizontal tail stabilizer. The stripping of the threads of the gimbal nut may have led to loss of control of the horizontal stabilizer and the resultant crash. Both the lubricants and the maintenance procedures are suspect.

C. Design and Manufacturing Problems

Problems associated with design and manufacturing arise because of the presence of geometric stress raisers such as oil holes, excessive machining of the case, undercutting, the characteristics of the gear teeth, and dimensional changes due to heat treating. In addition, the grinding process usually leaves the surface in a state of residual tensile stress, and may also cause grinding checks and burns.

VIII. CASE STUDY: FRICTION AND WEAR; BUSHING FAILURE

In the course of a 2000-hour overhaul, a small plane had the two steel bushings and the steel shaft that held a carburetor butterfly valve replaced in the fall of the year. The bushings were short, thin hollow cylinders used as guides and non-roller bearings. After the overhaul, the plane was not flown until the following spring. After only a few hours of flying, the pilot, in making an approach for a landing at an airport, found that the control linkage to the carburetor would not operate, and he began to lose altitude. He therefore attempted a landing on a highway, but unfortunately the plane struck a car, killing an occupant. In the course of the accident investigation, it was discovered that the carburetor shaft was frozen in one of the bushings, thereby accounting for the mishap. The only lubrication provided to the shaft-bushing system came from the fuel vapors of the carburetor.

In a subsequent investigation of the frozen shaft-bushing system, it was found that a great deal of force was required to separate the two as they had to be pressed apart, that is, the combination was truly frozen. Examination of the shaft showed that an abrasive scratch ran for some distance from the bushing along the length of the shaft. Evidently, when the shaft was inserted in the bushing, a work-hardened burr on the bushing had scratched the shaft, and the plowed-out material from the shaft had lodged in the interface between the bushing and shaft. During the winter, this debris had oxidized and hardened. This hardened mass caused further abrasive damage during flight operations, until the shaft froze in the bushing and the accident occurred.

It was next discovered that the manufacture of the parts by a firm now bankrupt and out of business had involved a process known as reverse engineering, a term that indicates that a part originally designed and made by one manufacturer had been copied by another. In this reverse engineering process, every detail of the original part save one had been faithfully replicated. Both bushings and shaft were of the same stainless steel (ASI 416: 0.15C, 12-14Cr, 1.25 Mn, 0.60 max Mo (optional), 0.06 max P, 0.15 min S, 1.00 max Si), and in fact had been machined from the same rod. The essential difference was that the original design called for the shaft to be heat treated to a hardness of R_C 24, with the bushings to be much softer at an R_B of 95. In the shaft-bushing system that seized, the hardness of both components was R_B 92. In the course of reverse engineering, this important difference had been overlooked.

Had the shaft been at the proper hardness, the burr on the bushing would not have caused the observed abrasive scratch, and the accident would not have occurred. Even without abrasive damage, the soft shaft wore excessively due to adhesive wear, as shown by examination of similar shafts that had seen 2000 hours of service. Had the shaft been at R_C 25, only the bushings would have worn, which is usually the desired situation.

In a final report issued by the NTSB on this matter, the board held the repair facility largely responsible, and placed little blame on the manufacturer, a decision that seems a bit strange given the circumstances.

IX. SUMMARY

This final chapter has reviewed curved beam theory with applications to crane hooks and coil springs. The role of contact stresses and wear in determining the useful lifetimes of roller bearings, bushings, and gears was also discussed. Case studies were used to illustrate a number of types of failure involving wear.

REFERENCES

(1) P. G. Laurson and W. J. Cox, Mechanics of Materials, Wiley, New York, 1938.
(2) A. M. Wahl, Mechanical Springs, 2nd ed., McGraw-Hill, New York, 1963.
(3) Tapered Roller Bearings, Section 1, Engineering J., The Timken Co., Canton, OH, 1972.
(4) C. Moyer, Fatigue and Life Prediction of Bearings, in ASM Handbook, vol. 19, Fatigue and Fracture, ASM, Materials Park, OH, 1996, pp. 355–362.
(5) D. W. Dudley, Fatigue and Life Prediction of Gears, in ASM Handbook, vol. 19, Fatigue and Fracture, ASM, Materials Park, OH, 1996, pp. 345–354.
(6) L. E. Alban, Systematic Analysis of Gear Failures, ASM, Materials Park, OH, 1985.
(7) M. Mackaldener and M. Olssen, Fat.and Fract. Eng. Mats and Structures, vol. 23, 2000, p. 283.
(8) S. Timoshenko, Theory of Elasticity, McGraw-Hill, New York, 1934.
(9) M. F. Ashby and D. R. H. Jones, Engineering Materials, Pergamon Press, Oxford, UK, 1980.

Concluding Remarks

The previous 14 chapters have provided a basic and, it is hoped, an interesting introduction to the subject of metal failures. For more extensive information, there is a large body of literature that deals with this subject. The ASM Metals Handbook Series provides additional detail, in particular Volume 19 on Fatigue and Fracture, Volume 11 on Failure Analysis and Prevention, and Volume 12 on Fractography. The weekly publication Aviation Week and Space Technology provides excellent coverage of aircraft accidents and safety-related matters. The Internet is now another source of information. For example, a one-year conference on Pipeline Integrity is available. Failures differ greatly in terms of loading conditions, environments, and material characteristics, and the competent investigator should be aware of such resources in carrying out a failure analysis.

Some additional sources of information concerning failures are listed below.

C. R. Brooks and A. Choudhury, Metallurgical Failure Analysis, McGraw-Hill, New York, 1993.

S.-I. Nishida, Failure Analysis in Engineering Applications, Butterworth Heinemann, Oxford, UK, 1992.

D. R. H. Jones, Materials Failure Analysis, International Series on Materials Science and Technology, Pergamon Press, Oxford, UK, 1993.

G. A. Lange, ed., Systematic Analysis of Technical Failures, DGM Informations Gesellschaft, Oberursel, Germany, 1986.

V. J. Colangelo and F. A. Heiser, Analysis of Metallurgical Failures, Wiley Interscience, New York, 1987.

D. J. Wulpi, How Components Fail, 2nd Ed., ASM, Materials Park, OH, 1999.

J. A. Collins, Failure of Materials in Mechanical Design, Wiley Interscience, New York, 1981.

C. E. Witherell, Mechanical Failure Avoidance, McGraw-Hill, NY, 1994.

Problems

1-1. This problem is related to the Comet (jet plane) disasters. The atmospheric air pressure in centimeters of mercury as a function of altitude in kilometers can be expressed as

$$p = 76 - 8.45h + 0.285h^2.$$

The Comet's cruising altitude was 10 km, about twice the cruising altitude of commercial propeller-driven transport planes. Assume the cabin pressure in the Comet was maintained at a pressure equivalent to that at an altitude of 2.37 km. The skin of the Comet was a high-strength aluminum alloy, 0.91 mm in thickness, t. The fuselage was approximately 3.7 m in diameter, D, and 33 m long. Standard atmospheric pressure is 1.013×10^5 Pa .

(a) At an altitude of 10 km, determine the differential between the external pressure and the cabin pressure. This differential is the pressure p that the fuselage must support.

(b) The circumferential (hoop) stress σ_h is given as $\sigma_h = pD/2t$. Determine the hoop stress in the fuselage at the cruising altitude.

(c) Assume that design details introduce a stress concentration of 3.0. Compute the maximum stress developed.

1-2. This problem is related to the Kansas City Hyatt balcony collapse. If the vertical rod in Fig. 1-3 carried 10^5 N between the lower and upper balconies, and if an additional load of 5×10^4 N were to be carried by the rod between the upper balcony and the overhead support, determine the maximum bearing load at the upper balcony for each of the cases shown in Fig. 1-3.

2-1. The von Mises criterion can be expressed in terms of nonprincipal stresses as

$$\frac{1}{\sqrt{2}}[(\sigma_x - \sigma_y)^2 + (\sigma_y - \sigma_z)^2 + (\sigma_z - \sigma_x)^2 + 6(\tau_{xy}^2 + \tau_{yz}^2 + \tau_{zx}^2)]^{1/2} = \sigma_o,$$

where σ_o is the yield stress in simple tension. In the case of a thin-walled cylinder that is subjected to both tension and torsion, this expression reduces to

$$[\sigma_x^2 + 3\tau_{xy}^2]^{1/2} = \sigma_o,$$

where the x-direction is taken along the axis of the cylinder.

(a) Plot the first quadrant of this relationship using σ_x/σ_o and τ_{xy}/σ_o as axes.

(b) Determine the Tresca equivalent to part (a), and compare with part (a). (Use Mohr circle to obtain σ_1 and σ_3. Plot σ_x and τ_{xy}, 0 and $-\tau_{xy}$.)

2-2. A thin-walled box beam was subjected to in-phase cyclic bending and torsional stresses in service. After a period of time, a crack at 30° to the transverse direction was discovered in the lower wall of the box beam. Use the Mohr circle for this situation to determine the ratio of the bending stress to the shear stress.

2-3. Consider three flat, identical 2-mm thick steel (E = 200 GNm^{-2} ksi, ν = 0.25) tensile specimens that contain a centrally located through-hole of radius a whose axis is in the thickness direction. The origin of the in-plane x, y coordinate system is at the center of the hole, with the y-direction being in the loading direction of the specimen. The specimens are each loaded in tension to a stress of 10 ksi. The stress developed at the point $\pm a$, 0 is 207 MPa, and the stress developed at 0, $\pm a$ is -69 MPa. Consider the stress in the thickness direction to be zero when the specimens are independently loaded (plane stress).

(a) What are the strains that develop at the coordinate positions given above?

(b) Next, consider that the three specimens are firmly bonded together along their flat faces to form a single tensile specimen. This specimen is then loaded to 69 MPa with the interfaces remaining plane. What are the stresses at mid-thickness at the coordinate positions given above?

2-4. The included angle of the notch in the standard Charpy V-notched specimen is 45°, and the radius of the notch tip is 0.25 mm.

(a) Assume the material of the Charpy bar is elastic-plastic (no strain hardening). Determine the maximum stress that can exist in a plastic zone ahead of the notch as a function of the uniaxial yield stress σ_Y, according to the von Mises criterion, and also according to the Tresca criterion.

(b) At what distance ahead of the notch does this stress first exist?

2-5. The hydrostatic stress in plane strain at initial yielding is numerically equal to k, the yield stress in shear. Draw the Mohr circles at yield in tension based upon both of the following:

(a) The Tresca criterion.

(b) The von Mises criterion.

2-6. Show that the dilatation Δ is equal to $\varepsilon_x + \varepsilon_y + \varepsilon_z$.

3-1. Based upon ASTM Designation E 399, Standard Test Method for Plane-Strain Fracture Toughness of Metallic Materials, the following expression gives the stress intensity factor K_I for the widely used compact specimen shown in Fig. 3-6 (the depth of the notch is $0.2W$):

$$K_I = \left(\frac{P}{BW^{1/2}}\right) f\left(\frac{a}{W}\right),$$

where $f(a/W)$ is given as

$$f\left(\frac{a}{W}\right) = \frac{(2 + a/W)(0.866 + 4.64a/W) - 13.32a^2/W^2 + 14.72a^3/W^3 - 5.6a^2/W^4}{(1 - a/W)^{3/2}}.$$

Plot the ratio of $K_I/(P/BW^{1/2})$ for values of a/W between 0.25 and 0.75.

3-2. K_I for a center-cracked panel is given as $K_I = \sigma\sqrt{\pi a}\sqrt{\sec \pi a/W}$. For the same B, compare the loads required to attain the same value of K_I for a compact specimen ($W = 50$ mm, $a/W = 0.5$) and a center-cracked panel ($W = 200$ mm, $a/W = 0.25$).

3-3. Determine the residual strength diagram (σ_{max} vs. a) for a center-cracked plate of an aluminum alloy ($\sigma_{YS} = 350$ MPa, $K_{IC} = 50$ MPa$\sqrt{\text{m}}$, $W = 10$ inches) for a/W values between 0.25 and 0.75.

3-4. The stress intensity factor for a semicircular flaw of depth a (equal to 0.01 m) in a proposed cylindrical pressure vessel is given as $0.71\,\sigma\sqrt{\pi a}$. The internal pressure is to be 15 MPa. The total volume of the pressure vessel is to be 1000 m^3. Three steels are available, and their characteristics are given below. Structural weight and safety are primary considerations. The maximum stress is to be $\sigma_{YS}/2$, and a safety factor ≥ 1.1 is called for based upon the ratio of the fracture stress to the working stress. Which is the preferred steel?

Steel	Thickness, m	σ_{YS}, MPa	K_{IC}, MPa$\sqrt{\text{m}}$
A	0.08	965	280
B	0.06	1310	66
C	0.04	1700	40

3-5. A sheet of material whose yield strength is 275 MPa is biaxially loaded with one principal stress being 300 MPa and the other 150 MPa. According to the

Tresca criterion, the material will yield. Will the material yield according to the von Mises criterion? Draw the Mohr circle at yield for each of these criteria.

4-1. The Hall-Petch relation (see Fig. 6-10):

(a) A side of the block represents grain size d.

(b) A shear stress results in an elastic displacement Δ.

(c) The shear strain $\gamma = \Delta/d$.

(d) Dislocations move on the dotted plane, and as a result, the elastic deformation is relaxed and is replaced by plastic deformation, that is, $\Delta = nb$, where n is the number of dislocations and b is their Burger's vector.

(e) $\gamma = \Delta/d = nb/d = \tau/G$; $n = \tau d/Gb$.

(f) Shear stress at the blocked end of the slip band is $n\tau = \tau^2 d/Gb$.

(g) The critical local shear stress required to initiate slip in the next grain is τ_c, and slip in the next grain occurs when $\tau^2 d/Gb = \tau_c$. Therefore, the shear stress τ required for slip is $\tau = (Gb\tau_c/d)^{1/2}$.

(h) With the inclusion of lattice friction τ_i, the expression for τ is $\tau = \tau_i + k_y\, d^{-1/2}$.

(i) In terms of normal stress, $\sigma_y = \sigma_i + k_y d^{-1/2}$.

(j) If $\sigma_i = 70.60$ MPa and $k_y = 0.74$ MN m$^{-3/2}$, plot the yield strength as a function of $d^{-1/2}$ (mild steel). Compare with aluminum for which typical constants are 15.69 and 0.07.

4-2. Why is the amount of recycled automobile body scrap limited in steel making?

4-3. Derive the lever rule and explain how it is used to determine the relative amounts of the phases present in a two-phase field.

5-1. A. In an SEM, the depth of field, z, can be expressed in microns as

$$z = \frac{2 \times 10^5 W}{MD},$$

where W is the working distance in millimeters, M is the magnification, and D is the diameter of the final aperture in microns. Determine the depth of field for the following cases:

(a) $M = 10^4$, $D = 100\ \mu$m, $W = 10$ mm.

(b) $M = 10^4$, $D = 300\ \mu$m, $W = 5$ mm.

(c) $M = 10^3$, $D = 100\ \mu$m, $W = 10$ mm.

(d) $M = 200$, $D = 100\ \mu$m, $W = 10$ mm.

B. Compare with the depth of field of an oil immersion lens where, at a magnification of 400×, $W = 1$ mm, and the numerical aperture, equivalent to $D/2W$, is 1.60.

5-2. A two-beam X-ray stress analysis is carried out on a high-strength steel component for which $E = 200$ GPa and $\nu = 0.25$. The part contains a biaxial residual stress system, and it is desired to measure the residual stress in a particular direction. The values of the (511) spacings for normal incidence and for incidence at $\psi = 45°$ are 0.0550 nm and 0.0552 nm, respectively. Determine the stress.

6-1. In Fig. 6-8, suppose the strain rate is fixed at 10^{-4}/sec. Assume a local fracture stress for ABS-C steel of 200 ksi. What is the highest temperature at which fracture will occur in a brittle manner for a Charpy V-notch specimen of this steel at the same strain rate according to the von Mises criterion?

6-2. The strength of a low-carbon steel is 622 MPa for ASTM grain size #2 and 663 MPa for ASTM grain size #8. What will the strength be for ASTM grain size #10? (Assume that the grains have a square shape.)

6-3. Plot the value of the hydrostatic stress in the neck of a tensile bar in terms of $\overline{\sigma}$ as a function of r/a for values of a/R of $\frac{1}{3}$, 1, and 2.

6-4. For a given alloy, the condition for necking $d\sigma/d\varepsilon = \sigma$ is met at a true stress of 350 MPa and a true strain of 0.50.

(a) Determine the corresponding values for the ultimate tensile strength and the engineering strain at necking.

(b) What is the work per unit of volume required to strain this alloy to the point of necking, assuming $\sigma = k\varepsilon^n$?

6-5. In a tensile test, what is the effect of the gauge length on the calculated strain:

(a) Before necking?

(b) After necking?

6-6. The stress concentration factor (based on net section stress) k for the notched-round bar in bending shown in Fig. 6-24 is given as

$$k = 3.04 - 7.236\left(\frac{2h}{D}\right) + 9.375\left(\frac{2h}{D}\right)^2 - 4.179\left(\frac{2h}{D}\right)^3,$$

where h is the notch depth, 3.175 mm, and D is the diameter of the bar. Determine k for this bar.

7-1. What problems might you anticipate with a ceramic coating that experiences a sudden rise in temperature such that the temperature rise in the metal beneath the coating is much less than that of the coating?

7-2. Consider the bimetallic strip figure shown in Fig. 7-1. Assume material 1 is steel ($\alpha = 0.0000065/°F$) and material 2 is aluminum ($\alpha = 0.00001/°F$). The yield stress of the aluminum is 69 MPa, that of the steel is 690 MPa. The total area of steel equals that of aluminum.

(a) What temperature rise is needed for the aluminum to reach its yield stress?

(b) Discuss what happens to the system if the temperature rises above that in part (a).

(c) If the yield stress of the aluminum is exceeded and 0.05% plastic deformation of the aluminum occurs as the temperature continues to rise, what is the state of stress in the strip when the temperature returns to its original value?

(d) Why is α higher for fcc aluminum than for bcc iron?

9-1. The bolts of an elevated temperature unit are tightened to 70 MPa at a temperature of 550°C where Young's modulus is 175 GPa. $\dot{\varepsilon}_{ss}$ at this stress and temperature is 5.0×10^{-8}/hr, and the power-law creep exponent is 4.0. Determine the stress in the bolts after one year.

9-2. A turbine engine operates at constant speed at 750°C, and the blades initially have a 1-mm clearance. The blades are proportioned such that, along a main section of 100 mm in length, the centrifugal stress is constant at 40 MPa. Tests have shown that, at the steady-state, power law creep rate at 800°C is 7.5×10^{-8}/sec and at 950°C it is 1.3×10^{-5}/sec. Estimate the time required for the tip clearance to be reduced to zero.

9-3. The power law creep relation for a Ni-Cr-Mo steel at 454°C is given as

$$\dot{\varepsilon}(\text{day}^{-1}) = 10 \times 10^{-20} \sigma^3$$

where σ is in pounds per square inch.

(a) Show that this relation can be written as $\dot{\varepsilon}(\text{hour}^{-1}) = 1.3 \times 10^{-14} \sigma^3$, when σ is in megapascals (1 MPa = 145 psi).

(b) It is desired that a 2-m long round tension member support a load of 4.48×10^4 N for 10 years at 454°C without exceeding 2.5 mm of creep deformation. What is the maximum stress that can be applied?

(c) If after three years it is decided to lower the total deformation allowed to 1.5 mm, what then is the maximum allowable stress?

(d) What cross section is required?

9-4. The 100-hour rupture lives of several nickel-base superalloys are shown in Fig. 9-4. For Udimet 700, use the Larson-Miller parameter with the constant equal to 20 to determine the life at 1500°F.

10-1. A thick-walled pressure vessel is made of steel for which $K_c = 50$ MNm$^{-3/2}$ (50 MPa $\sqrt{\text{m}}$). Nondestructive testing shows that the component contains semi-circular cracks of depth up to $a = 0.2$ mm. The crack growth rate under cyclic loading is given by $da/dN = A(\Delta K)^2$, where $A = 10^{-10}$ (MNm^{-2})$^{-2}$m^{-1}. The component is subjected to an alternating stress range $\Delta\sigma = 200$ MNm^{-2} ($R = 0$). Given that $\Delta K = 0.73 \Delta\sigma \sqrt{\pi a}$, determine the number of cycles to failure.

10-2. From Fig. 7-3, estimate the plastic strain range in the thermal cycle shown from the width of the hysteresis loop. Then use the Coffin-Manson law, $N_f^{1/2}$ $\Delta\varepsilon_p = C$, to determine the number of cycles to failure if the value of C is 0.6.

10-3. The Basquin law for an alloy tested $R = -1$ is $N_f^b(\Delta\sigma/2) = C$. At a $\Delta\sigma$ of 300 MPa, N_f is 10^5 cycles, and at a $\Delta\sigma$ of 200 MPa, N_f is 10^7 cycles.

(a) Determine the values of the constants b and C.

(b) What is N_f for a stress range of 250 MPa?

(c) If a specimen is cycled at 250 MPa for 5×10^5 cycles, how many additional cycles based upon the Palmgren-Miner rule will the specimen be able to sustain at a stress range of 200 MPa?

10-4. A critical component whose fracture toughness is K_{Ic} is cyclically loaded at $R = 0$. The component must withstand at least N cycles before failure. The stress range is $\Delta\sigma$ and the rate of fatigue crack growth in the part is given as $da/dN = A(\Delta K)^2$, where $\Delta K = Y(\Delta\sigma)\sqrt{\pi a}$ is range of the stress intensity factor for a fatigue crack or any initial flaws, if present. Assuming linear elastic behavior:

(a) Show that the magnitude of the proof stress to insure satisfactory service that must be applied before the component is put into service is given by $\sigma_{proof} = \sigma e^{(A/2)Y^2\pi\sigma^2 N}$. A procedure of this nature is used in safeguarding the wings of the F-111 swing-wing aircraft. If this were a pressure vessel, the working pressure would be raised by the same factor in conducting a proof test.

(b) If $A = 2 \times 10^{-9}$ $(\text{MPa})^{-2}$, $Y = 1.0$, $\sigma = 140$ MPa, and $N = 10{,}000$ cycles, determine the required proof stress level (259 MPa).

11-1. **(a)** The effect of the reinforcement angle at a butt weld in steel on the fatigue range $\Delta\sigma$ at 2×10^6 cycles at $R = 0$ is given as $\Delta\sigma = 120[2 - \cos(\theta - \pi/2)]$, MPa. Here θ is the reinforcement angle in radians. Plot the value of $\Delta\sigma$ as a function of θ for values of θ between $\pi/2$ and π radians.

(b) The ultimate tensile strength of the steel is 600 MPa. Use the Goodman relationship to determine the stress range $\Delta\sigma$ at 2×10^6 cycles for $R = -1$ loading conditions. For comparison purposes, plot as a function of θ on the same graph as in part (a).

12-1. The rate of stress corrosion crack growth expressed in meters per hour for an Al-Zn-Mg alloy in a saline solution is given as $da/dt = 8 \times 10^{-7}\ K^2$. The fracture toughness of the alloy is 30 MPa $\sqrt{\text{m}}$. Suppose the material is used in the form of a pipe of wall thickness 12 mm and diameter of 96 mm through which flows a saline solution at a pressure of 6 MPa. The pipe contains flaws on the interior surface which are 0.1 mm deep. Assume that $K = \sigma\sqrt{\pi a}$.

(a) How long will it take for the pipe to either burst or leak through?

(b) If it is desired that the pipe last for at least 10,000 hours, what is the maximum pressure that can be applied?

(c) What proof test pressure has to be applied to insure that the pipe lasts 10,000 hours?

(d) If the system is operated at a pressure of 6 MPa for 2000 hours, what is the life expectancy if the pressure is then dropped to 3.45 MPa?

13-1. A sound beam traveling in water impinges upon a steel plate. What percent of the energy of the beam is reflected? What percent is transmitted? The energy reflected, *IR,* is given by

$$IR = \left(\frac{Z_2 - Z_1}{Z_2 - Z_1}\right)^2, \ Z_{water} = 0.149 \text{ g/cm}^2\text{-}\mu\text{sec}, \ Z_{steel} = 4.68 \text{ g/cm}^2\text{-}\mu\text{sec}.$$

13-2. The longitudinal velocity of sound in lucite is 0.267 cm/μsec. In steel, the longitudinal wave velocity of sound is 0.594 cm/μsec, and the transverse wave velocity is 0.324 cm/μsec.

(a) Determine the first critical angle of incidence for sound transmission from the lucite to the steel.

(b) Determine the second critical angle for sound transmission from lucite into steel.

(c) There is no critical angle when sound goes from the steel into the lucite. Why?

13-3. A fatigue crack in a steel is completely filled with an impermeable substance whose magnetic properties are the same as the steel. Will the MT technique detect the crack? Will the dye penetrant method detect the crack? Will the eddy current method detect the crack? Will the UT method detect the crack?

14-1. The outer radius of a hollow steel axle is designated as a, the inner radius as b. The change Δa in a, due to an external pressure q resulting from a press fit of a bearing on the axle, is given as

$$\Delta a = \frac{-qa}{E}\left(\frac{a^2 + b^2}{a^2 - b^2} - \nu\right).$$

For a given Δa, plot the variation in q as a function of b/a. $E = 210$ GPa, $\nu = 0.25$, $a = 20$ cm.

14-2. **(a)** What is the life adjustment factor for a bearing reliability of 95%?

(b) Determine the L_5 life for a system containing four bearings whose rated lives are 12,000, 10,000, 9000, and 8000 hours?

14-3. What is the reliability of a system containing four bearings which requires a 5000-hour life when the rated lives of the bearings are 12,000, 10,000, 9000, and 8000 hours?

Index

Author Index